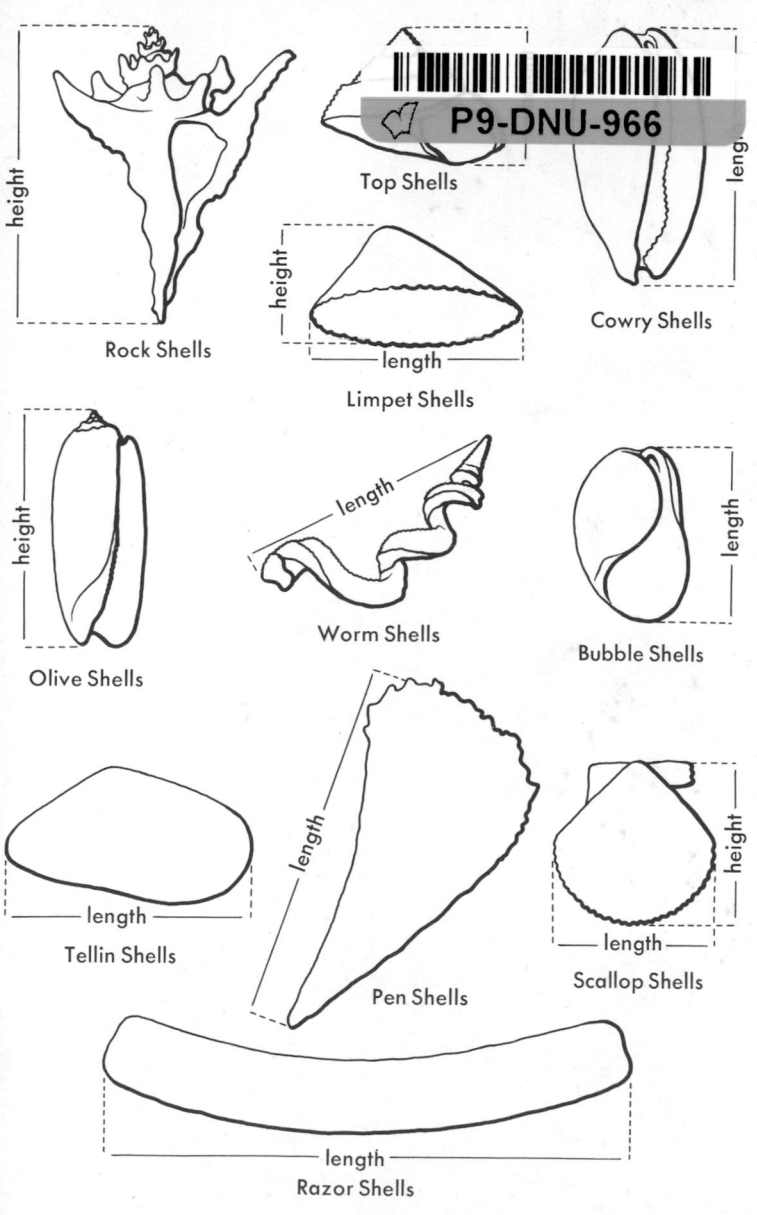

Top Shells

Rock Shells

Cowry Shells

height

length

length

height

Limpet Shells

Olive Shells

height

length

Worm Shells

Bubble Shells

length

Tellin Shells

length

Pen Shells

length

Scallop Shells

height

length

length

Razor Shells

SHELL MEASUREMENTS

A Field Guide to Shells

of the Atlantic and Gulf Coasts
and the West Indies

THE PETERSON FIELD GUIDE SERIES

EDITED BY ROGER TORY PETERSON

THE PETERSON FIELD GUIDE SERIES

A Field Guide to Shells

of the Atlantic and Gulf Coasts and the West Indies

BY PERCY A. MORRIS

Illustrated with Photographs

Third Edition
edited by
WILLIAM J. CLENCH

Sponsored by the National Audubon Society
and National Wildlife Federation

HOUGHTON MIFFLIN COMPANY BOSTON

Second Printing of the 3rd Edition
c

ISBN 0-395-16809-0 hardbound
ISBN 0-395-17170-9 paperbound
Library of Congress Catalog Card Number: 72–75612
Printed in the United States of America

Editor's Note

WHEN I was a boy half a century ago many lads my age collected birds' eggs, and in so doing launched their interest in natural history. Eggs were a popular outlet for that craving of all boys to collect. Finding nests was a kind of treasure hunt: the eggs were the gems and could be hoarded. With some boys the hobby became a mild form of kleptomania; others, more thoughtful, took only a set or two of each species and eventually became deeply interested in the birds themselves. Many ranking ornithologists among the older men can trace their involvement to this juvenile hobby. A few deplore the fact that egging is no longer allowed, but in the interests of conservation this was inevitable. With the advent of the binocular, the "list" has become the egg substitute.

Although butterflies are easily collectible, and so are minerals (more so than plants), none of nature's varied productions lend themselves more to the gratification of the collecting instinct than shells. Because of their calcareous texture and delicate colors, shells have some of the same tactile and visual appeal of jewels, jade, and fine porcelain. That is the attraction to the esthete. Others, aware that the shell is but the garment of a once-living animal, take the naturalist's point of view and concern themselves with classification and distribution. All too few inquire into the life of the living mollusk, its ecology and its habits.

Recognition, however, always comes first. That is why the Field Guide Series was launched—as a shortcut to recognizing and naming the multitude of living things that populate America. The first volume to appear, *A Field Guide to the Birds,* met with instant approval; then followed *A Field Guide to Western Birds;* the third was Percy Morris's admirable *Field Guide to the Shells* (1947), now presented in its new form.

Whether your interest in the seashore is that of a bird watcher (no seascape is ever devoid of birds), that of a surf fisherman casting for channel bass, or merely that of a beachcomber, you cannot overlook the shells. From the rocky headlands of Maine to the long sandy beaches of the Texas coast stretch thousands of miles of good collecting grounds. Rock, sand, and mud offer radically different environments. Some specimens are found at the line of high tide, where the sea has cast them up. Others occur commonly in tide pools. Still others can only be obtained in deep water. And here we add a cautionary conservation note: some mollusks are so rare that we do not encourage the collecting of the living animal except for scientific purposes. Commercialization of such rarities could eliminate low-level populations. Among the mollusks there are "endangered species" as vulnerable as certain birds or mammals.

Like many other books in the Field Guide Series, this guide has evolved considerably from its earlier beginnings. The Second Edition (1951) added 112 species of mollusks not covered in the original book. Six new plates of photographs were added and seven of the earlier plates were rebuilt with new photographs. The use of common or popular names was extended and much of the text was rewritten.

In this, the Third Edition, the area covered has been enlarged to include the West Indies and the number of plates increased from 45 to 76 (68 in black and white, 8 in color). The black and white plates have been rebuilt and the color plates are new. Common names have now been adopted for all shells and these are given priority in the species headings.

Percy Morris died shortly after preparing the manuscript of this new edition, and it fell to Dr. William J. Clench, former Curator of Mollusks at the Museum of Comparative Zoology, Harvard University, to act as technical editor, checking the nomenclature and also the many details that would have fallen to Mr. Morris had he lived to see his book through to publication. To Dr. Clench we are much in debt for his scholarly assistance.

We are also most grateful to three members of the staff of the Peabody Museum of Natural History at Yale University for their help in securing the eight new color plates: to William K. Sacco for his superb photography and to Dr. Willard D. Hartman and Mrs. Suzanne Stevenson for their cooperation in making available for photographing the shells originally chosen by Percy Morris.

Take this book with you whenever you go to the shore. Do not leave it home on your library shelf; it is a *Field Guide,* meant to be used.

ROGER TORY PETERSON

Acknowledgments

THE AUTHOR wishes to acknowledge his indebtedness to many who have aided materially in the production of this book. Carl O. Dunbar and Stanley C. Ball of the Peabody Museum of Natural History, Yale University, originally gave constructive criticism and advice during the writing of the First Edition. Earl H. Reed of the Springfield Science Museum, Springfield, Massachusetts, has been of inestimable help in matters of identification and distribution while this Third Edition was being prepared. William J. Clench and Ruth D. Turner of the Museum of Comparative Zoology, Harvard University, have been ever ready to answer questions regarding nomenclature and classification throughout the writing of the three editions. Others who have aided in many ways include: Willard D. Hartman of the Peabody Museum of Natural History, Yale University, Heathcote M. Woolsey of Kent, Connecticut, and J. Brookes Knight of Sarasota, Florida, both of whom are no longer living, and members of the Connecticut Shell Club. The writer wishes to express his sincere gratitude to all of these and at the same time point out that any errors that may creep into the book are the sole responsibility of the author.

Roger Tory Peterson, Editor of the Peterson Field Guide Series, has been ever helpful, and his many suggestions are deeply appreciated. Finally, the author's sincere thanks go to Paul Brooks and Helen Phillips of Houghton Mifflin Company for their patience and their untiring efforts in seeing the work through the press.

PERCY A. MORRIS

New Haven, Connecticut

Contents

Illustrations

Introduction

OF THE MANY amusements for the wanderer by the seashore, one of the most rewarding is the search for shells. None but those who have given a little time to shell collecting can conceive of the multitude of varieties which are discovered when practice sharpens the eyes.

A well-organized collection can be a joy to own. Properly cleaned shells will never be marred by insects; they are most durable, and the beautiful colors and patterns that many of them exhibit are practically permanent. While it is true that a shell will fade and lose its luster if exposed to the bright sun for a long period (the bleached examples to be seen on any beach prove this), it is also true that if kept in a cabinet or drawer away from direct sunlight the same shells will retain their beauty for years. In our great museums one can find shells that were collected more than a century ago, and they are just as colorful and fresh-appearing as specimens that are living along our shores today.

Collecting your shells: For collecting one should of course wear old clothes and sturdy shoes. Never collect barefooted, because many of our shells, buried in the sand, have razor-sharp edges. You will need a bag or sack in which to put specimens, a jar or two of seawater in which to put fragile specimens until you can care for them later, and a few small vials for minute forms. A stout bar of some sort for probing and overturning stones is useful; a shovel or trowel is a necessity. A pair of tweezers for handling tiny mollusks is helpful, and a small hand lens makes critical examination on the spot possible.

One scarcely needs to be told where to find shells. The beginner usually starts by walking along the shore and keeping a sharp eye for specimens that have been washed in by the tides. While many excellent shells may be obtained in this way, a large majority will be wave-worn, eroded, and broken, or will be otherwise imperfect. Bivalve shells will in most cases be represented by single valves, and the lips of univalves will show broken edges, worn sculpture, or damaged spires.

The young collector soon learns that to build up a really fine collection he must look for living specimens. Just as in bird studies we look for certain kinds of birds in open fields, others in deep woods, and still others in marshes, so it is with the mollusks. We find that some—in fact most—show decided preferences for particular types of marine territory. A vast assemblage of different forms live on rocky shores, clinging to or hiding under the rocks while the tide is out. Others prefer sandy flats, where they burrow into the sand during the ebb tide. Many like nothing better than a muddy bottom, and the blacker and stickier the better. Some will live only in the purest seawater, tolerating no brackishness at all, whereas others seem to like it better where the water is brackish.

A large and interesting part of our molluscan fauna lives in what is known as the *littoral zone,* that portion of the beach between high and low tides which is exposed to the sun twice each day. Others live out beyond the low-water level, and many species are found in deep water, even up to several hundred fathoms. Temperature controls the distribution of many varieties, and we shall find a different group in Maine from those found at Cape Hatteras. Yet some of the shells that may be collected close to shore in northern waters are to be found living in deeper water in the south.

On rocky coasts the receding tides leave pockets of water in all depressed areas, and in these tide pools one can find various kinds of molluscan life carrying on in a business-as-usual manner. It will reward the collector to scan these miniature aquariums carefully. The pilings and undersides of wharves are rich collecting grounds too, the logs generally being covered with a heavy growth of marine algae that conceals many of the smaller mollusks.

The best time to collect is of course at low tide, when one can get out to the low-water limits. Any stone, plank, or section of driftwood should be turned over; a host of marine creatures, including mollusks, take refuge under such objects to await the return of the next tide. A good practice is to roll the stone or log back the way it was after you are through searching. On sand and mud flats the presence of bivalves can usually be detected by small holes that reveal the location of the clam's siphons. As you trudge along the mollusk quickly draws in its siphons, commonly ejecting a thin squirt of water as it does so. The snails generally burrow just beneath the surface and produce small telltale mounds. By energetic digging in likely places while the tide is out, one should come up with a good representative collection of the species inhabiting that particular beach, all of them "taken alive" specimens, far superior to the old and empty shells picked up along the shoreline.

Nevertheless, the collecting of dead shells is not to be neglected altogether, since many fine examples may be so obtained. This is particularly true of the minute forms. On nearly every exposed beach you will find a long wavy line of flotsam and jetsam — known as sea wrack, beach litter, or beach drift — which marks the high-tide limit. It will pay you

to examine this debris carefully. Here you will find large pieces of seaweeds that have been torn from their moorings and washed up on the shore. Concealed in the folds and among the roots you may find scores of tiny mollusks that live out in deeper waters. If you gather freshly washed-up material, such as on a day following a violent storm, most of the mollusks will be live ones. Sections of sponges should be pulled apart and examined, and pieces of driftwood broken up to reveal piddocks and shipworms. Dead crustaceans and sea urchins are worth looking over for parasitic mollusks. The fine material in the beach litter should be run through a sieve to eliminate most of the sand grains, and the shelly particles remaining searched for tiny pelecypods and gastropods. Most collectors take home small bags of this material, commonly called "scratch," to work on at evening or on rainy days.

Try collecting at night with a strong flashlight. Several of our snails are nocturnal, and certain species that have to be searched for under stones and in crevices during the daylight hours will be found crawling over the rocks and the sandbars after dark. It is suggested that any night collecting be done with a companion or two, for safety reasons.

One of the most exciting as well as rewarding types of collecting is dredging. We cannot all afford the boats, winches, and heavy gear to do this on a large scale, much as we would like to, but it does not cost very much to rig up a small iron frame, attach a netting or fine wire bag, and drag it along the bottom from a rowboat in 15 to 20 feet of water. Low tide is the best time. Try all sorts of bottoms — sandy, muddy, gravelly, and so on. After dragging your scoop for several minutes, pull it to the surface, jiggle it a bit to wash out most of the mud and silt, then haul it over the side of the boat and empty it into a basin or tub of water. Separating your catch in this way is easier, and small fragile shells are less likely to be injured than if you dumped the contents out onto the deck.

Preparing your shells: While the collecting of shells is sheer pleasure, the necessary cleaning and preparing comes under the label of work; but it is work that has to be done if we are to gather living material. First of all get rid of the animal in the shell. It is common practice to place the shells in a pan of water and boil them 5 to 15 minutes. After that the bivalves present no problem: the valves gape open and the soft parts can be removed with no difficulty, though the muscle scars generally require a slight scraping. Valves with an external ligament will remain attached in pairs. After the interior has been thoroughly cleaned the two valves can be closed and the whole shell wrapped with strong thread until the ligament has dried. In each tray of specimens you will want a pair of valves that are separated to show the interiors.

With the univalves, the animal will have to be drawn out of

its shell by a hook or bent wire. After being boiled the s
parts generally slip out without too much trouble, but som
times the small end will break off and stay in the uppe
whorls. Soaking in plain water for several days often will rot
and soften this part so that it can be flushed out. If this fails,
the usual practice is to put the shell in a 70 percent solution of
alcohol for a few hours to harden the animal matter. Then the
shell can be dried in a shady place, and after this there should
be no offensive odor. Very tiny snails are generally kept in the
70 percent alcohol solution for several days and then dried.

Many of the snails will have an *operculum* attached to the
body in such a manner that when the "foot" is withdrawn it
serves as a door, effectively blocking the aperture (see left endpaper). Opercula may be horny, leathery, or calcareous. This
structure must be cut away from the flesh and its reverse side
scraped clean. When the shell is prepared and ready for the
cabinet, the operculum is glued to a tuft of cotton and then
inserted in the aperture and positioned as it would be in life. In
cleaning your gastropods take special care to see that each operculum is kept with its own shell.

Some of the larger bivalves, particularly of the genera *Spondylus* and *Chama,* usually have their shells more or less coated
with worm tubes, barnacles, calcareous algae, and other foreign
growths. They can be cleaned only by careful and painstaking
work with small chisels, needles, and scrapers. Some collectors
paint the unwanted encrustations with a weak solution of muri-
atic acid, but although the acid will dissolve all limy deposits
it will also attack the shell itself, so its use is not recommended.

The *periostracum* (or noncalcareous covering that protects
the outside of many shells) can be removed by soaking the shell
in a solution of caustic soda, about one pound to a gallon of
water. Most collectors, however, prefer their specimens in a
natural state, even though some — such as the cone shells
(*Conus*) — do not reveal their beauty until the periostracum has
been removed. It is ideal to have two or three examples of each
species in a tray, one denuded of its periostracum and the
others as they were in life.

After your shells are cleaned, washed, and dried and are
ready for their trays or the cabinet, they may be lightly rubbed
with a small amount of mineral oil applied with a tuft of cotton. This will impart a slight luster to the surface, make them
fresh-looking, and aid somewhat in preserving the delicate coloring. Some workers use olive oil, but that may in time become
rancid and it also may attract insects. Never coat your shells
with lacquer, varnish, or shellac.

Arranging your shells: A collection's value is proportional
to its labeling. The precise locality and date are far more important than having the shell correctly named. Any competent
malacologist can put the right name on your shell at any time,

but you and *only you* can provide the exact location and date. Adopt some form of label that provides space for a number (which will be entered in your catalog and written in waterproof ink on each specimen), the name, date, and locality. This last should be as exact and detailed as possible. "West Indies" and "southern Florida" do not mean very much, but "½ mile south of Mayagüez, Puerto Rico" and "Cape Romano, s. Florida" tie the specimen down accurately. The collector's name should be included, and most collectors can be depended upon not to neglect that. Under the entry number in your catalog you may include notes on the tide conditions, the weather, whether or not the mollusk was a living specimen or an empty shell and, if the former, what type of bottom it was living on, its relative abundance at that time and place, what it was doing, and anything else that may seem pertinent. Below is a sample of the label used by the writer:

> **YALE PEABODY MUSEUM**
> **Division of Invertebrate Zoology**
>
> YPM No. *26971*
>
> *Nassarius obsoletus* (Say)
>
> Mud flats, 1 mile W. /6602/
> of New Haven, Conn.
> Coll. H. M. Woolsey Date 7-21-1954

The number in the small box is an accession number applying to all shells from a certain collection. For example, if we spend an hour or a day or a week at some particular locality all the shells gathered on that trip can have the same accession number; or if we purchase a complete collection from some source we may give an accession number to the whole lot. Then, although each species will eventually have its catalog number, the number in the box instantly associates each shell with a certain group. In the accessions catalog you will have detailed information about the trip or the purchase.

Names of shells: The beginner is apt to be confused and at first annoyed by some of the "jawbreaking" names that have been given to shells. This is true in any branch of natural history but is perhaps more apparent in the molluscan field, since many of our species are rarely seen by the average person and only a few kinds have become well enough known to have acquired common or popular names. Perhaps this is a good thing, because English names are very likely to be too local in character. Thus the same shell may be called one thing in one place and go under a completely different name a few miles down the coast, and in three widely separated localities the same name may be applied to three totally different shells.

Scientists have agreed upon Latin as best for the naming of animals (and plants) for very good reasons. It is a dead language and therefore not subject to change. Serious scholars all over the world, regardless of their individual nationalities, are familiar with it, so it is about as close as one can get to an international tongue. The name "Crown Conch" means something to a resident of Florida but it means nothing to a Frenchman or a Japanese, or to most Californians. Use the snail's scientific name of *Melongena corona,* however, and every shell enthusiast, whatever his country, knows what shell we are talking about.

In this book common names are given for all the shells. However, the serious collector is strongly urged to learn his treasures by their authentic names. Scientific names are not difficult after we become familiar with them. We use the words alligator, boa constrictor, and gorilla commonly enough without stopping to realize that they are perfectly good scientific names. In botanical reference we all speak or write of our iris, forsythia, geranium, crocus, delphinium, and scores of others with the greatest of ease, just because we are familiar with them. Who can say that chrysanthemum or rhododendron is easier to say than *Murex, Tellina,* or *Littorina?*

It has been found convenient to give each animal two names: first, a general, or generic, name (always capitalized) to indicate the group (*genus*) to which it belongs; second, a special, or specific, name (not capitalized) to apply to that animal alone as the *species* name. In older works the specific name was capitalized if a proper name, but that practice has been discontinued for many years.

Take the well-known Junonia, that gaily spotted shell hopefully looked for by nearly all collectors on Florida beaches. This was named *junonia* by Lamarck in 1804; Lamarck placed it in the genus *Voluta.* Since the name of the person who first describes a species, known as the *author,* follows the scientific name, our snail became listed officially as *Voluta junonia* Lamarck. If later research proves that a species belongs in a genus different from the one to which the author has allocated it, it is removed to the proper genus but the species name is retained and parentheses are placed around the author's name to indicate that a change has been made from the original allocation. In this case the gastropod was put in the genus *Scaphella,* so our name now reads *Scaphella junonia* (Lamarck). The parentheses are a technicality of importance chiefly to specialists;

many books for nonspecialists do not use them, although the International Rules of Zoological Nomenclature insist that they are a valid part of the name. A list of authors whose names appear in this *Field Guide* can be found at the back as Appendix B.

Your fellow collectors: One of the many joys connected with "shelling" is meeting others with the same interests. One cannot search the sandflats or pry under driftlogs and stones for any length of time without running into someone similarly occupied; many lasting friendships have been initiated on the beach. Shell people like to get together, and shell clubs have been organized in many localities throughout the country. In fact, their growth during the past decade has been little short of phenomenal. Most are affiliated with the American Malacological Union. It is strongly urged that anyone seriously collecting shells consider joining this society. The only requirement is an interest in mollusks. The membership is about one-third professional malacologists and two-thirds enthusiastic amateurs. The corresponding address of this society varies with that of the acting secretary (who now is Mrs. Robert Hubbard, 3957 Marlow Court, Seaford, New York 11783), and any inquiry about the association will receive prompt attention.

Some of the eastern shell clubs are listed in Appendix D. If you are within commuting distance of any, by all means get in touch with the Secretary and plan to attend one of the meetings. Here you will find kindred souls, perhaps enrich your own collection by exchanges, enjoy the good fellowship of local field trips, and obtain help in the identification of puzzling material. Many of these shell clubs issue a publication for the benefit of club members, possibly a bimonthly newsletter or monthly bulletin that varies from being a mimeographed list of members with program notes and local items to including short articles about collecting and classification or presenting discussions on a particular group or genus. Some clubs issue a regularly printed journal that publishes reports of work done by some of the country's leading scientists.

The Phylum Mollusca: This group is one of the major branches of the Animal Kingdom, including as it does the clams, oysters, scallops, the snails and slugs, and the chitons, squids, octopuses, and some others. About 100,000 living forms are known throughout the world, besides many thousands of fossil species, for the group is represented in the most ancient fossil-bearing rocks. The mollusks are divided into the 6 major classes that follow.*

---

* The order of listing here follows that of J. E. Morton in *Molluscs,* 4th ed. (London, 1967), whereas Percy Morris follows Thiele's *Handbuch* in general for his systematic chapters. Ed.

Monoplacophora: Gastroverms

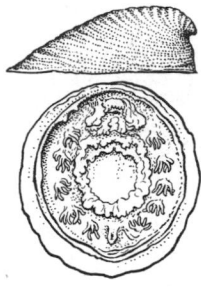

Amphineura: Chitons

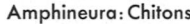
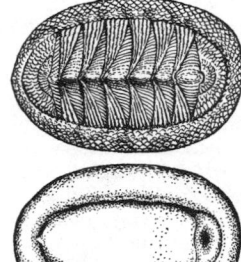

Gastropoda: Univalves
Snails, periwinkles, whelks, conchs

Bivalvia (Pelecypoda): Bivalves
Clams, oysters, mussels, scallops, cockles

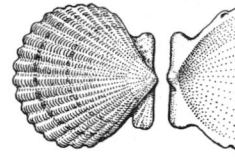

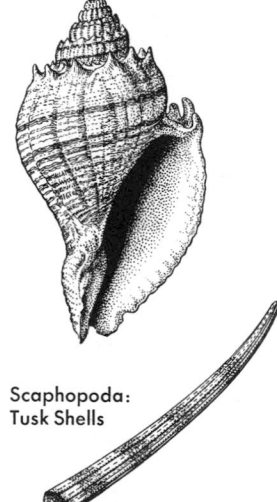

Cephalopoda: Squids,
Octopuses, Nautiluses

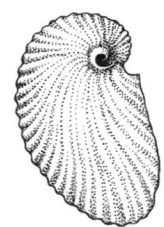

Scaphopoda:
Tusk Shells

THE SIX CLASSES OF MOLLUSKS

1. The *Monoplacophora* are a limpetlike primitive group with some paired organs. They were known only as ancient fossils and were believed to have become extinct about 250 million years ago until 1957, when living specimens were first dredged from abyssal depths off Guatemala. The first one found was named *Neopilina galatheae* Lemche.

2. The *Amphineura* are primitive mollusks of very sluggish habits, mostly preferring shallow water close to shore. The typical chiton (the common name) is an elongate, depressed mollusk bearing a shelly armor of 8 saddle-shaped plates, or valves, arranged in an overlapping series along the back. The foot is broad and flat and serves as a creeping sole or sucking pad, by which the creature clings to the rocks. The ancestral mollusk, from which all existing forms have been evolved, is believed to have been very similar to the present-day chiton.

3. The *Gastropoda* are numerically the largest division of the Mollusca. Examples occur in marine and fresh waters, and also as terrestrial air-breathing animals. There is but a single valve, usually spiral or caplike, and a few are shell-less. There is a distinct head, often with eyes, and the mollusk is provided with a toothed radula, or lingual ribbon, by means of which it shreds its food. Gastropods may be either herbivorous or carnivorous.

4. The *Scaphopoda* are elongate mollusks enclosed in a tapering conical shell open at both ends and slightly curved. From the larger end project the foot and several slender filaments. Scaphopods live in clean sand, from shallow water to considerable depths. They are commonly called tusk or tooth shells.

5. The *Bivalvia* (*Pelecypoda*) are entirely aquatic and predominantly marine, with a pair of valves joined by a hinge and held together by strong muscles within. The animal has no head and feeds upon microscopic plant and animal matter drawn into the mantle cavity through the siphon. There is commonly a muscular foot for burrowing, although some live permanently attached to rocks or coral.

6. The *Cephalopoda* are highly specialized mollusks, keen of vision and swift in action. The head is armed with a sharp parrotlike beak, and is surrounded by long flexible tentacles studded with sucking disks. Besides the well-known devilfishes (octopuses), this class includes the delicate Paper Argonaut and the beautiful Pearly Nautilus of the western Pacific. *Spirula,* a delicate white chambered shell frequently washed up on our southern beaches, belongs to this class, and the spectacular *Ammonites* found in rocks of Jurassic and Cretaceous periods were cephalopods too.

About This Book

IN THIS third edition of the Atlantic Coast shell guide the scope has been expanded to include the more common shells of the West Indies, so the book will be useful to anyone visiting the Caribbean area. It has been completely rewritten, much new material has been added, and old material has been brought up to date. The text now lists information under boldface subheadings of range, habitat, description, and remarks, in line with the more recent members of the Peterson Field Guide Series, and these should make the book much easier to use than the previous editions. Another departure from the earlier editions is to be found in the main species entries, where the English common name is listed first and the Latin scientific name second. There now are 76 plates instead of the former 45. The black and white plates have been entirely rebuilt, with many new photographs, and the number increased from 37 to 68. The 8 color plates also are new. It is hoped that this revised edition will prove useful to the growing army of enthusiasts who are embracing shell collecting as a hobby.

A total of 1035 species and subspecies are described and all but 2 are illustrated by one or more photographs. It should be understood that the coverage is by no means complete. Some families contain hundreds of described species, many of them very rare or often differing from each other so slightly that only a specialist can tell them apart — and not always is there complete agreement even between specialists. The author has tried to include all of the common mollusks that the collector is likely to discover on trips to beaches, and it is hoped that the selection herein presented will prove adequate for building a foundation of knowledge about one of the most interesting animal groups in nature. For more advanced identification work, the reader is referred to some of the technical works listed in Appendix C, the bibliography.

Classification: Unlike books on birds or minerals or flowers, shell books rarely agree on the arrangement of families — which in other words means the classification of mollusks. The accepted plan begins with the most primitive forms and works up to the most advanced or complex forms. This is determined largely by the mollusk's anatomy. The early conchologists were concerned chiefly with the characters of the shell rather than with the soft parts. Modern workers have arrived at a fairly satisfactory listing, certainly better, bringing together those forms most closely related, and the classification is constantly being improved as more anatomical work is done. For example, until a few years ago practically all writers began their gastropod sections with *Patella* and *Acmaea,* the true limpets, followed by the keyhole limpets, *Fissurella, Diodora,* and so on. It is now recognized that the keyhole limpets are more primitive than the true limpets, so in modern listings they precede them. As more intensive research is done it can occasionally be demonstrated that whole genera should be taken out of one family and placed in another. We can hope that someday the classification will be stabilized, but there may never be complete agreement between equally competent malacologists.

The arrangement in this book is taken from Johannes Thiele's *Handbuch der systematischen Weichtierkunde,* Volumes 1 and 2 (Jena, 1929–35), with some modifications. Within families the arrangement follows that adopted by the editors of *Johnsonia,* published by the Museum of Comparative Zoology, Harvard University. The listing of species within a genus is alphabetical in this book.

Nomenclature: The nomenclature of any branch of natural science is never static but always subject to change. Although name changes are often annoying to the amateur collector, they are necessary if we are to abide by the rules and eventually arrive at something resembling stability. The International Commission on Zoological Nomenclature is a body set up to formulate rules and pass judgment on such matters; and perhaps the most important rule is the law of priority — the first published name is to be the valid one, even if it is misspelled or misleading. A common clam from our East Coast carries the name of *Lucina pensylvanica,* with one *n,* and a helmet shell from the Caribbean is listed as *Cassis madagascarensis.*

Let us consider the reasons for name changes in a well-known eastern freshwater shell. An abundant ram's horn snail of ponds and streams was named *Planorbis bicarinatus* by Thomas Say in 1816. In 1834 Timothy Conrad described the same snail as *Planorbis antrosum.* Of course Say's name had priority, and *antrosum* became a synonym. Later it was discovered that Lamarck had used the name *bicarinatus* for a Paris Basin planorbid, before Say's publication, so *bicarinatus* became unavailable for our American shell, and the next in line, Conrad's *antrosum,* became official. In 1840 William Swainson erected the subgenus *Helisoma* for the planorbids of the New World, and in 1928 F. C. Baker elevated it to full generic rank, so our snail became *Helisoma antrosum* (Conrad). It was eventually discovered that the same mollusk had been described in a German publication by Karl Menke in 1830. Menke named it

Planorbis anceps, based upon the illustration in a work by Martin Lister, and since his is the earliest naming our common pond snail is now correctly known as *Helisoma anceps* (Menke).

The nomenclature in this *Field Guide* is believed to be as up to date as possible before time of publication, but in the years ahead, possibly in the months ahead, there undoubtedly will be additional name changes.

In many listings you will find species, subspecies, and varieties. A *variety* is simply a color form, or perhaps an unusually slender form, occupying the same geographic range as the *typical species,* and it has no taxonomic validity. A *subspecies,* on the other hand, differs slightly but constantly from the typical species; more important, it occupies a different geographic range, so that, except where the two forms meet and perhaps overlap a bit, there can be no intergrades. However, this law is not absolute, for we have what is termed a *sympatric subspecies.* This is a form slightly but demonstrably different from the typical, which though occupying the same geographic range is confined to a particular ecologic niche and ordinarily does not come in contact with the typical species.

Measurements: The *length of a bivalve* shell is the distance, in a straight line, from the anterior to the posterior end. The *height* is a straight line from the dorsal margin to the ventral margin (see right endpaper). It should be noted that with some shells, such as *Lima,* the height may be greater than the length. The *anterior* end of a bivalve is usually (but not always) the shorter; that is, it is closer to the beaks than the *posterior* end, and the beaks generally incline forward. In some cases, especially with orbicular shells, one has to see the interior to be sure. The foot protrudes from the anterior end and the siphons from the posterior. The pallial sinus, when present, opens toward the posterior end. Oysters, irregular in form but commonly elongate, are usually measured in length from the beaks to the opposite end.

Bivalves have *right* and *left* valves. When a clam is held with the dorsal margin up, and the anterior end away from the observer, the right valve is on the right and the left valve on the left. With sessile bivalves such as oysters and jewel boxes, one usually refers to the valves as upper and lower. The same is true of some of the scallops, particularly where one valve is deeply cupped (lower) and its mate is more or less flat (upper).

With *gastropods* it is always a question whether to give the distance between the apex and the opposite end as height or length. Since most of our snails customarily carry their shells in a horizontal or semihorizontal position, it is probably more correct anatomically to regard this measurement as length; but because virtually all illustrations of gastropods show them with the apex uppermost, that dimension will be given in this book as height. As with the bivalves, there will be some exceptions. With the *limpets* the height is the distance of the apex, or summit, above the place of the snail's attachment, and the length is a straight line from the anterior to the posterior end. *Cowries,* which crawl about on the apertural area and thus have an upper and lower surface, are measured from end to end as length, and length is also used for the maximum distance from apex to aperture in the elongate worm shells (see right endpaper).

Measurements in this book refer to average, mature shells. The collector will find many young examples that are smaller than the stated dimensions, and may obtain occasional specimens that exceed them.

Range and habitat: The ranges cited are based on records compiled over many years by collectors and marine laboratories of the region. Ranges may be extended from time to time as new material is collected and evaluated. For the purposes of this book the term *shallow water* ranges from the tidal area to a depth of about 30 feet, *moderately shallow water* from 30 to about 80 feet, *moderately deep water* from 80 to about 200 feet, and *deep water* anything beyond 200 feet.

Except for cases where space limitation prevented, the color of each shell is given on the legend page for each black and white plate. Where the periostracum is an important factor in the shell's color that fact is shown.

A Field Guide to Shells

of the Atlantic and Gulf Coasts
and the West Indies

Pelecypods

Veiled Clams: Family Solemyacidae

IN THIS FAMILY the valves are equal and considerably elongated. There is a tough and glossy periostracum (noncalcareous covering) extending well beyond the margins of the shell. These are rather uncommon bivalves, but the family is nearly worldwide in distribution and is represented on our East Coast from Canada to the West Indies.

Genus *Solemya* Lamarck 1818

NORTHERN VEILED CLAM   Pl. 9
Solemya borealis Totten
  Range: Nova Scotia to Connecticut.
  Habitat: Shallow water; burrows in sand and mud.
  Description: Fragile, attains a length of 3 in. Oblong; beaks scarcely elevated. Weak radiating ribs, nearly concealed by strong, shiny periostracum that overhangs margin of shell as a ragged fringe. Color tan to greenish brown; interior lead-colored.
  Remarks: Like its more common relative, Veiled Clam (below), it is a good swimmer, progressing by extending and rapidly withdrawing its foot, forcing water from the shell by spurts and traveling by a series of sudden dashes.

WEST INDIAN VEILED CLAM   Pl. 9
Solemya occidentalis Desh.
  Range: Florida to West Indies.
  Habitat: Moderately shallow water.
  Description: Length about ¼ in. An oval shell, rounded and gaping at both ends. Beaks inconspicuous, hinge toothless. Surface smooth, covered by shiny brown periostracum.

VEILED CLAM *Solemya velum* Say   Pl. 9
  Range: Nova Scotia to Florida.
  Habitat: Shallow water; burrows in sand and mud.
  Description: Fragile, 1 to 1½ in. long. Shape oblong, ends rounded. Surface smooth, with about 15 slightly impressed

double lines, most conspicuous on posterior margin. Greenish-brown periostracum firm and elastic, shiny; projects far beyond the shell itself, where it is slit at each of the radiating lines and gives edges a ragged, fringed appearance. Interior bluish white.

Remarks: Also known by such popular names as Swimming Clam and Awning Clam, this bivalve is easily recognized by the tough covering that overhangs the edge of the shell like a veil. Examples of *Solemya* are exceedingly fragile when dry and should not be kept loose in trays. Attach them by a drop of adhesive to a card cut to fit the tray, or place them on soft cotton. Specimens free to shift about as a drawer is opened and closed soon become mere shell fragments.

Nut Shells: Family Nuculidae

THREE-CORNERED or oval, small shells, with no pit for the ligament between the umbones (beaks). There is a series of tiny but distinct teeth on each side of the beak cavity. The inside is polished, often pearly, and the inner margins are crenulate. Distributed in nearly all seas but commonest in cool waters.

Genus *Nucula* Lamarck 1799

CANCELLATE NUT SHELL   Pl. 9
Nucula atacellana Schenck
    Range: Cape Cod to Virginia.
    Habitat: Moderately deep water.
    Description: Slightly under ¼ in. long. Shell oval, yellowish brown, inside only somewhat pearly. Surface cancellated by minute radiating and concentric lines.
    Remarks: Formerly listed as *N. cancellata* Jeffreys and *N. reticulata* Jeffreys.

DELPHINULA NUT SHELL   Pl. 9
Nucula delphinodonta Mighels & Adams
    Range: Labrador to New Jersey.
    Habitat: Moderately shallow water; muds.
    Description: Length less than ¼ in. Obliquely triangular, beaks near posterior end. Posterior end short, anterior sloping to rounded point. Hinge with 3 teeth behind and 7 before beaks. Surface smooth, color greenish olive; interior moderately pearly.
    Remarks: This tiny bivalve is usually common in the stomach of cod.

INFLATED NUT SHELL *Nucula expansa* Reeve   **Pl. 9**
Range: Arctic Ocean to Maine.
Habitat: Moderately shallow water; muds.
Description: Grows to length of nearly ½ in., although average is less. A swollen, oval shell with full beaks and rounded margins, the posterior end much longer than anterior. Hinge has 10 teeth in anterior series, 15 in posterior. Surface smooth and shiny, dark brown in tone; interior lead-colored, with some iridescence.

NEAR NUT SHELL *Nucula proxima* Say   **Pl. 9**
Range: Maine to Florida.
Habitat: Sheltered bays and harbors; muds.
Description: Slightly more than ¼ in. long and white, with a thin olive-green periostracum (commonly missing on beach specimens). Shell triangular, beaks somewhat elevated. Surface marked with scarcely perceptible striae (fine lines). Interior pearly and highly polished, margins crenulate. Teeth of hinge sturdy, the posterior series very distinct and regular.
Remarks: Our most abundant nut shell close to shore. Common in the stomachs of marine fishes; it is also eaten by various bottom-feeding ducks.

THIN NUT SHELL *Nucula tenuis* (Montagu)   **Pl. 9**
Range: Circumboreal; Labrador to Maryland.
Habitat: Moderately deep water.
Description: Thin and fragile, length ¼ in. Outline more oval than triangular. Hinge teeth usually number 4 or 5 before and about 8 behind the beaks. Color pale yellowish green; interior silvery white but not pearly.

VERRILL'S NUT SHELL *Nucula verrilli* Dall   **Pl. 9**
Range: Massachusetts to Mexico.
Habitat: Deep water.
Description: Length less than ¼ in. Sharply triangular, beaks pointed. Teeth robust. Color pale brown; interior glossy white but not pearly.
Remarks: A rare shell discovered by Professor Verrill in 1885 and named *N. trigona*. However, that name proved to have been used previously for another species, and in 1886 Dall renamed the bivalve *verrilli* in honor of the distinguished Yale zoologist. Should not be confused with another shell named Verrill's Nut Shell which is *Nuculana verrilliana* (p. 6).

Nut Shells and Yoldias: Family Nuculanidae

THESE were formerly classed with the Nuculidae. The shell is more or less oblong, usually rounded in front and prolonged into an angle behind. Margins are not crenulate. There are 2 lines of hinge teeth, separated by an oblique pit for the ligament. Distributed in nearly all seas, but chiefly in cool waters.

Genus *Nuculana* Link 1807

POINTED NUT SHELL *Nuculana acuta* (Conrad) **Pl. 9**
  Range: Massachusetts to West Indies.
  Habitat: Sandy mud just offshore.
  Description: Length ½ in. Anterior end broadly rounded, posterior prolonged to an acute tip. Shell sculptured with distinct concentric grooves. Hinge composed of a number of V-shaped teeth, interrupted by a central pit for the ligament. Color white, with a very thin brownish or greenish periostracum; interior white, highly polished. See Verrill's Nut Shell, *N. verrilliana* (below), for similarity.
  Remarks: A moderately uncommon shell in the north, it occurs rather frequently in the south, where living examples can often be obtained by sifting the bottom material just offshore.

FAT NUT SHELL *Nuculana buccata* (Steens.) **Pl. 9**
  Range: Circumpolar; Greenland to Maine.
  Habitat: Moderately deep water.
  Description: A stout little shell about ¾ in. long. Beaks moderately prominent, hinge teeth robust. Shell fairly thick, sculptured by closely spaced concentric lines. Color pale brown; interior white with little or no iridescence.

CARPENTER'S NUT SHELL **Pl. 9**
Nuculana carpenteri Dall
  Range: N. Carolina to West Indies.
  Habitat: Moderately deep water.
  Description: A narrow, elongate shell slightly less than ½ in. long. Anterior end broadly rounded, posterior drawn out to a small square tip. Dorsal line concave, the beaks closer to front end. Hinge teeth delicate and few. Surface quite shiny, with faint concentric lines. Color greenish gray.
  Remarks: A graceful small bivalve; not common and considered a prize in most collections.

TAILED NUT SHELL *Nuculana caudata* (Donovan) **Pl. 9**
Range: Maine to Virginia.
Habitat: Moderately deep water.
Description: This species is quite large for the genus, measuring a full ¾ in. Anterior end regularly rounded, posterior drawn out to a square tip. Weak concentric lines. About 10 teeth in front series, more than 20 behind. Color grayish green; interior not pearly.

CONCENTRIC-LINED NUT SHELL **Pl. 9**
Nuculana concentrica (Say)
Range: Florida to Texas; also Pleistocene.
Habitat: Moderately deep water.
Description: Slightly under ½ in. long, the shell thin and not inflated. Posterior end double the length of anterior and narrowed to a nearly acute tip. Surface with a series of concentric regular, equidistant, rounded striations. Hinge teeth angulated toward beaks. Color pale yellowish brown; interior white.

MINUTE NUT SHELL *Nuculana minuta* (Fabr.) **Pl. 9**
Range: Labrador to Maine; also Pacific Coast from Arctic Ocean to California.
Habitat: Moderately deep water.
Description: A small brown species only about ¼ in. long; interior white. Anterior end broadly rounded; posterior slightly prolonged, and narrowed at the tip, where it is abruptly truncate. Very fine concentric lines present on the valves.

MÜLLER'S NUT SHELL *Nuculana pernula* (Müller) **Pl. 9**
Range: Circumpolar; Greenland to Massachusetts.
Habitat: Moderately deep water.
Description: Length about ¾ in. Rather thin, with a prolonged but only slightly narrowed posterior end. Hinge teeth sharp. Sculpture consists of rather weak concentric lines. Color olive-green; interior white.

SULCATE NUT SHELL **Pl. 9**
Nuculana tenuisulcata (Couthouy)
Range: Gulf of St. Lawrence to Rhode Island.
Habitat: Shallow water; muds.
Description: Averaging about ½ in. long, this shell is quite thin, the valves scarcely inflated at all. Posterior end double the length of anterior and narrowed to a truncate tip. Surface with strong, concentric, closely spaced grooves. Color pale yellowish brown; interior white.
Remarks: Commonest of *Nuculana* on the New England

coast. Specimens are not often found on the beach, but individuals may sometimes be obtained by dragging in the soft mud well out beyond the low tide limits. A common shell in the stomach of cod.

VERRILL'S NUT SHELL *Nuculana verrilliana* Dall **Pl. 9**
Range: Off Florida Keys and Gulf of Mexico; West Indies.
Habitat: Deep water.
Description: A small shell, slightly more than ¼ in. long. Shaped much like Pointed Nut Shell (above), but less pointed at posterior end, and concentric lines far less conspicuous. Color white. Should not be confused with *Nucula verrilli,* also called Verrill's Nut Shell (p. 3).

Genus *Ledella* Verrill & Bush 1897

MESSANEAN NUT SHELL **Pl. 10**
Ledella messanensis (Seguenza)
Range: Massachusetts to West Indies.
Habitat: Moderately deep water.
Description: A tiny shell, usually less than ¼ in. long. Shape oval, moderately elongated; both ends rounded. Shell well inflated; full beaks only slightly closer to front end. A row of prominent teeth on each side of beak cavity. Surface smooth, color pale brown.
Remarks: Some malacologists regard *Ledella* as a subgenus of *Nuculana,* and list this species as *Nuculana (Ledella) messanensis.*

Genus *Yoldia* Möller 1842

ARCTIC YOLDIA *Yoldia arctica* (Gray) **Pl. 10**
Range: Greenland to Gulf of St. Lawrence.
Habitat: Moderately deep water.
Description: A rather squarish or oblong shell, well inflated, attaining length of nearly 1 in. Beaks quite prominent, located closer to anterior end, which is regularly rounded. Posterior truncate, basal margin sloping upward to form a blunt point. From 12 to 14 teeth in each series. Color lead-gray, with thin greenish periostracum; interior white.
Remarks: Fossil shells (Pleistocene) are to be found in New Brunswick and Maine.

IRIS YOLDIA *Yoldia iris* (Verrill & Bush) **Pl. 10**
Range: Gulf of St. Lawrence to N. Carolina.
Habitat: Moderately deep water.
Description: A minute shell, thin and glossy, less than ¼ in.

long. Shell rather short for height, rounded in front and bluntly pointed behind. About 12 teeth on each side of the small ligament pit. Color greenish, with brilliant iridescence.
Remarks: See Shining Yoldia (below).

FILE YOLDIA *Yoldia limatula* (Say) **Pl. 1**
Range: Gulf of St. Lawrence to New Jersey; West Coast from Alaska to San Diego.
Habitat: Moderately shallow water; muds.
Description: A handsome, shiny, light green clam 2 to 2½ in. long. Shell oval, elongate, and thin; beaks nearly central but low. Anterior and basal margins regularly rounded, posterior end drawn out to a pointed and partially recurved tip. Teeth along hinge prominent, about 20 on each side. Interior bluish white.
Remarks: This streamlined bivalve may be recognized by its length (more than twice as great as its height) and by the peculiarly upturned, snoutlike posterior tip. It is an active mollusk, living in mud but capable of swimming and leaping to an astonishing height.

SHINING YOLDIA *Yoldia lucida* (Lovén) **Pl. 10**
Range: Gulf of St. Lawrence to N. Carolina.
Habitat: Moderately deep water.
Description: This species is very much like Iris Yoldia, discussed above. It is perhaps a bit smaller, and a little less blunt. The color is more yellowish, but this shell also reflects brilliant prismatic colors. About 12 teeth in each series.
Remarks: Not likely to be found on the beach. It is shown here, along with Iris Yoldia, because it may be common to those who search the stomachs of bottom-feeding fishes for unusual shells.

OVAL YOLDIA *Yoldia myalis* (Couthouy) **Pl. 10**
Range: Labrador to Massachusetts.
Habitat: Moderately shallow water; muds.
Description: Smooth, elongate-oval, averages slightly less than 1 in. Beaks low, situated close to middle of shell. Both ends slightly rounded, posterior one somewhat narrower. About 12 teeth in each series. Color yellowish olive, and there is a thin periostracum that in fresh specimens is often arranged in alternate dark and light zones. Interior glossy yellowish white.

SHORT YOLDIA *Yoldia sapotilla* (Gould) **Pl. 10**
Range: Labrador to Massachusetts.
Habitat: Moderately shallow water; muds.
Description: Length 1 in. Oval, somewhat elongate, thin and fragile, green. Anterior end regularly rounded, posterior

end narrowed. Interior white, with about 50 teeth on each side of beak cavity.

Remarks: Found commonly in stomachs of fishes taken off New England coast; single valves are often found among the beach litter that marks the high-tide limits.

AX YOLDIA *Yoldia thraciaeformis* (Storer) **Pl. 10**
Range: Arctic Ocean to N. Carolina; West Coast from Arctic Ocean to Oregon.
Habitat: Moderately shallow water; muds.
Description: Length averages about 2 in. A strong, firm, squarish shell, broadest behind and gaping at both ends. An oblique fold extends from the beaks to posterior 3rd of basal margin, giving exterior a wavy appearance. Hinge with a spoon-shaped cavity for the ligament, about 12 robust teeth on each side. Color yellowish brown; interior pure white, not shiny.
Remarks: The general shape strongly suggests the blade of an ax. With its squarish outline, wavy surface, and salient rows of V-shaped teeth along the hinge line, this species is not likely to be confused with any other shell.

Genus *Malletia* Desmoulins 1832

DILATED NUT SHELL *Malletia dilatata* Phil. **Pl. 10**
Range: Off Florida.
Habitat: Deep water.
Description: Length ½ in. Valves well inflated, shape almost boxlike. Beaks somewhat swollen. Anterior end acutely rounded, posterior longer and abruptly truncate. About 10 strong teeth in forward series, about 16 in the rear. Color glistening white, inside and outside, and there is a sculpture of narrow but distinct concentric lines.

BLUNT NUT SHELL *Malletia obtusa* (Mörch) **Pl. 10**
Range: Massachusetts to N. Carolina.
Habitat: Deep water.
Description: Oblong, length ¾ in. Beaks very small. Anterior end short and evenly rounded, posterior longer and abruptly rounded. Dorsal line from the beaks to posterior tip is straight, the line in front of beaks gradually curved. About 16 teeth in front series, more than 30 in rear. Color greenish yellow in live specimens, pale straw color in dead shells.

OVATE NUT SHELL **Pl. 10**
Malletia subovata Verrill & Bush
Range: Nova Scotia to N. Carolina.
Habitat: Moderately deep water.

Description: About ½ in. long, a rather short and high shell. Beaks full, anterior end broadly rounded, posterior bluntly pointed. Basal margin presents a gentle curve. Teeth small. Color shiny yellowish green; interior of shell white.

Genus *Tindaria* Bellardi 1875

LOVABLE NUT SHELL *Tindaria amabilis* Dall   **Pl. 10**
Range: Florida and West Indies.
Habitat: Deep water.
Description: A solid, rather plump shell about ½ in. long. Both ends rounded, the posterior slope slightly longer and more pronounced. About 12 teeth in front series, nearly 30 in posterior; they are robust at the two ends and grade gradually to a very fine series under the beaks. Color dull yellowish; interior polished white.

Ark Shells: Family Arcidae

IN THIS GROUP the shell is rigid, strongly ribbed or cancellate, the hinge line bearing numerous teeth arranged in a line on both valves. Usually a heavy, commonly bristly periostracum (non-calcareous covering). There is no siphon. Some prefer to attach themselves by a silky byssus (threadlike anchor). Worldwide, from shallow water to considerable depths (1800 ft.).

Genus *Arca* Linnaeus 1758

MOSSY ARK *Arca imbricata* Brug.   **Pl. 10**
Range: N. Carolina to West Indies.
Habitat: Attached to rocks in shallow water.
Description: Some 2 in. long, boxlike and elongate. The Mossy Ark is well named: its shell is purplish white inside and outside, but the mollusk is almost completely covered with a dark brown, mossy periostracum. Beaks prominent and widely separated, with a broad area between them often scored with a geometric pattern. Anterior end rounded, the longer posterior end sharply carinated (keeled). Valves gape at ventral margin. Surface of shell irregularly cross-ribbed by growth lines. Margins smooth, not scalloped or crenulate.
Remarks: Formerly listed as *A. umbonata* Lam.

TURKEY WING *Arca zebra* Swain.   **Pl. 3**
Range: N. Carolina to West Indies; Bermuda.
Habitat: Attached to rocks in shallow water.

Description: From 2 to 4 in. long, the shell is sturdy, oblong, and gaping at both ends. Beaks slightly incurved and rather widely spaced, a broad and flat area between them. Hinge line perfectly straight, with about 50 small teeth. A colorful clam, fresh specimens yellowish white, irregularly streaked with reddish brown; interior pale lavender. On living examples the colors are fairly well concealed by a thick and shaggy periostracum.

Remarks: Formerly listed as *A. occidentalis* Phil.

Genus *Barbatia* Gray 1847

HAIRY ARK *Barbatia cancellaria* (Lam.) **Pl. 10**
Range: Florida to West Indies.
Habitat: Moderately shallow water.
Description: Another shaggy shell, from 1 to 2 in. long and purplish brown. Differs from the Mossy Ark, *Arca imbricata* (above), by being rounded at both ends and in lacking flattened area between beaks. Valves decorated with numerous fine, closely set radiating lines, each faintly beaded throughout its length. Hinge typically comblike, the teeth small and few. A heavy, hairy periostracum.
Remarks: Formerly listed as *Arca barbata* Linn., but this name belongs to a Mediterranean species.

BRIGHT ARK *Barbatia candida* (Helb.) **Pl. 10**
Range: N. Carolina to Texas; West Indies.
Habitat: Attached to stones in shallow water.
Description: Length 2 to 3 in. Shell compressed, rounded at anterior end and bluntly pointed at posterior. Surface sculptured with fine ribs and concentric growth lines, imparting a somewhat beaded appearance. Notch at ventral margin for byssus. Hinge teeth quite small. Color white, with soft and shaggy yellowish-brown periostracum.

RETICULATE ARK *Barbatia domingensis* (Lam.) **Pl. 10**
Range: Cape Hatteras to West Indies; Bermuda.
Habitat: Shallow water; under coral and stones.
Description: About 1 in. long. Strong and rugged, moderately inflated. Posterior end abruptly pointed, with a distinct ridge running from the posterior point to neighborhood of the beaks. Surface with strong radiating ribs that cut across stronger concentric beads and produce a distinctive network pattern. Hinge line short, teeth medium. Margins of shell crenulate. Color yellowish white; periostracum yellowish brown.
Remarks: Formerly listed as *Arca reticulata* Gmelin.

STOUT ARK *Barbatia tenera* (C. B. Adams) **Pl. 11**
Range: S. Florida to West Indies.
Habitat: Moderately shallow water.
Description: About 1½ in. long. Beaks swollen, front end the shorter. Shell thin and considerably inflated. Sculpture of fine radiating lines. Color white, with a thin brownish periostracum.

Genus *Arcopsis* Koenen 1885

ADAMS' ARK *Arcopsis adamsi* (E. A. Smith) **Pl. 11**
Range: Cape Hatteras to Brazil.
Habitat: Shallow water; under stones.
Description: Nearly ½ in. long, shape oblong. Both ends rounded, anterior end broadly and posterior more abruptly. Shell relatively thick, decorated with numerous very fine beaded lines. Beaks well elevated. Hinge line bears teeth that are moderately large but few. Inner margins smooth. Color yellowish white.

Genus *Anadara* Gray 1847

INCONGRUOUS ARK *Anadara brasiliana* (Lam.) **Pl. 11**
Range: N. Carolina to Brazil.
Habitat: Moderately shallow water; gravelly bottoms.
Description: Length 1½ to 2 in., height nearly the same. Inequivalve, one valve slightly overhanging the other. Considerably inflated. Beaks high and not well separated, so that in many cases there are specimens with the beaks showing signs of wear (frequently flattened). About 25 broad ribs with narrow grooves between them, the ribs crossed by equidistant lines, less conspicuous toward posterior end. Row of comblike teeth is graduated, the smaller teeth at center. Color white. See Chemnitz's Ark for similarity.
Remarks: Maxwell Smith has pointed out the interesting fact that when looking directly at the basal ventral margin, with the beaks directed away from the viewer, that margin presents a gentle curve instead of a straight line as in most bivalves. This species used to be known as *Arca incongrua* Say.

CHEMNITZ'S ARK *Anadara chemnitzi* (Phil.) **Pl. 11**
Range: S. Florida to West Indies.
Habitat: Moderately shallow water; gravelly bottoms.
Description: About 1½ in. long, this species appears at first sight quite like the Incongruous Ark. It is slightly inequi-

valve, short and high, and well inflated. There are some 25 rounded or flattened ribs, crossed by distinct grooves. Beaks more widely spaced and considerably higher than in Incongruous Ark, and ventral line lacks characteristic curve of that species. Color white.

Remarks: This is chiefly a West Indian shell, uncommon in Florida.

CUT-RIBBED ARK   Pl. 11
Anadara lienosa floridana (Conrad)
    Range: N. Carolina to Texas; West Indies.
    Habitat: Moderately shallow water.
    Description: From 3 to 5 in. long, a sturdy white shell with a brownish periostracum. Beaks low, slightly incurved, hinge line straight. About 35 radiating ribs that widen as they approach the margin, each rib bearing a deep central groove for most of the length.
    Remarks: Formerly listed as *Arca secticostata* Reeve. It is now recognized as a subspecies of *Anadara lienosa* (Say) — a fossil shell from Florida (Pleistocene) — so our clam has to be burdened with three names.

EARED ARK   *Anadara notabilis* (Röding)   Pl. 11
    Range: Bermuda; Florida to Brazil.
    Habitat: Shallow water; grassy and muddy bottoms.
    Description: Thick, heavy, sturdy, rather oblong, attains length of 3½ in. Beaks well elevated. Anterior end short and rounded, posterior longer and squarish. Hinge teeth numerous and small. About 25 strong radiating ribs crossed by thread-like lines. Color white, with a silky brown periostracum.
    Remarks: Known as the Eared Ark from the thin dorsal edge of the posterior tip. Formerly listed as *Arca auriculata* Lam., but that species comes from the Red Sea. It was also called *Arca deshayesi* Hanley, but Röding's name has priority.

BLOOD ARK   *Anadara ovalis* (Brug.)   Pl. 11
    Range: Massachusetts to West Indies.
    Habitat: Shallow water; sandy bottoms.
    Description: About 2 in. long, shell thick and solid, with prominent beaks terminating in points nearly in contact, so that viewed from the end the shell is heart-shaped. About 35 radiating ribs. Color white, but lower half of shell covered with a thick and bristly greenish-brown periostracum.
    Remarks: This clam derives its popular name from being one of the very few mollusks that have red blood. It used to be listed as *Arca pexata* Say, *A. campechiensis* Gmelin, and *A. americana* Wood.

TRANSVERSE ARK   *Anadara transversa* (Say)   Pl. 11
  Range: Massachusetts to Texas; West Indies.
  Habitat: Shallow water; sandy bottoms.
  Description: Length about 1 in. Shell transversely oblong, the beaks incurved and set closer to anterior end. Valves slightly unequal, so margin of one commonly passes a little beyond that of its mate. About 12 radiating ribs. Color white, the hairy periostracum dark brown.

Genus *Noetia* Gray 1857

PONDEROUS ARK   *Noetia ponderosa* (Say)   Pl. 11
  Range: Virginia to Texas.
  Habitat: Shallow water; sandy bottoms.
  Description: Length 2½ in. Shell somewhat oblique, quite thick and sturdy. Beaks prominent, inclined to turn backward. Some 30 strong ribs covered near the margins by deeply chiseled lines. The comblike teeth are larger at the ends, smaller at the center. Color white. In life the shell is covered with a heavy, velvety periostracum that is nearly black.
  Remarks: Single valves are now and then washed up on the beaches at Cape Cod, Nantucket, and Long Island, but these are believed to be fossil shells, and the species is not thought to be living north of Virginia at present.

Genus *Bathyarca* Kobelt 1891

BALL-SHAPED ARK   *Bathyarca glomerula* (Dall)   Pl. 11
  Range: N. Carolina to West Indies.
  Habitat: Deep water.
  Description: About ¼ in. long, the height nearly the same. Full beaks almost central in position. Whole margin of shell rounded, with a slight lean toward the posterior. Teeth relatively large. Surface with minute radiating and concentric lines. Color white, with brownish periostracum.

SCALLOPLIKE ARK   Pl. 11
  Bathyarca pectunculoides (Scacchi)
  Range: Gulf of St. Lawrence to Cape Cod.
  Habitat: Deep water.
  Description: Nearly ½ in. long, fairly well inflated. Anterior end short and abruptly rounded, posterior end curves out and back to the dorsal line. Beaks full, hinge teeth graduated — smallest at center. Yellowish brown, the surface with tiny striations.

Genus *Bentharca* Verrill & Bush 1898

DEEP-SEA ARK **Pl. 11**
Bentharca profundicola (Verrill & Smith)
 Range: Virginia to S. Carolina.
 Habitat: Deep water.
 Description: Small, elongate, somewhat oblique, length nearly ½ in. Anterior end short and oblique, posterior long and expanded. Beaks small and full. Shell covered with a rather coarse dark brown periostracum that rises into conspicuous scales, forming a fringe beyond the posterior margin. Surface bears weak concentric ridges under the periostracum. Color yellowish brown.

Limopsis: Family Limopsidae

THESE are chiefly small oval clams with a soft, furry periostracum. The beaks are nearly central, and the toothed hinge line is slightly curved. They are inhabitants of deep waters as a rule.

Genus *Limopsis* Sasso 1827

GREGARIOUS LIMOPSIS *Limopsis affinis* Verrill **Pl. 12**
 Range: Massachusetts to Florida.
 Habitat: Deep water.
 Description: Length nearly ½ in., height a trifle more. An oblique shell, moderately swollen. Valves rather plump. Inner margins crenulate. The shell is covered with a soft, yellowish-brown, hairy periostracum that forms a fringe at the margins, but after drying out the hairs are easily rubbed off, revealing a grayish surface.

EARED LIMOPSIS *Limopsis aurita* (Brocchi) **Pl. 12**
 Range: Gulf of Mexico.
 Habitat: Deep water.
 Description: Length ½ in. Obliquely oval, valves somewhat compressed. Sculpture of distinct concentric lines, especially marked on young shells, but during life concealed by a pale brown hairy periostracum. Beneath this shaggy coat the shell is white.

MINUTE LIMOPSIS *Limopsis minuta* (Phil.) **Pl. 12**
 Range: Newfoundland to Florida.
 Habitat: Deep water.

Description: Just under ½ in. long. Outline oblique like the others of this genus, the valves well compressed. It bears a velvety, furry periostracum that is brownish and underneath the surface has a weakly cancellate sculpture.

FLAT LIMOPSIS *Limopsis plana* Verrill **Pl. 12**
Range: Virginia to Florida.
Habitat: Deep water.
Description: One of the largest members of its genus, attaining a length of ¾ in. Shell well compressed and decidedly oblique. Dorsal margin short and straight; inside are 4 or 5 prominent teeth on each side of a conspicuous cartilage pit. Periostracum pale brown, bearing numerous rows of fine slender hairs that are crowded at the margins.

SULCATE LIMOPSIS **Pl. 12**
Limopsis sulcata Verrill & Bush
Range: Massachusetts to West Indies.
Habitat: Deep water.
Description: Length about ½ in., shape obliquely oval. Posterior margin prolonged and obtusely rounded, dorsal margin short and straight. Beaks small but prominent. About 18 strong teeth, the posterior series curving downward. Surface sculptured with rather strong concentric grooves marked by faint vertical lines. Margins crenulate. Color grayish, with brownish, furry periostracum.

Bittersweet Shells: Family Glycymeridae

THESE are rather solid, roundish, and well-inflated shells with strong hinges bearing curved rows of teeth. Living chiefly in warmer waters, they are also popularly known as button shells.

Genus *Glycymeris* Da Costa 1778

AMERICAN BITTERSWEET **Pl. 12**
Glycymeris americana Defrance
Range: Virginia to Texas.
Habitat: Moderately shallow water.
Description: About 1½ in. long in the northern section of its range, it grows to nearly 5 in. in the southern section. Shell circular, rather compressed, the beaks centrally located. There is a curving row of hinge teeth, becoming feeble or ab-

sent beneath the beaks. Surface bears numerous, fine striations; margins crenulate. Color grayish tan, sometimes weakly mottled with yellowish brown.
Remarks: A relatively rare species.

GROOVED BITTERSWEET Pl. 12
Glycymeris decussata (Linn.)
   Range: S. Florida, West Indies, and south to Brazil.
   Habitat: Moderately shallow water.
   Description: About 1¼ in. long. Outline orbicular, moderately inflated. Shell solid. Numerous radiating lines. Cream-colored, variously blotched with chestnut-brown. See Lined Bittersweet (below) for similarity.
   Remarks: Formerly listed as *G. pennaceus* (Lam.).

COMB BITTERSWEET Pl. 12
Glycymeris pectinata (Gmelin)
   Range: N. Carolina to West Indies.
   Habitat: Shallow water on a sand and gravel bottom.
   Description: Generally less than 1 in. long. Shell rather compressed, with about 20 well-rounded radiating ribs crossed by minute growth lines. About 10 teeth on each side of beak cavity. Color white to yellowish white, with spotted bands of yellowish brown.
   Remarks: This is a pretty shell when obtained fresh and in good condition. Specimens rolled about in the surf and left drying on the beach lose much of their delicate color.

LINED BITTERSWEET *Glycymeris undata* (Linn.) Pl. 12
   Range: N. Carolina to West Indies.
   Habitat: Moderately shallow water.
   Description: About 1¾ in. long. Shell solid, moderately inflated. Surface appears smooth but there are numerous fine radiating lines. Color pale cream to white, variously blotched with brown.
   Remarks: Easily confused with Grooved Bittersweet (above), but its radiating lines are finer and the beaks point toward each other; in *G. decussata* the beaks are inclined to point backward. It used to be known as *G. lineata* Reeve.

Mussels: Family Mytilidae

SHELLS with equal valves, the hinge line very long and the umbones (beaks) sharp. Some members burrow in soft mud, clay, or wood, but the majority are fastened by a byssus (threadlike anchor). Most mussels are edible, though seldom eaten in this country. In Europe the mussel is farmed much the same as the oyster is here, and it forms an important item of European seafood. Worldwide, and best represented in cool seas.

Genus *Mytilus* Linnaeus 1758

BLUE MUSSEL *Mytilus edulis* Linn. **Pl. 12**
 Range: Greenland to S. Carolina; Alaska to California; Europe.
 Habitat: Intertidal.
 Description: Length about 3 in. Shell roughly an elongate triangle, the beaks forming the apex. Anterior margin generally straight, posterior margin broadly rounded. Surface bears many fine concentric lines and is covered with a shining periostracum. Color bluish to bluish black; interior white, the margins violet. Young specimens are usually brighter-colored, and may be greenish or even banded or rayed.
 Remarks: Acres of blue mussels are exposed at low tide all along the Atlantic Coast to the Carolinas. We find them attached to stones and pebbles where the water is clear, on pilings of wharves, and in rocky places generally. They are attached by a series of strong byssal threads but are capable of moving about to some extent. This clam is a tasty morsel, much relished by those who have tried it; for some reason — probably the abundance of larger bivalves — it has never become very popular here, although it is eaten by the ton in Europe.

Genus *Modiolus* Lamarck 1799

TULIP MUSSEL *Modiolus americanus* (Leach) **Pl. 12**
 Range: N. Carolina to West Indies.
 Habitat: Moderately shallow water.
 Description: From 2 to 4 in. long, the anterior end is short and narrow, posterior end broadly rounded. Beaks anterior, not terminal. Shell rather thin in substance and moderately inflated. Surface smooth, periostracum glossy. Color yellowish brown; interior purplish.
 Remarks: Formerly listed as *M. tulipa* Lam.

RIBBED MUSSEL *Modiolus demissus* (Dill.) **Pl. 1**
 Range: Nova Scotia to Florida.
 Habitat: Intertidal.
 Description: Length 2 to 4 in. Shell moderately thin, oblong-oval, and much elongated. Surface ornamented with numerous radiating, somewhat undulating ribs that occasionally

branch. Color yellowish green to bluish green. Interior silvery white, often iridescent.

Remarks: The common mussel on muddy flats and in brackish waters, it seems to thrive best in partially polluted situations. Formerly listed as *M. plicatulus* Lam. Some authorities place this species in the genus *Brachidontes*. Florida specimens are sometimes listed as a subspecies, *M. demissus granosissimus* (Sby.).

HORSE MUSSEL *Modiolus modiolus* (Linn.) **Pl. 12**
Range: Arctic Ocean to n. Florida; Arctic to California.
Habitat: Moderately deep water.
Description: Length of adults 4 to 6 in. Shell heavy and coarse, oblong-oval, the beaks placed slightly to one side. Anterior end short and narrow, posterior broadly rounded. Surface marked by lines of growth, and sometimes a few radiating lines. Color bluish black, with a thick and leathery periostracum.
Remarks: This large mussel is generally considered unfit for food. The empty valves, noticeable on account of their size, are thrown up on nearly every beach exposed to the open sea. Bleached shells often turn pale lavender or reddish.

Genus *Brachidontes* Swainson 1840

YELLOW MUSSEL *Brachidontes citrinus* (Röding) **Pl. 12**
Range: S. Florida to West Indies.
Habitat: Moderately shallow water.
Description: About 1¼ in. long and fan-shaped on one side; this gives the outline a peculiarly lopsided appearance. Surface ornamented with numerous fine wavy riblets. Color yellowish brown; interior purplish, often with a metallic sheen.

SCORCHED MUSSEL *Brachidontes exustus* (Linn.) **Pl. 12**
Range: Cape Hatteras to West Indies.
Habitat: Moderately shallow water.
Description: About 1 in. long, this species is not as elongate as Yellow Mussel. Shell thin, fan-shaped on one side. Surface ribbed, strongest near margins. Color bluish gray, with a yellowish-brown periostracum.
Remarks: Commonly washed ashore in clusters attached to other shells and seaweeds.

BENT MUSSEL *Brachidontes recurvus* (Raf.) **Pl. 12**
Range: Cape Cod to West Indies.
Habitat: Shallow water.
Description: Length 1 to 2 in., triangular shell moderately solid, slightly inflated, strongly and obliquely curved. Surface

decorated with a pattern of fine, elevated lines that often divide as they approach posterior end. Color bluish black; interior polished, purplish with a whitish margin.

Remarks: Formerly listed as *Mytilus hamatus* Say. Living specimens have been taken in Long Island Sound and at Cape Cod, but it is believed that these were imported with young oysters from more southern waters, and that this mussel is unable to live through our northern winters.

Genus *Gregariella* Monterosato 1884

RIDGED MUSSEL Pl. 13
Gregariella coralliophaga (Gmelin)
  Range: West Indies.
  Habitat: Bores into rocks and corals.
  Description: Length ¾ in. Anterior end extremely short. Beaks prominent. Posterior long and with pronounced ridge, making that end sharply sloping. Hinge weak, dorsal margin with fine teeth. Color brownish to blackish; interior pearly.

ARTIST'S MUSSEL *Gregariella opifex* (Say) Pl. 13
  Range: N. Carolina to West Indies.
  Habitat: Bores into rocks and other shells.
  Description: A tiny mussel, usually less than ½ in. long. Outline somewhat cylindrical, the tips drawn out to a point and often frayed. Color reddish brown; interior highly iridescent.
  Remarks: This diminutive mussel sometimes bores into other shells or stone but more often it attaches itself to some support and becomes encased in a small round mound of cemented sand grains. Thomas Say, the discoverer of this pelecypod, wrote in 1826: ". . . on a single valve of *Pecten nodosus* were several elevations that, on a cursory glance, presented an appearance not unlike the *Balanus* [barnacle]. On a more particular inspection, each elevation proved similar to the others . . . and composed of fine dark sand agglutinated together, attached by a broad base to the surface of the *Pecten,* and rising in the shape of a very low cone around an intruded shell . . . with its byssus very firmly affixed to the supporting surface."

Genus *Amygdalum* Mühlfeld 1811

PAPER MUSSEL *Amygdalum dendriticum* Mühl. Pl. 13
  Range: West Indies.
  Habitat: Moderately shallow water.
  Description: Length to 1½ in. An elongate shell, very thin

and fragile. Front end short, posterior long and widening to a rounded tip. Hinge weak. Surface glossy, with delicate pale green periostracum.

Genus *Musculus* Röding 1798

DISCORDANT MUSSEL *Musculus discors* (Linn.) **Pl. 13**
Range: Labrador to Long Island Sound.
Habitat: Moderately deep water.
Description: A rather plump little mussel about 1 in. long, oblong-oval, and slightly produced at the posterior end. Weak radiating lines are discernible on both ends of the shell, and there is a slightly excavated channel across the middle of each valve. Color brownish black; interior bluish white and somewhat iridescent.

LATERAL MUSSEL *Musculus lateralis* (Say) **Pl. 13**
Range: N. Carolina to West Indies.
Habitat: Moderately shallow water.
Description: Length nearly ½ in. An oblong shell, moderately inflated, with radiating lines at both ends; the center shows only concentric growth lines. Color pale brown; interior slightly iridescent.

LITTLE BLACK MUSSEL *Musculus niger* (Gray) **Pl. 13**
Range: Arctic Ocean to N. Carolina.
Habitat: Moderately shallow water.
Description: Shell thin and oval, 1 to 2 in. long. Beaks close to front end. Rather prominent radiating lines at both ends, a relatively smooth area at center of each valve. Color deep brownish black, with a rusty brown periostracum; interior pearly.
Remarks: This mussel is rather more active than most of the others, and easily moves from place to place, using its foot as a prehensile organ and spinning a new byssus when a satisfactory situation has been found.

Genus *Crenella* Brown 1827

LITTLE BEAN MUSSEL *Crenella faba* (Müller) **Pl. 13**
Range: Greenland to Nova Scotia.
Habitat: Moderately shallow water.
Description: Length ¼ to ½ in. Oval, with a small portion of the hinge line straight. Anterior acutely rounded, posterior broadly rounded. Sculpture consists of numerous distinct radiating lines that make the margins crenulate. Color yellowish brown; interior lead-colored but polished.

GLANDULAR BEAN MUSSEL   **Pl. 13**
Crenella glandula (Totten)
  Range: Labrador to N. Carolina.
  Habitat: Moderately shallow water.
  Description: Slightly more than ½ in. long. Shape oval,
  beaks close to anterior end. Surface bears fine but distinct
  radiating lines crossed by even finer concentric lines. Color
  yellowish brown; interior bluish white.

Genus *Botula* Mörch 1853

DUSKY MUSSEL   *Botula fusca* (Gmelin)   **Pl. 13**
  Range: N. Carolina to West Indies.
  Habitat: Shallow water; bores into rocks and wood.
  Description: An oddly shaped shell, about ¾ in. long. Beaks
  situated at one end, which they sometimes overhang, producing a hooked effect. Shell rather cylindrical and somewhat
  curved. Surface smooth but coarsely wrinkled by growth
  lines. Dark chestnut-brown, with a shiny periostracum.

Genus *Lioberus* Dall 1898

CHESTNUT MUSSEL   *Lioberus castaneus* (Say)   **Pl. 13**
  Range: Florida to West Indies.
  Habitat: Shallow water.
  Description: About ¾ in. long, beaks close to front end.
  Shell slightly cylindrical. Color light brown; usually the pos-
  terior half is shiny and the posterior part is covered by a
  grayish, feltlike periostracum; interior lavender.
  Remarks: These mussels may be found living in clusters,
  attached to stones and dead shells.

Genus *Lithophaga* Röding 1798

GIANT DATE MUSSEL   **Pl. 13**
Lithophaga antillarum (Orb.)
  Range: S. Florida to West Indies.
  Habitat: Moderately shallow water; bores into limestone.
  Description: Length 2 to 4 in. Shell thin, elongate, cylindrical; wedge-shaped when viewed from above. Beaks low and
  insignificant, hinge line without teeth. Surface with numerous
  concentric furrows, most pronounced on posterior end. When
  young this species suspends itself from rocks by a byssus;
  when adult it forms a cavity in limestone and other moderately soft rocks which corresponds to the shape of its valves.
  Color brown, the thin periostracum also brown.

SCISSOR DATE MUSSEL Pl. 13
Lithophaga aristata (Dill.)
Range: S. Florida to West Indies and Brazil.
Habitat: Moderately shallow water; bores into soft rocks.
Description: A small rock borer about 1 in. long. Pale brown and cylindrical. This species is instantly recognized by the extension of the extremities of the valves, which cross each other at the posterior end of the shell.

TWO-FURROWED DATE MUSSEL Pl. 13
Lithophaga bisulcata (Orb.)
Range: N. Carolina to West Indies.
Habitat: Moderately shallow water; bores into soft rocks.
Description: About 1 in. long. A smooth and somewhat polished shell bearing a pair of radiating furrows from beaks to posterior end of shell. Anterior end bluntly rounded, posterior abruptly tapering. Color pale brown, but surface generally covered by a calcareous encrustation.

BLACK DATE MUSSEL *Lithophaga nigra* (Orb.) Pl. 13
Range: S. Florida to West Indies.
Habitat: Moderately shallow water; bores into coral.
Description: Length 1 to 2 in. Color dark brown or black. Cylindrical, surface with strong concentric growth lines, crossed by prominent vertical striations. Heavy, glossy periostracum.
Remarks: Fleshy parts of the animal are luminous in the dark.

Genus *Dacrydium* Torell 1859

GLASSY TEARDROP *Dacrydium vitreum* (Möller) Pl. 12
Range: Greenland to Florida.
Habitat: Deep water.
Description: This tiny bivalve is only slightly more than ⅛ in. long. Shape somewhat triangular, with a short and acutely rounded anterior end and a deep and broadly rounded posterior end. Valves thin, considerably inflated, the color a vitreous white — clean and shining.

Genus *Idasola* Iredale 1915

SILVER CLAM *Idasola argenteus* (Jeffreys) Pl. 13
Range: Labrador to Massachusetts.
Habitat: Deep water.
Description: Oblong shape, less than ¼ in. long. The ap-

pearance has been described as like a miniature *Arca,* but the hinge is toothless. Surface smooth, color pale brown.

False Mussels:
Family Dreissenidae

SMALL BIVALVES, these superficially resemble the mussels. There is a shelflike platform under the beaks for the attachment of the anterior muscle. Found in shallow waters, under brackish conditions.

Genus *Congeria* Partsch 1835

PLATFORM MUSSEL   **Pl. 13**
Congeria leucopheata (Conrad)
   Range: New York to Texas.
   Habitat: Brackish waters.
   Description: An elongate, partially curved shell about 1 in. long. Valves considerably inflated and bluntly keeled. Beaks terminal, the beak cavity containing a small plate, or platform, where the muscle is attached. Color grayish brown; interior dull white, not polished.

Genus *Mytilopsis* Conrad 1857

FALSE MUSSEL   *Mytilopsis domingensis* Récluz   **Pl. 20**
   Range: West Indies.
   Habitat: Brackish water.
   Description: Length to 1 in. Shaped very much like a *Mytilus:* elongate-oval, beaks rather sharp. Surface smooth, color grayish with thin brown periostracum.

Purse Shells:
Family Isognomonidae

THESE are greatly compressed shells, characterized by a hinge with vertical parallel grooves. The interior is pearly. Inhabitants of warm seas around the world.

Genus *Isognomon* Solander 1786

FLAT TREE OYSTER   *Isognomon alatus* (Gmelin)   **Pl. 13**
  Range: S. Florida to West Indies.
  Habitat: Mangrove roots and submerged brush near shore.
  Description: Length about 3 in. Shell greatly compressed, the right valve flat and the left but little inflated. Hinge line short, perpendicular grooves on inside. Surface may be smooth or scaly; color brown, black, or purplish; juvenile examples often rayed. Inside with pearly layer that does not extend all the way to the margins.
  Remarks: This genus may be found in old books under the names *Perna* Brug. or *Pedalion* Dill.

LISTER'S TREE OYSTER   **Pl. 13**
Isognomon radiatus (Anton)
  Range: S. Florida to West Indies.
  Habitat: Shallow water; attached to rocks.
  Description: Length to 3 in. Elongate, irregular in outline. Valves compressed, hinge with vertical grooves. Surface wrinkled, color greenish brown, commonly with paler rays; interior pearly.
  Remarks: Formerly listed as *I. listeri* (Hanley).

Pearl Oysters: Family Pteriidae

IN this group the shells are very inequivalve, the right valve with an opening under its wing for the passage of a byssus (threadlike anchor). This family contains the valuable pearl oysters. Found generally on rocks, sea fans, and other firm objects.

Genus *Pteria* Scopoli 1777

WINGED PEARL OYSTER   **Pl. 15**
Pteria colymbus (Röding)
  Range: N. Carolina to West Indies.
  Habitat: Shallow water; attached to sea fans.
  Description: Moderately solid, 1½ to 3 in. long. Hinge line straight, posterior margin broadly rounded, the wing strongly notched. Surface wrinkled; young specimens usually covered with prickly spines. Color brownish purple, with radiating lines of paler brown; interior pearly.
  Remarks: Formerly known as *Avicula atlantica* Lam., this is

a member of the pearl oyster group. The valuable shells of this group, however, are found in Ceylon, Australia, and the Persian Gulf, although some pearl fishing is done off Baja California, Panama, and n. S. America.

Genus *Pinctada* Röding 1798

SCALY PEARL OYSTER Pl. 15
Pinctada radiata (Leach)
    Range: S. Florida to West Indies.
    Habitat: Shallow water; attached to rocks.
    Description: About 2 in. long; larger in the southern part of its range. Valves flattish and nearly equal in size. Hinge line straight; byssal notch under right wing. Surface sculptured with scaly projections arranged concentrically, although these may be lacking in some specimens. Color variable, generally some shade of brown or green; interior very pearly.

Pen Shells: Family Pinnidae

LARGE wedge-shaped bivalves, thin and fragile, and gaping at the posterior end. They are attached by a large and powerful byssus, and live partially buried in mud or sand. Articles of wear such as gloves have been woven from the byssus of a Mediterranean species. There are large muscles in both valves. Occurring in warm seas, some grow to a length of more than 2 ft., and occasional specimens contain black pearls.

Genus *Pinna* Linnaeus 1758

FLESH PEN SHELL *Pinna carnea* Gmelin   Pl. 14
    Range: S. Florida to West Indies.
    Habitat: Shallow water; attached to pebbles in gravelly sand.
    Description: Thin, fragile, and wedge-shaped, more than 1 ft. long when fully grown, but most specimens are less than that. Posterior end rounded and gaping; shell tapers to a point at the other end, where the beaks are situated, so that one might almost say there is no front end. Valves decorated by a number of undulating, radiating folds with smaller lines in between. Color pale yellowish orange.
    Remarks: In the genus *Pinna,* on the inside there is a narrow groove down the center separating the iridescent portion into 2 lobes. In the genus *Atrina* (below) this groove is absent and the iridescent part is undivided.

Genus *Atrina* Gray 1842

STIFF PEN SHELL *Atrina rigida* (Sol.)   **Pl. 14**
   Range: N. Carolina to West Indies.
   Habitat: Shallow water; gravelly mud.
   Description: Large but delicate, growing to a length of 1 ft.
Shell wedge-shaped in outline — dorsal margin straight, ventral rounded, and posterior gaping. Valves decorated with about 15 rounded and slightly elevated ribs, which fade near beaks. Highly elevated tubular spines adorn the ribs, particularly near margins. There is a strong silky byssus. Color of shell purplish black.
   Remarks: The Spiny Pen Shell, *A. seminuda* (Lam.), shown on Plate 14, occupies about the same range and cannot be distinguished from the Stiff Pen Shell externally, although it is usually a more tan purple; differences are in the soft parts and muscle scars.

SAW-TOOTHED PEN SHELL   **Pl. 14**
Atrina serrata (Sby.)
   Range: N. Carolina to Florida.
   Habitat: Shallow water; sandy mud.
   Description: A yellowish-brown shell from 8 to 12 in. long. Shaped like Stiff Pen Shell but more delicately sculptured. Ribs closely set, with smaller but much more numerous scales.
   Remarks: Large "scallops" offered for sale in southern markets are likely to be cut from the muscles of one of these big pen shells.

Scallops: Family Pectinidae

THE VALVES are commonly unequal, the lower strongly convex, the upper flat or even concave. Surface usually ribbed and margins scalloped. Juvenile specimens may be attached by a byssus, but adults generally are free-swimming. The bivalve progresses through the water by a series of jerks and darts produced by rapidly opening and closing its valves, forcing a jet of water out from between the wings so that the mollusk travels with the hinge backward. This manner of locomotion is powered by a single large muscle — the only part of the scallop we eat. A row of tiny eyes fringes the mantle, each eye complete with cornea, lens, and optic nerve. Scallops are found in all seas, from shallow water to great depths. Shells of this family have always had a certain artistic appeal. The Crusaders employed the shell as a badge of honor, using a Mediterranean species (*Pecten jacobaeus* Linn.), and at present a scallop shell is the well-known trademark of a great oil company. All of the scallops used to be placed in the genus *Pecten;* the great diversity within the group, however, has led to the erection of many genera and subgenera.

Genus *Pecten* Müller 1776

RAVENEL'S SCALLOP *Pecten raveneli* Dall   **Pl. 15**
    Range: N. Carolina to West Indies.
    Habitat: Moderately shallow water.
    Description: About 2 in. long, a pinkish to purplish and occasionally golden-yellow scallop. Upper valve is quite flat, and deeply colored with irregular dark markings. Very convex lower valve decorated with about 25 strong, grooved ribs with wide spaces between them. Hinge line straight, the wings unequal; basal margin forms an almost perfect semicircle.
    Remarks: A rather uncommon species, sometimes confused with the Zigzag Scallop, but it is a smaller shell, more lightly colored, and the ribs on the flat valve are less crowded.

ZIGZAG SCALLOP *Pecten ziczac* Linn.   **Pl. 15**
    Range: Bermuda; N. Carolina to West Indies.
    Habitat: Moderately shallow water.
    Description: Length 2 to 4 in. This species has a flat upper valve, often with a central concavity, and a lower valve that is deep, cup-shaped, and overlaps the upper. Lower valve has low radiating ribs, so low that the surface seems rather smooth, and the color of this valve is a mottled brown, verging on reddish. The flat, heavily blotched upper valve bears crowded ribs and is ornamented with zigzag lines of black.
    Remarks: Sometimes mistaken for Ravenel's Scallop (see above).

Genus *Chlamys* Röding 1798

LITTLE KNOBBY SCALLOP   **Pl. 16**
Chlamys imbricata (Gmelin)
    Range: S. Florida to West Indies.
    Habitat: Moderately shallow water.
    Description: Length about 1½ in., the height somewhat more. Unusually flat, the valves scarcely arched at all. 9 or 10 stout ribs, each with a series of regularly spaced hollow knobs. Wings unequal size. Color mainly white, sometimes variegated with pinkish; interior yellowish, the margins and hinge area purplish.

ICELAND SCALLOP *Chlamys islandica* (Müller)   Pl. 16
Range: Greenland to Massachusetts.
Habitat: Moderately deep water.
Description: Length to 4 in. Shell oval, upper valve slightly more convex than the lower. Surface with about 50 narrow, crowded, radiating ribs bearing numerous small, erect scales; ribs frequently grouped, forming a number of unequal ridges. Wings very unequal, the posterior one shorter. Pale orange to reddish brown, lower valve paler tone. Interior glossy white, muscle scar large and shallow.

MINIATURE SMOOTH SCALLOP   Pl. 16
Chlamys nana (Verrill & Bush)
Range: N. Carolina to West Indies.
Habitat: Moderately deep water.
Description: About ½ in. long. A nearly orbicular shell, thin and delicate, wings about equal. Surface smooth, but there are microscopic radiating and concentric lines. Color translucent white; interior polished.

ORNATE SCALLOP *Chlamys ornata* (Lam.)   Pl. 16
Range: S. Florida to West Indies.
Habitat: Shallow water.
Description: About 1 in. long. Approximately 20 strong ribs, a few of which are usually unspotted, so they stand out as plain white lines. Ribs studded with short but sharp spines, especially near margins. Wings very unequal, one scarcely discernible. Color white, spotted with red and purple.

THORNY SCALLOP *Chlamys sentis* (Reeve)   Pl. 16
Range: S. Florida to West Indies.
Habitat: Shallow water.
Description: A small species 1 to 1½ in. long. Wings very unequal. Surface with numerous crowded ribs, each with tiny scales. Often bright scarlet, but may be brown, purple, or even white. Both valves generally colored alike.

Genus *Placopecten* Verrill 1897

DEEP-SEA SCALLOP   Pl. 16
Placopecten magellanicus (Gmelin)
Range: Labrador to N. Carolina.
Habitat: Moderately deep water.
Description: Commonly from 6 to 8 in. long. This scallop's shell is large, orbicular, somewhat higher than long, and moderately thick and solid. Lower valve nearly flat, upper but slightly convex. Surface sculptured with a multitude of narrow radiating lines and grooves, wings about equal in size.

Upper valve reddish or pinkish brown, sometimes rayed with white, lower valve pinkish white; interior glossy white, with very prominent muscle scar.

Remarks: The convex valves were commonly used as dishes by the Indians, and today visitors to our northern shores nearly always take home a shell or two to be used as ashtrays. This species has been listed as *Pecten tenuicostatus* Mighels and *P. grandis* Sol.

Genus *Lyropecten* Conrad 1862

ANTILLEAN SCALLOP Pl. 17
Lyropecten antillarum (Récluz)
  Range: S. Florida to West Indies.
  Habitat: Shallow water.
  Description: Small, rather flattish, not quite 1 in. long. Valves thin, about equal in shape and convexity. About 15 low ribs, with relatively wide spaces in between. Color buffy yellow, pale orange, or light brown, sometimes mottled with white.

LION'S PAW *Lyropecten nodosus* (Linn.) Pl. 2
  Range: Cape Hatteras to West Indies.
  Habitat: Moderately deep water.
  Description: Shell very robust and heavy, 4 to 6 in. long. Valves equal in size and shape and but little arched. Surface with numerous closely spaced radiating ribs, and about 10 broad folds, the crests of which are marked at regular intervals with blunt raised knobs. Interior has channels corresponding to the outside folds. Color reddish brown to bright orange; glossy interior generally some shade of pink or salmon.
  Remarks: A handsome shell eagerly sought by collectors and seldom found on shore. Single valves can be picked up at Sanibel Island after a storm, but most of the fine specimens seen are brought in by sponge fishermen.

Genus *Aequipecten* Fischer 1886

CALICO SCALLOP *Aequipecten gibbus* (Linn.) Pl. 2
  Range: N. Carolina to West Indies.
  Habitat: Shallow water.
  Description: Length 1 to 2 in., most specimens about 1½. Shell inflated, with about 20 rounded radiating ribs marked by numerous growth lines that give the shell a somewhat rough surface when not wave-worn. Wings about equal. Color patterns of this little clam are numerous; various com-

binations of mottled white, rose, brown, purple, and orange-yellow, the colors most striking on the upper valve.
Remarks: See Nucleus Scallop (below) for similarity.

BAY SCALLOP *Aequipecten irradians* (Lam.) **Pl. 17**
Range: Nova Scotia to Florida and Texas.
Habitat: Shallow water.
Description: Shell roughly round, well inflated, length 2 to 3 in. Wings nearly equal-sized, covered with small radiating ridges. Valves about equally convex. This species is divided into 3 subspecies, each illustrated on Plate 17. (1) Bay Scallop, *A. i. irradians* (Lam.), is found from Nova Scotia to Long Island; it has 17 or 18 rounded radiating ribs; color grayish brown, more or less mottled, and the valves are nearly the same color. (2) Southern Scallop, *A. i. concentricus* (Say), is found from New Jersey to Florida; it has 20 to 21 ribs that are somewhat squarish in cross section; slightly more inflated than the typical form and commonly more brightly colored, being orange-brown to bluish gray, occasionally with concentric bands; lower valve usually uncolored. (3) Gulf Scallop, *A. i. amplicostatus* (Dall), occurs in Gulf of Mexico along Texas coast; it is the stoutest of the 3 subspecies, with only 12 to 17 ribs; color mottled gray and black, the lower valve usually pure white.
Remarks: This is the common scallop of our East Coast; tons are dredged annually for the markets. It prefers to live among eelgrass. The unfortunate disappearance of eelgrass in many localities has led to a corresponding scarcity of scallops. Fishermen obtain them by dragging a rakelike affair through this grass.

ROUGH SCALLOP *Aequipecten muscosus* (Wood) **Pl. 2**
Range: N. Carolina to West Indies.
Habitat: Shallow water.
Description: About 2 in. long, the shell is sturdy, with wings of unequal size. Surface with about 20 strong ribs, each composed of a bundle of smaller ribs. Lower portions studded with erect sharp scales, giving the surface a rough, spiny appearance. Pinkish red to deep reddish brown; some individuals are bright lemon-yellow.
Remarks: Formerly listed as *Pecten exasperatus* Sby. The yellow examples are eagerly sought by collectors (see Plate 2).

NUCLEUS SCALLOP *Aequipecten nucleus* (Born) **Pl. 17**
Range: S. Florida to West Indies.
Habitat: Shallow water.
Description: About 1 in. long, well inflated. Very similar to Calico Scallop (above) but slightly more globose and with a few more ribs. Colors the same, except that the Nucleus

Scallop does not show the bright shades of red or orange.
Remarks: This scallop is sometimes listed as a subspecies,
A. gibbus nucleus Born.

SPATHATE SCALLOP Pl. 17
Aequipecten phrygium (Dall)
Range: Cape Cod to West Indies.
Habitat: Deep water.
Description: Length 1 in. Shell fan-shaped, rather flattish.
Both valves sculptured with about 18 ribs; interspaces nearly
equal. Each rib has a sharp median keel, made up of tiny
scales. Wings unequal. Color greenish gray, with irregular
bands of dull pink.

Genus *Amusium* Röding 1798

PAPER SCALLOP *Amusium papyraceus* (Gabb) Pl. 15
Range: Gulf of Mexico to West Indies.
Habitat: Moderately deep water.
Description: A rather flattish, smooth and glossy shell, some
2 in. long, each valve only slightly arched. Exterior polished
and without ribs of any kind, but there are distinct radiating
ribs on the inside. Upper valve reddish brown, lower one pure
white, sometimes with a yellow border. Inside, this valve
nearly always has a rim of bright yellow.
Remarks: Many authorities regard *Amusium* as a subgenus
and would list this bivalve as *Pecten (Amusium) papyraceus*
Gabb. Similar and larger species are to be found in tropic
seas around the world.

Genus *Pseudamussium* Mörch 1853

FRAGILE SCALLOP Pl. 17
Pseudamussium fragilis (Jeffreys)
Range: Off Cape Hatteras.
Habitat: Deep water.
Description: A small, paper-thin, roundly oval shell, the
length less than 1 in. Surface decorated with broad concen-
tric ridges. Color silvery white.
Remarks: An arctic bivalve, originally described from Euro-
pean material. It has been taken off our coast at a depth of
more than 9000 ft.

STRIATE SCALLOP Pl. 16
Pseudamussium striatus (Müller)
Range: Northern seas; off Massachusetts.
Habitat: Deep water.
Description: Small and delicate, about ¾ in. long. Valves

quite round, thin, and moderately inflated, the wings slightly unequal. Surface appears smooth but is decorated with numerous very fine radiating lines (striae). Color white, variously marked with pinkish or rose.

TRANSPARENT SCALLOP   Pl. 17
Pseudamussium vitreus (Gmelin)
   Range: Labrador to Massachusetts.
   Habitat: Deep water.
   Description: A nearly round shell, 1 in. across. Valves very thin and translucent and but little inflated. Wings unequal. Surface seems to be smooth and glossy, but under a lens it shows numerous very fine radiating lines and several rows of tiny beads arranged concentrically near the margins. Color silvery gray.

Genus *Propeamussium* Gregorio 1884

DALL'S SCALLOP   Pl. 17
Propeamussium dalli (E. A. Smith)
   Range: Florida to West Indies.
   Habitat: Moderately deep water.
   Description: About ½ in. long. Extremely thin and fragile, orbicular and considerably compressed. Outside smooth, with several low but distinct concentric ribs near margin, the interior bearing a number of prominent radiating ribs. Wings small and equal. Color silvery white.

File Shells: Family Limidae

THESE shells are obliquely oval and usually winged on one side. The ends gape and the hinge is toothless, with a triangular pit for the ligament. Color is generally white. Members of this family are most often called file shells. They are as competent in swimming as the scallops, but they dart about with the hinge foremost instead of backward, often trailing a long sheaf of filaments. Some of them build nests of broken shells and coral, held together by byssal threads, which act as an anchor. There are many fossil forms.

Genus *Lima* Bruguière 1797

SPINY FILE SHELL   *Lima lima* Linn.   Pl. 18
   Range: S. Florida to West Indies.
   Habitat: Shallow water.

Description: About 1½ in. long, sometimes slightly longer, with a white, moderately thick shell. Its rasplike surface does remind one of a file. Shell obliquely oval, ends only slightly gaping. Surface bears about 20 broad ribs, each with many closely set erect scales.

Remarks: This file shell occurs in one of its forms almost everywhere around the world in warm seas. *L. tetrica* Gould is found in the Gulf of California and *L. squamosa* Lam. is from the Mediterranean Sea.

INFLATED FILE SHELL Pl. 18
Lima pellucida C. B. Adams
 Range: N. Carolina to West Indies.
 Habitat: Shallow water.
 Description: Length 1 to 1½ in. Shell thin but sturdy, oblique, considerably inflated, and gaping at both ends, so that the valves are in contact only at the hinge and basal margin. Surface sculptured with fine lines, often with tiny lines in between. Color pure white.
 Remarks: Formerly listed as *L. inflata* Lam., which is a South American shell. It lives in crevices and under stones, attached by a byssus, and frequently constructs a crude nest of byssal threads, plastered with bits of seaweed and pebbles; but the mollusk can cast all this aside and go zigzagging off with speed and dispatch when the occasion demands action.

ROUGH FILE SHELL *Lima scabra scabra* Born Pl. 18
 Range: S. Florida to West Indies.
 Habitat: Shallow water.
 Description: About 2 in. long, but attains height of nearly 4 in. A moderately thick and robust shell, oval and not as oblique as most of this group. Valves rather compressed, gaping somewhat near hinge, and decorated with closely set ridges covered with small pointed scales. White in color, it usually bears a thin, yellowish-brown periostracum.
 Remarks: The species formerly known as *L. tenera* Sby. is now regarded as only a form of *L. scabra* and is listed as *L. s. tenera,* Delicate File Shell (Plate 18). It has the same shape, generally is slightly smaller, and the surface bears many very fine radiating lines notched by small, sharp scales, so the shell has a satiny luster. This form has the same range as *L. scabra.*

Genus *Limatula* Wood 1839

SMALL-EARED FILE SHELL Pl. 17
Limatula subauriculata Montagu
 Range: Greenland to West Indies.

Habitat: Moderately deep water.
Description: Length ½ in. Elongate-oval, hinge at top. Sculptured with very fine radiating lines. 2 lines, or ribs, along middle of shell are slightly stronger and more conspicuous, and these can usually be seen on the inside of the shell. Color white.

Genus *Limea* Bronn 1831

OVATE FILE SHELL *Limea subovata* (Jeffreys)   **Pl. 17**
Range: Greenland to Maine.
Habitat: Deep water.
Description: A very small shell, its length about ⅛ in. Rather strongly ribbed. Beaks at dorsal margin; ventral margin rounded. Valves fairly well inflated. Color pure white.

Spiny Oysters:
Family Spondylidae

THESE bivalves are attached to some object, commonly coral, by their right valves. The surface is ribbed or spiny and the hinge consists of interlocking teeth in each valve. Many species are highly colored. They are confined to warm seas.

Genus *Spondylus* Linnaeus 1758

SPINY OYSTER *Spondylus americanus* Hermann   **Pl. 18**
Range: Florida to West Indies.
Habitat: Moderately deep water; on coral.
Description: Heavy and strong-shelled, from 3 to 5 in. long. Color varies from white to brown and purplish, and sometimes bright yellow or red. Shell attached by its right valve, which has a broad triangular hinge area; crowded conditions often produce irregularly shaped individuals. Surface with many radiating ribs and numerous scattered spines, some short and needlelike and some long, broad, and blunt. Interlocking teeth are present in each valve, so a pair can be opened only partway without being fractured.
Remarks: Also known as Thorny Oyster and Chrysanthemum Shell, the latter because of the frilled, varihued appearance. Shells are sometimes found partially imbedded in chunks of coral washed ashore, and upper valves, badly worn,

are not uncommon on Florida beaches. The display specimens, however, are generally obtained by divers working with hammer and chisel on old wrecks and coral growths.

Cats' Paws: Family Plicatulidae

SMALL, trigonal, thick-shelled bivalves with broad radiating ribs or folds. They live attached to rocks and coral (by either valve) in warm seas.

Genus *Plicatula* Lamarck 1801

CAT'S PAW   *Plicatula gibbosa* Lam.   **Pl. 19**
  Range: N. Carolina to West Indies.
  Habitat: Intertidal.
  Description: About 1 in. long, although most individuals found on beach are smaller. Shell solid and fan-shaped, with 6 or 7 broad folds radiating from beaks. Right valve is larger of the two. Color white, often with gray or reddish lines.
  Remarks: Very common in the drift along shore, but the delicate pencil-like coloring fades rapidly when the shell is exposed to the sun's rays, and most specimens picked up will be a lusterless white.

SPINY CAT'S PAW   **Pl. 19**
Plicatula spondyloidea (Meus.)
  Range: Florida to West Indies.
  Habitat: Intertidal.
  Description: A larger shell than Cat's Paw, growing in clusters, one individual attached to another in crowded fashion. Hinge teeth sturdy, the valves not easily separated, so that one occasionally discovers clusters washed ashore that are intact.
  Remarks: Some modern authorities believe that this bivalve is simply a large variety of the abundant Cat's Paw, *P. gibbosa,* and not entitled to specific rank.

Oysters: Family Ostreidae

SHELL irregular, with unequal valves, and is often large and heavy. The lower valve usually adheres to some solid object; the upper valve generally is smaller. Distribution is worldwide in temperate seas as well as warm seas. Probably the most valuable of the food mollusks belong to this family.

Genus *Ostrea* Linnaeus 1758

CRESTED OYSTER   *Ostrea equestris* Say   Pl. 19
> **Range:** N. Carolina to West Indies.
> **Habitat:** Moderately shallow water.
> **Description:** About 2 in. long. Shell roughly oval, with raised and fluted margins. Color yellowish gray; interior dull gray with a greenish tinge. Muscle scar uncolored.
> **Remarks:** This species seems to prefer saltier waters than the commercial Common Oyster, *Crassostrea virginica* (below).

COON OYSTER   *Ostrea frons* Linn.   Pl. 19
> **Range:** N. Carolina to West Indies.
> **Habitat:** Mangrove roots and submerged brush.
> **Description:** Length 1½ to 2 in. Shell moderately thin and curved, and there is a broad longitudinal midrib with coarse folds from it to the margins. The attached valve bears several processes that clutch the stems of sea plants or tree roots. Color rosy brown to deep brown; interior white, usually with violet margins.
> **Remarks:** Coon Oysters are found growing together in huge masses sometimes larger than a bushel basket. They derive their name from the fact that raccoons delight in feeding upon them.

SPONGE OYSTER   *Ostrea permollis* Sby.   Pl. 19
> **Range:** N. Carolina to West Indies.
> **Habitat:** Moderately shallow water; in sponges.
> **Description:** A small, rather soft-shelled oyster, in the majority of cases living imbedded in masses of the "bread sponge." Slightly over 1 in. long, it is a golden-brown shell outside and bluish white inside. Shell compressed, variable in outline, with a narrow hinge. Lower valve flat, upper but slightly convex. Surface wrinkled by irregular wavy ridges, and covered by a periostracum that is thick but soft.

Genus *Crassostrea* Sacco 1897

CARIBBEAN OYSTER   Pl. 19
Crassostrea rhizophorae (Guild.)
> **Range:** West Indies.
> **Habitat:** Mangrove roots.
> **Description:** Length to 6 in. Very much like the Common

Oyster of our northern coasts. Shape variable, generally a well-cupped lower valve and flat upper valve. Color grayish; interior white with purple muscle scar.

COMMON OYSTER   Pl. 19
Crassostrea virginica (Gmelin)
Range: Gulf of St. Lawrence to Florida.
Habitat: Moderately shallow water.
Description: From 6 to 10 in. long and lead-gray. The rough and heavy shell is generally narrow, elongate, gradually widening and moderately curved, but it varies in surface and shape according to the position in which it has lain during growth. Upper valve smaller and flatter than lower and moves forward as the shell advances in age; growth of the ligament leaves a lengthening groove along beak of the adhering valve. Interior dull white, the muscle scar nearly central and deep violet.
Remarks: This well-known shellfish is our most important commercial bivalve. The tiny young (called spat) are free-swimming for a short period before they settle upon some hard object and become sessile for life. Those individuals that chance to settle in the mud perish, and to minimize the annual loss oystermen spread tons of broken shells (called clutch) over the beds each year. The oyster's chief enemies are the starfish and various snails, especially the Oyster Drill (*Urosalpinx cinerea,* see p. 194). The starfish wraps its 5 arms or rays around the unlucky oyster and exerts a steady pull that may last for hours, until the bivalve's muscles are exhausted and it is forced to gape a little; then the starfish is rewarded with a good meal. The Oyster Drill bores a neat round hole through the shell by means of its sandpaperlike tongue (radula), and feeds upon the succulent parts within. Contrary to popular belief, valuable pearls are not likely to be found in the valves of the Common Oyster. The shell lining is not pearly but is smooth and dull, so that any pearl formed is porcelaneous and without luster or iridescence.

Jingle Shells: Family Anomiidae

THESE are thin, translucent clams, usually pearly inside. They are attached to some solid object by a stalklike byssus (thread-like anchor) that passes through an opening in the lower valve; the byssus becomes calcified and the bivalve is permanently attached. The cuplike upper valve is the one that is washed ashore after the animal dies, and the perforated lower valve is not so often found. Native to warm and temperate seas.

Genus *Anomia* Linnaeus 1758

PRICKLY JINGLE SHELL   **Pl. 18**
Anomia aculeata Gmelin
    Range: Maine to N. Carolina.
    Habitat: Moderately shallow water.
    Description: About ¾ in. long and yellowish gray. Shell rounded, upper valve convex, lower thin and flat. Surface of upper valve covered with minute prickly scales, commonly arranged in radiating rows; lower valve with perforation near beaks for passage of byssus. Interior purplish white.

JINGLE SHELL   *Anomia simplex* Say   **Pl. 18**
    Range: Nova Scotia to West Indies.
    Habitat: Shallow water.
    Description: About 1 to 2 in. long and varies in color, which ranges from sulphur-yellow to coppery red, and many specimens silvery gray or black. Shell circular, and variously distorted according to object on which it is attached. Margins sometimes undulating or jagged. Surface minutely scaly and of a waxy luster. Upper valve convex; lower smaller, flat, with subcircular hole for passage of a fleshy byssus, by which the mollusk adheres.
    Remarks: These are greatly admired by children at the seashore, and perhaps are the most abundant and familiar shells on many beaches. It used to be a popular custom to string these shells on cords and hang them in an open window or doorway at a shore cottage, where each passing breeze produced a pleasing tinkle.

Genus *Pododesmus* Philippi 1837

FALSE JINGLE SHELL   *Pododesmus rudis* (Brod.)   **Pl. 18**
    Range: Florida to West Indies.
    Habitat: Moderately shallow water.
    Description: Length 1 to 2 in., sometimes longer. A moderately solid shell, the upper valve decorated with very coarse and irregular radiating ribs. Lower valve has a much larger hole than true jingle shells have. Pale yellowish white; interior greenish.
    Remarks: Formerly listed as *P. decipiens* Phil.

Arctic Hard-shelled Clams: Family Arcticidae

SHELLS large and thick, almost circular, the periostracum thick and wrinkled. There is no lunule (depressed area in front of beaks). Native to cold seas.

Genus *Arctica* Schumacher 1817

BLACK CLAM   *Arctica islandica* (Linn.)   **Pl. 19**
  Range: Arctic Ocean to Cape Hatteras.
  Habitat: Moderately deep water.
  Description: A large and robust pelecypod, 4 in. long when fully grown. Shell thick and heavy, roughly circular. Beaks elevated and turned forward, nearly in contact. Periostracum black or deep brown, coarse, shiny, and roughened with crowded and loose wrinkles. Interior white.
  Remarks: Also known as the Ocean Quahog. Large colonies are dredged south of Cape Cod and taken to the Cape to be cleaned, diced, and quick-frozen for shipment to hotels and restaurants.

Marsh Clams: Family Corbiculiidae

THESE are mollusks of brackish or semifresh waters. The shell is somewhat oval, and there is a rough periostracum, often eroded in places. They inhabit warm and temperate seas.

Genus *Polymesoda* Rafinesque 1820

CAROLINA MARSH CLAM   **Pl. 19**
Polymesoda caroliniana (Bosc)
  Range: Virginia to Texas.
  Habitat: Brackish water.
  Description: Oval, considerably swollen, length 1½ in. Valves more or less corroded in neighborhood of beaks, as a rule, owing to the mollusk's preference for living in brackish waters where acids are apt to be present. Shining greenish periostracum.

Remarks: At first glance this species appears like a typical river clam. It is generally abundant in tidal marshes and river-fed lagoons.

Genus *Pseudocyrena* Bourguignat 1854

FLORIDA MARSH CLAM   Pl. 19
Pseudocyrena floridana (Conrad)
  Range: Florida to Texas.
  Habitat: Brackish water.
  Description: About 1 in. long, shell thin but sturdy. Outline oval, beaks moderately inflated. Anterior end rounded, posterior prolonged. Hinge structure weak. Thin periostracum. Color purplish white, darker at margins. Interior may be white with purple margin or it may be solid purple.
  Remarks: Generally abundant in mangrove swamps.

Astartes: Family Astartidae

SMALL brownish pelecypods, usually sculptured with concentric furrows. The ligament is external, lunule distinct. Soft parts commonly bright-colored. There are many species, distributed chiefly in cool seas. This is a confusing group. The species making up the genus *Astarte* all are much alike but at the same time show some variation; many of them have been named and renamed, with the result that a shell could be listed under several different names in early conchological literature. Take for example the species *borealis:* this pelecypod was named by Schumacher in 1817; it was named *semisulcata* by Leach in 1819, *veneriformis* by Wood in 1828, *withami* by Smith in 1839, *richardsonii* by Reeve in 1855, *lactea* by Broderip in 1874, *producta* by Sowerby in 1874, *placenta* by Mörch in 1883, and *rhomboidalis* by Leche in 1883. Its correct name is *A. borealis* Schum.

Genus *Astarte* Sowerby 1816

NORTHERN ASTARTE   *Astarte borealis* Schum.   Pl. 20
  Range: Circumpolar; Greenland to Massachusetts.
  Habitat: Moderately shallow water.
  Description: Shell solid and oval, about 1¼ to 2¾ in. long. Beaks rather low, centrally located. Surface bears distinct concentric furrows on upperpart of each valve, but they fade toward the margins. Color deep brown; interior white.

CHESTNUT ASTARTE *Astarte castanea* Say **Pl. 20**
Range: Nova Scotia to Massachusetts.
Habitat: Moderately shallow water.
Description: Shell small but solid, rather compressed, about 1 in. long. Somewhat kidney-shaped, beaks nearly central and considerably elevated. Surface bears numerous concentric wrinkles but it lacks the deeper furrows so characteristic of most of this group, or has them only weakly defined. Shell covered with a thick chestnut-brown periostracum, often eroded near the beaks. Interior shiny white.
Remarks: In life the foot of the animal is bright vermilion, and when seen protruding from the partly open valves in shallow water it presents an extremely colorful sight.

STRIATE ASTARTE *Astarte striata* Leach **Pl. 20**
Range: Greenland to Massachusetts.
Habitat: Moderately shallow water.
Description: About ½ in. long. Shell oval-triangular, moderately stout. Beaks prominent, pointing forward, and the lunule (depressed area in front of beaks) is broad and deeply excavated. Surface marked by numerous closely spaced concentric ridges that are but slightly elevated. Periostracum dark brown; interior of shell white.

LENTIL ASTARTE *Astarte subaequilatera* Sby. **Pl. 20**
Range: Labrador to Florida.
Habitat: Water 80 to 180 ft. deep.
Description: Slightly more than 1 in. long. Anterior slope a bit concave, posterior end broadly rounded. Some 15 squarish concentric ridges, more or less obsolete toward posterior end. Margin finely crenulate within. Periostracum yellowish brown.
Remarks: Formerly listed as *A. lens* Stimpson.

WAVED ASTARTE *Astarte undata* Gould **Pl. 20**
Range: Labrador to Maryland.
Habitat: Moderately shallow water.
Description: Length 1¼ in. Shell robust and roughly triangular. Posterior slope rather straight. Beaks elevated and pointed. Surface decorated with about 15 strongly developed concentric ridges and furrows, widest and strongest at center of shell and vanishing at each end. A thick and glossy reddish-brown periostracum; interior of shell polished white.

Crassatellas:
Family Crassatellidae

SHELLS thick and solid, equivalve, and often rostrate (beaked) posteriorly. Strong hinge structure. Occur in shallow to moderately deep water.

Genus *Eucrassatella* Iredale 1924

GIBB'S CLAM   *Eucrassatella speciosa* (A. Adams)   **Pl. 20**
  Range: N. Carolina to West Indies.
  Habitat: Moderately shallow water.
  Description: Strong and robust, length 2 in. Anterior end rounded, posterior partially truncate and bearing a weak ridge. Beaks near center, not high. Surface sculptured with closely spaced concentric ridges. Inside, the hinge is very strong and the muscle scars deep. Color yellowish brown, with a thin periostracum; interior generally pinkish.
  Remarks: Formerly listed as *Crassatella gibbesii* Tuomey & Holmes, but the genus *Crassatella* is now restricted to fossil forms.

Genus *Crassinella* Guppy 1874

LUNATE CRASSINELLA   **Pl. 20**
Crassinella lunulata (Conrad)
  Range: Florida, Bahamas, West Indies.
  Habitat: Shallow water.
  Description: Length ¼ in. Shell somewhat triangular, posterior and anterior sharply sloping, basal margin rounded. Beaks pointed. Sculpture consists of relatively large and thick concentric ribs. Color white or pinkish.

THICKENED CRASSINELLA   **Pl. 20**
Crassinella mactracea (Linsley)
  Range: Massachusetts to New York.
  Habitat: Shallow water.
  Description: A small but rugged shell about ½ in. long. Shape triangular, beaks forming the apex. Anterior and posterior slopes pronounced, basal margin rounded. Hinge thick and sturdy. Surface has a few undulating concentric waves, often rather indistinct, and very minute radiating lines (striae). Color yellowish green; interior often brownish.
  Remarks: Formerly listed as *Gouldia mactracea* Linsley.

Carditas: Family Carditidae

SMALL, generally solid shells, equivalve, and usually strongly ribbed. There is an erect, robust tooth under the umbones (beaks). These pelecypods are found in warm, temperate, and cold seas.

Genus *Cardita* Bruguière 1792

DOMINGO CARDITA   *Cardita dominguensis* Orb.   **Pl. 20**
    Range: N. Carolina to Florida.
    Habitat: Shallow water.
    Description: An oval shell about ¼ in. long. Beaks set closer to front end, and valves moderately inflated. Sculpture consists of strong radiating ribs, some of which are weakly beaded, so surface has a rough appearance. Color pinkish white.

BROAD-RIBBED CARDITA   **Pl. 20**
Cardita floridana Conrad
    Range: Florida to Texas.
    Habitat: Shallow water.
    Description: About 1 in. long, sometimes a little more. Shell heavy and solid, bluntly oval. About 20 robust, scaly, radiating ribs. Yellowish white, blotched with purple and brown. Old individuals may be unspotted. Interior porcelaneous and white.
    Remarks: Many thousands of shells are used annually in the manufacture of shell novelties.

WEST INDIAN CARDITA   *Cardita gracilis* Shutt.   **Pl. 20**
    Range: West Indies.
    Habitat: Shallow water.
    Description: Length to 1½ in. An elongate shell, sturdy and strong. Anterior end rounded and short; posterior long and broadly rounded. Surface with strong, flattened radiating ribs, scaly on posterior. Color grayish brown; interior white, heavily stained with purplish brown.

Genus *Venericardia* Lamarck 1801

NORTHERN CARDITA   **Pl. 20**
Venericardia borealis (Conrad)
    Range: Arctic Ocean to Cape Hatteras.
    Habitat: Moderately shallow water.

Description: Length about 1 in. Shell thick and solid, beaks elevated and incurved. About 20 radiating ribs, wider than the spaces between them. Grayish white, with thick and shaggy brownish periostracum; interior glossy white.

FLAT CARDITA *Venericardia perplana* (Conrad)   **Pl. 20**
Range: N. Carolina to Florida.
Habitat: Moderately shallow water.
Description: Length about ¼ in. Shaped much like Northern Cardita, but smaller and less inflated. Color pinkish brown, often more or less mottled, interior white. No periostracum.

THREE-TOOTHED CARDITA   **Pl. 20**
Venericardia tridentata (Say)
Range: N. Carolina to Florida.
Habitat: Moderately shallow water.
Description: Length ¼ in. Trigonal in shape, considerably inflated. Beaks at apex of triangle and point forward. Hinge sturdy, dominated by 3 teeth. About 15 strongly beaded ribs. Grayish brown; interior white.

Diplodons: Family Diplodontidae

MOSTLY small, thin-shelled pelecypods, orbicular and well inflated. They generally are white, and live in the sands from shallow water to moderate depths.

Genus *Diplodonta* Bronn 1831

FLAT DIPLODON   **Pl. 20**
Diplodonta notata Dall & Simpson
Range: Florida to West Indies.
Habitat: Shallow water.
Description: Length about ½ in. Shell nearly orbicular, compressed. Beaks about central, fairly prominent. Sculpture of weak concentric lines, surface appears pitted. Color white.

WAGNER'S DIPLODON   **Pl. 20**
Diplodonta nucleiformis Wagner
Range: Florida to West Indies.
Habitat: Shallow water.
Description: Length ½ in. An orbicular shell, beaks somewhat swollen. Surface shiny, with minute concentric lines. Color white.

ATLANTIC DIPLODON Pl. 20
Diplodonta punctata (Say)
Range: N. Carolina to West Indies.
Habitat: Moderately shallow water.
Description: Length about ½ in. Very inflated, its shape almost globular. Surface appears smooth, but under a lens you can see fine concentric lines. Color pure white.
Remarks: Formerly in the genus *Taras* Risso 1826.

PIMPLED DIPLODON Pl. 20
Diplodonta semiaspera (Phil.)
Range: S. Florida to West Indies.
Habitat: Shallow water.
Description: Length about ½ in. A nearly orbicular shell, beaks about central. Well inflated. Sculpture is distinctly pustulate; color is chalky white.

VERRILL'S DIPLODON *Diplodonta verrilli* (Dall) Pl. 20
Range: Massachusetts to N. Carolina.
Habitat: Moderately deep water.
Description: A thin-shelled clam, highly inflated, with full beaks swollen at the umbones. Length ¼ in. Outline nearly round, surface quite smooth. Color pure white.
Remarks: Formerly in the genus *Taras* Risso 1826.

Cleft Clams: Family Thyasiridae

THESE are small bivalves with a furrow, or cleft, on the posterior portion of each valve. Usually white, the shell fragile. Chiefly from cool seas.

Genus *Thyasira* Lamarck 1818

GOULD'S CLEFT CLAM *Thyasira gouldi* (Phil.) Pl. 21
Range: Labrador to Cape Hatteras.
Habitat: Moderately deep water.
Description: Length ¼ in. Roundish outline, posterior slope with a weak cleft from beaks to the tip. No hinge teeth; shell very delicate and fragile. Color white, with thin yellowish periostracum.

CLEFT CLAM *Thyasira insignis* (Verrill & Bush) Pl. 21
Range: Newfoundland to Massachusetts.
Habitat: Moderately deep water.
Description: An unusually large and thick shell for this genus, the length commonly more than 1 in. Beaks moder-

ately prominent and pointed. Front end short and sharply rounded; posterior longer, broadly rounded, and bearing a shallow fold running from beaks to that margin. Color white, with thin yellowish periostracum.

ATLANTIC CLEFT CLAM   **Pl. 21**
Thyasira trisinuata (Orb.)
   Range: Nova Scotia to West Indies.
   Habitat: Moderately deep water.
   Description: An oblong shell ½ in. long. Shell quite delicate, hinge weak. Posterior slope bears a weak cleft. Surface smooth, grayish white.

Lucines: Family Lucinidae

SHELLS generally round, compressed, and equivalve, with small but definite beaks. Members of this family are distributed mostly in tropical and subtropical seas and are usually white.

Genus *Lucina* Bruguière 1797

DECORATED LUCINE   *Lucina amiantus* Dall   **Pl. 21**
   Range: N. Carolina to West Indies.
   Habitat: Moderately deep water.
   Description: Length about ½ in. Nearly round, moderately inflated. Very ornate; about 12 broad radiating ribs with narrow furrows between them, and these are cut by concentric ridges that are bladelike in character. A row of tiny nodes decorate the posterior slope. Pure white.

NORTHERN LUCINE   *Lucina filosus* Stimpson   **Pl. 21**
   Range: Newfoundland to Florida.
   Habitat: Moderately deep water.
   Description: Length 2 in. Shape nearly circular, hinge line nearly straight. Valves compressed, beaks small but prominent. Sculpture of sharp concentric ridges, rather widely spaced. Color white.
   Remarks: Unlike most of this group, it is a cold-water lover, and is found at increasingly greater depths as it passes southward. Southern specimens are usually not as large. Formerly in the genus *Phacoides* Gray 1847.

FLORIDA LUCINE   *Lucina floridana* Conrad   **Pl. 21**
   Range: W. Florida to Texas.
   Habitat: Shallow water.
   Description: Length about 1½ in. Valves thick and solid,

circular in outline and considerably compressed. Beaks small but prominent, inclined to turn forward. Fold from beak to posterior margin less conspicuous than for many of this group. Surface with fine concentric lines. White, with thin yellowish periostracum.

SULCATE LUCINE *Lucina leucocyma* Dall **Pl. 21**
Range: N. Carolina to West Indies.
Habitat: Moderately deep water.
Description: Small but sturdy, length slightly more than ¼ in. Shape somewhat triangular, with prominent beaks at apex. Shell with 5 broad folds, producing a lobed effect at the margins. Sculpture of fine concentric lines. Color white. Interior margins finely crenulate.

MANY-LINED LUCINE **Pl. 21**
Lucina multilineata Tuomey & Holmes
Range: N. Carolina to Florida.
Habitat: Shallow water.
Description: Length ¼ in. Beaks nearly central, shell nearly circular and well inflated. Surface with numerous fine, crowded, radiating lines. Color white.

FANCY LUCINE *Lucina muricata* Spengler **Pl. 21**
Range: S. Florida to West Indies.
Habitat: Moderately shallow water.
Description: A very pretty shell, circular and about 1 in. long. Valves well compressed, decorated with sharp concentric and radiating lines that cover the surface with a fine network of sharp ridges. Color pure white. See Woven Lucine for similarity.
Remarks: Formerly in the genus *Phacoides* Gray 1847.

WOVEN LUCINE *Lucina nassula* Conrad **Pl. 21**
Range: N. Carolina to Florida and n. Gulf Coast.
Habitat: Shallow water.
Description: Less than ½ in. long. Closely resembles Fancy Lucine but is smaller and slightly more inflated. Sculpture of fine but distinct radiating lines. Where the lines meet there is a sharp point; consequently the surface is almost filelike. Color white.
Remarks: Formerly in the genus *Phacoides* Gray 1847.

JAMAICA LUCINE *Lucina pectinatus* Gmelin **Pl. 21**
Range: N. Carolina to West Indies.
Habitat: Shallow water.
Description: Length about 2 in. Heavy and solid, roughly circular but little inflated. The characteristic fold is conspicu-

ous on the posterior end. Surface with many fine concentric ridges, rather widely spaced. Color pale yellowish white.
Remarks: Formerly in the genus *Phacoides* Gray 1847.

PENNSYLVANIA LUCINE   **Pl. 21**
Lucina pensylvanica (Linn.)
   Range: N. Carolina to West Indies.
   Habitat: Shallow water.
   Description: Length to 2 in. Nearly circular and moderately well inflated. A deep fold extends from the beak to posterior margin, giving impression of one shell cupped within another. Surface marked with widely separated concentric ridges. Color white, with pale yellow or brownish periostracum.
   Remarks: In the original description of this species in 1758, the name Pennsylvania was misspelled; so according to the rules of scientific nomenclature the specific name of this clam must be spelled with only one *n*.

Genus *Anodontia* Link 1807

BUTTERCUP   *Anodontia alba* Link   **Pls. 6, 21**
   Range: N. Carolina to West Indies.
   Habitat: Shallow water.
   Description: About 2 in. long, dull white, the interior bright yellow to orange. Shell strong, considerably inflated, with rounded margins; beaks low but prominent. Though surface appears smooth there are numerous very faint growth lines. Ligament bright red in living specimens.
   Remarks: Formerly listed as *Loripinus* or *Lucina chrysostoma* Phil. The Chalky Buttercup resembles the Buttercup but its interior is uncolored.

CHALKY BUTTERCUP   **Not illus.**
Anodontia philippiana (Reeve)
   Range: N. Carolina to West Indies; Bermuda.
   Habitat: Moderately shallow water.
   Description: 2 to 4 in. long. Shell strong and quite globose, with well-rounded and centrally placed beaks. Surface with fine concentric lines. Dull chalky white; interior uncolored.
   Remarks: Formerly listed as *Loripinus schrammi* Crosse. Similar to the Buttercup, but that species has brightly colored interior.

Genus *Myrtea* Turton 1822

LENTICULAR MYRTEID   **Pl. 21**
Myrtea lens (Verrill & Smith)
   Range: Massachusetts to Brazil.

Habitat: Moderately deep water.
Description: Length ½ in. A roundish shell with low beaks and very compressed valves. Posterior slope somewhat flattened. Surface rather smooth. Color dull white, with thin greenish periostracum.
Remarks: Originally described as *Loripes lens* Verrill & Smith.

Genus *Codakia* Scopoli 1777

COSTATE LUCINE   *Codakia costata* (Orb.)   **Pl. 22**
Range: N. Carolina to West Indies.
Habitat: Moderately shallow water.
Description: About ½ in. long, sometimes slightly longer. Valves more inflated than in Great White or Little White Lucines, and the lunule (depressed area in front of beaks) is proportionally smaller. Radiating ribs commonly arranged in pairs and crossed by fine concentric lines. Color white, sometimes tinged with yellow.

GREAT WHITE LUCINE   **Pl. 22**
Codakia orbicularis (Linn.)
Range: Florida to West Indies.
Habitat: Moderately shallow water.
Description: A handsome and showy shell about 3 in. long. Orbicular, solid, and but little inflated. Lunule small, heart-shaped. Beaks sharp and prominent, hinge teeth large and sturdy. Surface marked with many narrow radiating ribs crossed by elevated growth lines which give the shell a cross-ribbed appearance. Color white; sometimes with a border of pink or lavender on the inside.
Remarks: Also popularly known as the Tiger Lucine, after its former listing of *Lucina tigrina* (Linn.).

LITTLE WHITE LUCINE   **Pl. 22**
Codakia orbiculata (Montagu)
Range: N. Carolina to West Indies.
Habitat: Moderately shallow water.
Description: Length 1 in. Very much like a small edition of the Great White Lucine, but the lunule is elongate instead of heart-shaped. Shell sturdy, circular, and surface bears distinct radiating ribs crossed by numerous fine concentric lines. Pure white inside and outside.

Genus *Divaricella* Martens 1880

TOOTHED LUCINE   *Divaricella dentata* (Wood)   **Not illus.**
Range: West Indies.

Habitat: Moderately shallow water.
Description: Very like Crosshatched Lucine, but generally slightly larger, sometimes 1½ in. long, and having the inner margin smooth instead of crenulate.

CROSSHATCHED LUCINE Pl. 21
Divaricella quadrisulcata (Orb.)
    Range: Massachusetts to West Indies.
    Habitat: Moderately shallow water.
    Description: An odd shell, ivory-white and about 1 in. long. Shell moderately solid, circular, and rather plump. Surface sculptured quite unlike any similar shell, with prominent grooves bent obliquely downward at both ends. Inner margins of valves minutely crenulate.
    Remarks: See Toothed Lucine for differentiation.

Jewel Boxes: Family Chamidae

THICK, heavy, and irregular shells with unequal valves. They are attached to some solid object, the fixed valve being the larger and more convex. Native to tropical and subtropical seas.

Genus *Chama* Linnaeus 1758

CORRUGATED JEWEL BOX Pl. 22
Chama congregata Conrad
    Range: N. Carolina to West Indies.
    Habitat: Shallow water.
    Description: Usually less than 1½ in. long. Surface marked with wavy corrugations. Upper valve, the one not attached, often spiny. Color white, more or less mottled or streaked with purple. Interior generally reddish.

JEWEL BOX *Chama macerophylla* Gmelin Pl. 22
    Range: N. Carolina to West Indies.
    Habitat: Moderately deep water.
    Description: A thick and ponderous oysterlike bivalve 1 to 3 in. long. Shell irregularly rounded in outline, the adhering valve the larger of the two and also deeper-cupped. Surface sculptured with many distinct scalelike foliations. Pallial line extends from the outer edges of the muscle scars. Inner margin finely crenulate. Color varies from pink and rose to yellow.
    Remarks: These mollusks are usually firmly attached, so a hammer and chisel are necessary parts of the collector's equip-

ment. See Smooth-edged and Left-handed Jewel Boxes (below) for similarities.

LITTLE JEWEL BOX *Chama sarda* Reeve   **Pl. 22**
  Range: S. Florida to West Indies.
  Habitat: Shallow water.
  Description: About 1 in. long. Robust for its size. Surface with many wavy, tubular scales arranged like shingles and often frondlike in character. Lower valve white; upper sometimes with reddish rays.

SMOOTH-EDGED JEWEL BOX   **Pl. 22**
Chama sinuosa Brod.
  Range: S. Florida to West Indies.
  Habitat: Moderately deep water.
  Description: Length 2 to 2½ in. Almost identical with Jewel Box (above), but differing in a few ways: inner margins not crenulate; pallial line runs only to anterior and posterior muscle scars instead of past them as in the common Jewel Box. Color white; interior often greenish.

Genus *Pseudochama* Odhner 1917

LEFT-HANDED JEWEL BOX   **Pl. 22**
Pseudochama radians (Lam.)
  Range: S. Florida to West Indies.
  Habitat: Moderately deep water.
  Description: Length 1 to 3 in. Almost the same as Jewel Box (above), except that the adhering valve turns the other way. Members of the genus *Chama* are attached by their left valves, but in this genus it is the right valve that is anchored to a solid object. Color whitish or grayish; interior often brownish.

Genus *Echinochama* Fischer 1887

CARIBBEAN SPINY JEWEL BOX   **Pl. 22**
Echinochama arcinella (Linn.)
  Range: West Indies to S. America.
  Habitat: Shallow water.
  Description: About 1 in. long. Very much like the next species but usually smaller. Strong attachment scar on right valve. Color yellowish white, often marked with rosy purple, especially on inside.
  Remarks: This name was for many years applied to the common Spiny Jewel Box of Florida, which is correctly named *E. cornuta* (Conrad).

SPINY JEWEL BOX Pl. 22
Echinochama cornuta (Conrad)
    Range: N. Carolina to Florida and Texas.
    Habitat: Shallow water.
    Description: Length about 1½ in. Shell robust and solid, beaks curved forward. 7 or 8 strong ribs on each valve, spread fanwise, each with erect tubular spines throughout the length. Surface between ribs covered with beadlike pustules. Color white; interior often shows splashes of red.
    Remarks: It is attached to some solid object in its youth and later in life it becomes free, but the attachment scar (in form of a smooth area) is always present and visible just in front of the umbo on the right valve. Formerly listed as *E. arcinella* (Linn.), but that name is now reserved for a closely related species from the Caribbean region (see above).

Leptons: Family Leptonidae

THESE are small, fragile, generally inflated shells that often live attached to other marine organisms, commonly lobsters and crabs. Others live on vegetation, and some are free crawlers.

Genus *Mysella* Angas 1877

FLAT LEPTON *Mysella planulata* (Stimpson) Pl. 20
    Range: Nova Scotia to West Indies; Texas.
    Habitat: Eelgrass.
    Description: Only ⅛ in. long, an oval shell with both ends rounded, the anterior end the shorter of the two. Surface smooth and polished. Color white, with thin brownish periostracum.

Genus *Kellia* Turton 1822

RED KELLIA *Kellia rubra* (Montagu) Pl. 20
    Range: Bermuda, s. Florida.
    Habitat: Commonly nestles in barnacles.
    Description: Slightly less than ¼ in. long. Shape roundly oval, beaks full, valves well inflated. Shell fragile, color pinkish white.

Genus *Kelliella* Sars 1870

SHINING KELLIELLA *Kelliella nitida* Verrill Pl. 20
    Range: Massachusetts to Delaware.

Habitat: Moderately deep water.
Description: Length ¼ in. Considerably swollen, the shell quite plump. Outline rounded, beaks high and prominent. Shell fragile, white.
Remarks: The shell shown on Plate 20 is the type specimen described by Verrill in 1885.

Cockles: Family Cardiidae

SHELLS equivalve and heart-shaped, frequently gaping at one end. Margins of the shell are serrate or scalloped. Native to all seas, this family contains the cockles, or heart clams. In Europe these bivalves are regularly eaten and cockle-gathering is a recognized seaside activity, but they are not used as food to any extent in this country. For many years all of the shells belonging to this family were assigned to the genus *Cardium* Linn. 1758, and those found on our Atlantic Coast are listed in old books as *Cardium muricatum, C. islandicum,* etc. It is now recognized, however, that the genus *Cardium* should be restricted to those forms living in the e. Atlantic, and so our species have had their subgeneric names elevated to full generic rank.

Genus *Trachycardium* Mörch 1853

CHINA COCKLE Pl. 23
Trachycardium egmontianum (Shutt.)
    Range: N. Carolina to Florida.
    Habitat: Shallow water.
    Description: From 2 to 2½ in. long, valves well inflated and rather thin and oval. Moderately prominent beaks nearly central. Surface sculptured with deeply chiseled radiating ribs, the ends of which are studded with sharp recurving scales, most pronounced on anterior and posterior slopes. Yellowish to creamy white; interior salmon to salmon-purple, the anterior part whitish.
    Remarks: Very popular with seaside visitors. The graceful lines and delicate colors make it suitable for all sorts of articles, such as pincushions, ashtrays, shell flowers, etc. It used to be listed as *T. isocardia* (Linn.), but that name belongs rightfully to a slightly larger West Indian species (next).

PRICKLY COCKLE Pl. 8
Trachycardium isocardia (Linn.)
    Range: West Indies.
    Habitat: Shallow water.

Description: Length to 3 in. An elongate-oval shell, beaks well elevated. Sculpture of 30 or more radiating ribs that are set with sharp scales, almost becoming spines on posterior slope. Color creamy yellow, with irregular blotches of brown; interior coppery salmon.

COMMON COCKLE   **Pl. 23**
Trachycardium muricatum (Linn.)
    Range: N. Carolina to West Indies.
    Habitat: Moderately shallow water.
    Description: Length 2 in. Yellowish white, sometimes lightly speckled with brownish red, especially on the umbones. Interior yellow. Shell roundish and inflated, valves equal in size, and heart-shaped when viewed endwise. From 20 to 40 pronounced ribs, about a dozen of the central ribs almost or quite smooth over the umbonal region, the others crossed by erect, sharp scales. Margins of valves serrate and interlocking.

Genus *Dinocardium* Dall 1900

GREAT HEART COCKLE   **Pl. 23**
Dinocardium robustum robustum (Sol.)
    Range: Virginia to Texas.
    Habitat: Moderately shallow water.
    Description: Our largest member of this group, the length averaging between 3 and 5 in. Shell large and considerably inflated, with the posterior area flattened, dark, and polished. Beaks strongly rounded. About 35 robust flat ribs, regularly arranged. Margins of valves serrate. Color yellowish brown, irregularly spotted with chestnut and purplish marks; posterior slope brownish purple. Interior salmon-pink.
    Remarks: A form living on the western coast of Florida is generally larger, more triangular in outline, and brighter in color. This subspecies has been named Vanhyning's Heart Cockle, *D. r. vanhyningi* Clench & Smith (see Plate 23).

Genus *Trigoniocardia* Dall 1900

ANTILLEAN STRAWBERRY COCKLE   **Pl. 23**
Trigoniocardia antillarum Orb.
    Range: Florida to West Indies.
    Habitat: Moderately shallow water.
    Description: About ½ in long. Anterior end regularly rounded, posterior squarish. Surface with pronounced radiating ribs, crossed by scales near the outer margins. Color whitish, often blotched with reddish brown.

STRAWBERRY COCKLE   **Pl. 23**
Trigoniocardia ceramidum (Dall)
Range: West Indies.
Habitat: Shallow water.
Description: About ½ in. long. Solid, inflated, squarish posterior end with distinct slope. Sculpture of robust radiating ribs, knobby toward margins, which are scalloped. Color white.

Genus *Americardia* Stewart 1930

ATLANTIC STRAWBERRY COCKLE   **Pl. 23**
Americardia media (Linn.)
Range: Cape Hatteras to West Indies.
Habitat: Moderately shallow water.
Description: Length 1 in., sometimes a bit more; creamy white, more or less checkered with buff and purple. Shell solid, somewhat triangular; anterior margin regularly rounded and posterior margin partially truncate, forming a distinct slope on that end. Sculpture of strong, rounded radiating ribs.

Genus *Microcardium* Thiele 1934

EASTERN MICROCOCKLE   **Pl. 23**
Microcardium peramabile Dall
Range: Rhode Island to West Indies.
Habitat: Moderately deep water.
Description: A plump shell about ½ in. long. Shape quite round, with shell well inflated. Sculpture of numerous fine radiating lines, those on posterior slope spiny. Color white, at times somewhat mottled on posterior end.

Genus *Cerastoderma* Poli 1795

ELEGANT COCKLE   **Pl. 23**
Cerastoderma elegantulum (Beck)
Range: Circumpolar; Labrador.
Habitat: Moderately deep water.
Description: About ¾ in. long. Color white or yellowish white. About 25 ribs, strongly roughened by closely spaced arched and overlapping scales. Valves fairly well inflated.

LITTLE COCKLE   **Pl. 23**
Cerastoderma pinnulatum (Conrad)
Range: Labrador to N. Carolina.

Habitat: Moderately deep water.
Description: Length ½ in. Shell small, rather fragile, nearly orbicular. A blunt ridge passes from beaks to posterior tip. About 25 rounded ribs, each with a series of scales, most pronounced near the margin. Color dingy white; interior flesh-colored in fresh specimens.
Remarks: This little clam is an active one, said to scamper over the gravelly bottom with surprising ability by making expert use of its recurved, extensible foot.

Genus *Clinocardium* Keen 1936

ICELAND COCKLE *Clinocardium ciliatum* (Fabr.)   **Pl. 23**
Range: Greenland to Massachusetts.
Habitat: Moderately deep water.
Description: Shell large and well inflated, about 2½ in. long. Front end a little shorter and narrower than posterior. Surface bears about 38 sharp-edged radiating ribs, the furrows between them rounded and slightly wrinkled by lines of growth. Margins scalloped. Dull white; periostracum stiff and fringelike, brown. Interior straw-colored.
Remarks: Formerly listed as *Cardium islandicum* Brug.

Genus *Laevicardium* Swainson 1840

EGG COCKLE *Laevicardium laevigatum* (Linn.)   **Pl. 23**
Range: N. Carolina to West Indies.
Habitat: Moderately shallow water.
Description: Length about 1½ in. Shell thin and inflated, beaks rather small. Anterior end curved slightly more than posterior. Surface smooth and polished, but there are slight traces of ribs. Interior margin crenulate. Color ivory-white, usually with concentric bands of brownish orange. Thin brownish periostracum.
Remarks: Occasional specimens are found that when quite fresh are very brilliantly colored, but most beach specimens are plain white. The leaping ability of this clam is well established. One collector reports that a captive specimen made a successful getaway by using its powerful foot to leap from the boat. Formerly listed as *L. serratum* (Linn.).

MORTON'S EGG COCKLE   **Pl. 23**
Laevicardium mortoni (Conrad)
Range: Nova Scotia to Brazil.
Habitat: Shallow water.
Description: Height seldom exceeds 1 in., usually somewhat smaller; length ¾ in. Shell thin, inflated, obliquely oval. Surface weakly pebbled. Inner margins crenulate. Color yel-

lowish white, generally a little streaked with orange. Interior generally tinged with yellow, somewhat blotched with purple on the posterior side.

RAVENEL'S EGG COCKLE   Pl. 23
Laevicardium pictum Rav.
Range: Florida to West Indies.
Habitat: Moderately shallow water.
Description: Length about ¾ in. A roundish shell, well inflated. Front end regularly rounded, posterior slightly sloping. Surface smooth, color creamy white, with zigzag bars of yellow-brown. Interior yellow, the pattern often showing through.

DALL'S EGG COCKLE   Pl. 23
Laevicardium sybariticum Dall
Range: Florida to West Indies.
Habitat: Shallow water.
Description: Length ¾ in. Shell inflated. General shape squarish with slight posterior slope. Surface shiny, color whitish with brown scrawls; umbones pinkish. Interior pale yellow, variously marked with lilac.

Genus *Papyridea* Swainson 1840

SPINY PAPER COCKLE   Pl. 23
Papyridea soleniformis (Brug.)
Range: N. Carolina to West Indies.
Habitat: Shallow water.
Description: Length about 1½ in. Shell thin, compressed, somewhat elongate, gaping at posterior end. Beaks low. Many fine radiating ribs, smooth in center of shell but provided with short spines toward extremities and sometimes overhanging the margins. White or pink, heavily mottled with rosy brown — inside as well as outside.
Remarks: Formerly listed as *P. hiatus* (Meus.) and *P. spinosum* (Meus.).

Genus *Serripes* Gould 1841

GREENLAND COCKLE   Pl. 24
Serripes groenlandicus (Brug.)
Range: Greenland to Cape Cod.
Habitat: Moderately deep water.
Description: A fairly large clam, averaging in length between 3 and 4 in. Shell thin in substance but moderately inflated. Outline somewhat triangular, with beaks centrally located and slightly incurved. Anterior end regularly

rounded, posterior partially truncate and widely gaping. Surface bears numerous concentric lines, and several radiating ridges (most pronounced on the two ends). Color drab gray; juvenile specimens sometimes show a few zigzag darker lines.

Hard-shelled Clams: Family Veneridae

THIS is one of the largest pelecypod families and it probably has the greatest distribution, both in range and in depth. Named for the goddess Venus, the shells of this group are noted for their graceful lines and beauty of color and sculpture. The shells are equivalve, commonly oblong-oval in outline, and porcelaneous in texture. The mollusks are burrowers just beneath the surface of sand or mud, and are never fixed in one place. They are native to all seas, and since ancient times many of them have been used by man for both food and ornament.

Genus *Periglypta* Jukes-Brown 1914

LISTER'S VENUS *Periglypta listeri* (Gray)   **Pl. 24**
  Range: S. Florida to West Indies.
  Habitat: Shallow water.
  Description: Shell thick and solid, broadly oval, 2 to 4 in. long. Sculpture of impressed lines, crossed by concentric ridges that are both heavy and sharp. On posterior end these ridges often expand into flaring, bladelike structures. Color grayish white.
  Remarks: Formerly in the genus *Antigona* Schum. 1817.

Genus *Ventricolaria* Keen 1954

RIGID VENUS *Ventricolaria rigida* (Dill.)   **Pl. 24**
  Range: S. Florida to West Indies.
  Habitat: Shallow water.
  Description: Length 2½ in. Front end short and abruptly rounded, with definite lunule. Posterior regularly rounded, slightly squarish at tip. Sculpture of strong concentric ribs, with smaller concentric lines in between. Color yellowish gray, with brown mottlings.

QUEEN VENUS *Ventricolaria rugatina* (Heilprin)   **Pl. 24**
  Range: N. Carolina to West Indies.
  Habitat: Moderately shallow water.
  Description: Thick-shelled, nearly round, and about 1½ in.

long. Posterior end regularly rounded, anterior sharply rounded. Surface decorated with rather strong concentric ribs, each with a pair of smaller riblets between it and the next rib. Color yellowish white, somewhat mottled or marked with pale brown.

Genus *Mercenaria* Schumacher 1817

QUAHOG *Mercenaria mercenaria* (Linn.)   **Pls. 1, 24**
  Range: Gulf of St. Lawrence to Florida.
  Habitat: Shallow water.
  Description: Length 5 or 6 in. when fully grown. Shell thick and solid, rather well inflated; beaks elevated and placed forward. Surface bears numerous closely spaced concentric lines, most conspicuous near the ends; central portion of the valves smoother. Around umbonal region the lines are rather widely spaced. Color dull grayish; interior white, often with a dark violet border.
  Remarks: This bivalve has many other popular names, among them Round Clam, Hard-shelled Clam, Cherrystone Clam, and Little-necked Clam. It will be found in many lists under its old name, *Venus mercenaria*. This is the chief commercial clam of the East Coast, ranking second only to the oyster in shellfish value. When young or half grown it is the delicious "cherrystone," said to have a flavor surpassing that of any other bivalve. When older it is less tender and is used extensively for bakes and chowders. That it was a favorite food of the coast Indians is attested by the many shell heaps of ancient vintage found scattered from Maine to Florida and by the name Quahog, by which this clam is known in New England. The noted purple wampum was made from the colored edge of this shell; the fact may explain the pelecypod's name of *mercenaria*. It is not uncommon to find pearls under the mantle of this bivalve, sometimes of fair size but of no commercial value.
  Some malacologists recognize a subspecies, *M. mercenaria notata* (Say), which is smaller, lacks the purple border as a rule, and bears a weak pattern of brownish zigzag marks on the outer surface. *M. campechiensis* (Gmelin) is a very closely related species that grows to a larger size and is found from Virginia to Texas.

Genus *Chione* Mühlfeld 1811

CROSS-BARRED CHIONE   **Pl. 25**
Chione cancellata (Linn.)
  Range: N. Carolina to West Indies.
  Habitat: Shallow water.

Description: About 1¼ in. long, cloudy yellowish white, decorated with zigzag or triangular patches of purplish brown, sometimes in the form of radiating bands. Shell small, thick, and solid, the beaks elevated and situated forward. Surface sculptured with a series of well-elevated rounded ribs, crossed by concentric ridges of the same size, forming a network of raised lines. Inner margins crenulate. Interior usually purple.

CLENCH'S CHIONE *Chione clenchi* Pulley   **Pl. 25**
Range: Texas coast.
Habitat: Moderately shallow water.
Description: About 1 in. long. Anterior end short and abruptly rounded, the posterior longer and more sharply pointed than for most of this genus. Sculptured with broad and distinct concentric ribs, the grooves between them quite small. Color gray, spotted and streaked with pale purplish brown; interior white.

BEADED CHIONE *Chione granulata* (Gmelin)   **Pl. 25**
Range: S. Florida to West Indies.
Habitat: Shallow water.
Description: Length 1 in. Roundish in outline, posterior end but little pointed. Sculptured with close radiating ribs that are somewhat scaly. Color gray, with darker mottling; interior generally purplish toward posterior end.

GRAY PYGMY CHIONE *Chione grus* Holmes   **Pl. 25**
Range: S. Carolina to Florida.
Habitat: Moderately shallow water.
Description: Length ½ in. Rather oblong, posterior much longer than anterior. Surface bears numerous fine radiating lines crossed by equally fine concentric lines. Color dull gray, sometimes with pinkish tinge; interior glossy white with purplish stain on hinge at both ends.

MOTTLED CHIONE *Chione intepurpurea* Conrad   **Pl. 25**
Range: N. Carolina to West Indies.
Habitat: Moderately shallow water.
Description: About 1½ in. long. Shell strong and solid, ventral margins strongly convex. Beaks prominent but low. Many crowded concentric ridges, generally wrinkled over posterior half. Inner surface smooth, margins crenulate. Color white to cream; interior with a violet splotch on posterior portion.

BROAD-RIBBED CHIONE   **Pl. 25**
Chione latilirata Conrad
Range: N. Carolina to Texas; West Indies.

Habitat: Moderately deep water.
Description: Attains a length of nearly 2 in. but averages a little more than 1 in. Shell very solid and sturdy, well inflated, irregularly triangular, and highly polished. Surface bears a series of large and broadly rounded concentric ribs, with deep furrows between them, the ribs not pinched out at ends. Color grayish white, irregularly marked with lilac and brown.
Remarks: Its broad rounded ribs, plus the high gloss, render this pelecypod easy to identify, but wave-worn specimens found on the beach are apt to have lost their polish.

KING CHIONE *Chione paphia* (Linn.) **Pl. 25**
Range: S. Florida to West Indies.
Habitat: Moderately deep water.
Description: May be confused with Broad-ribbed Chione (above), but it is not as heavy in build. Grows to about the same size and has about 12 rounded concentric ribs that are not as broad, and they tend to pinch out at each end. Grayish, rather heavily marked with lilac and brown; shell has a high polish.

OLD VENUS CHIONE *Chione pubera* Val. **Pl. 25**
Range: S. Florida to Texas.
Habitat: Moderately deep water.
Description: A giant among the East Coast *Chione* tribe, sometimes attaining a length of nearly 3 in. Outline rather round, beaks nearly central. Sculpture has both radiating and concentric ribs, the latter forming deep grooves around the shell. Color gray, sometimes irregularly mottled with purplish brown. Interior white, hinge violet.

WHITE PYGMY CHIONE *Chione pygmaea* Lam. **Pl. 25**
Range: S. Florida to West Indies.
Habitat: Shallow water.
Description: Small, somewhat elongate, length about ½ in. Surface with delicate radiating and concentric lines; rather scaly. Color white, the teeth purple only on posterior half of hinge.

Genus *Anomalocardia* Schumacher 1817

WEST INDIAN POINTED VENUS **Pl. 24**
Anomalocardia brasiliana (Gmelin)
Range: West Indies to Brazil.
Habitat: Shallow water.
Description: Length 1¼ in. This bivalve is sometimes con-

fused with the next one, but it is a larger shell, heavier in build, and considerably less elongate. Color variable, generally some shade of yellowish gray with tan and purple speckling.

Remarks: It does not occur in the U.S.

POINTED VENUS ～ **Pl. 24**
Anomalocardia cuneimeris Conrad
    Range: S. Florida to Texas.
    Habitat: Shallow water.
    Description: Length about ¾ in. Shell small, thin, wedge-shaped, decidedly pointed at posterior end. Beaks slightly elevated. Surface of shell glossy, decorated with rounded concentric lines. Color varies from grayish white to greenish or brownish, often with darker mottlings. Interior pale lavender, margins crenulate.
    Remarks: See West Indian Pointed Venus for similarity.

Genus *Tivela* Link 1807

ABACO TIVELA *Tivela abaconis* Dall **Pl. 24**
    Range: Bahamas to West Indies.
    Habitat: Moderately shallow water.
    Description: Length ½ in. Roughly triangular, moderately inflated. Beaks elevated. Both ends gently rounded. Surface polished, color yellowish; umbonal region (beaks) rosy lavender.

TRIGONAL TIVELA *Tivela mactroides* (Born) **Pl. 24**
    Range: West Indies.
    Habitat: Shallow water.
    Description: About 1 in. long. Outline triangular, beaks swollen and elevated. Both ends sloping, basal margin rounded. Surface smooth, color yellowish brown with radiating darker rays. Interior white; purple under beaks.

Genus *Cyclinella* Dall 1902

SMALL RING CLAM *Cyclinella tenuis* (Récluz) **Pl. 24**
    Range: N. Carolina to West Indies.
    Habitat: Moderately shallow water.
    Description: Length 1½ in. A rather thin shell, nearly orbicular. Valves moderately inflated, beaks small and inclined forward. Surface smooth but not polished. Color white.

Genus *Transennella* Dall 1883

CONRAD'S TRANSENNELLA Pl. 26
Transennella conradina Dall
  Range: S. Florida to West Indies.
  Habitat: Moderately shallow water.
  Description: About ½ in. long. Front end rounded, posterior sloping to a rounded point. Surface shiny, color yellowish gray, often with zigzag scrawls of brown.
  Remarks: A characteristic of this genus is the shell's inner ventral margin: it presents a corded appearance, with numerous microscopic threads twisted along the edge (see Plate 26).

CUBAN TRANSENNELLA Pl. 26
Transennella cubaniana (Orb.)
  Range: S. Florida to West Indies.
  Habitat: Moderately shallow water.
  Description: Length ½ in. Somewhat triangular, anterior end slightly shorter than posterior. Basal margin rounded. Surface with very faint concentric lines. Color white, flecked with brown; umbonal region rosy on inside.
  Remarks: See Remarks under Conrad's Transennella.

STIMPSON'S TRANSENNELLA Pl. 26
Transennella stimpsoni (Dall)
  Range: N. Carolina to Florida and Bahamas.
  Habitat: Moderately shallow water.
  Description: About ½ in. long, the shape more triangular than with *T. conradina* (above), its posterior end less elongate. Surface smooth and polished, color creamy white with radiating brown bands sometimes discernible. Interior commonly purplish.
  Remarks: See Remarks under Conrad's Transennella (above).

Genus *Pitar* Roemer 1857

WHITE VENUS *Pitar albida* (Gmelin) Pl. 26
  Range: West Indies.
  Habitat: Moderately shallow water.
  Description: About 1½ in. long. Rather plump, posterior somewhat longer than anterior. Beaks not prominent. Color pure white, rarely with faint traces of radiating bands.

WEST INDIAN VENUS Pl. 26
Pitar aresta Dall & Stimpson
  Range: West Indies.

Habitat: Shallow water.
Description: Length 2 in. A roundish shell, very much inflated. Beaks prominent and curved forward. The only sculpture consists of fine, somewhat wavy growth lines. Color grayish white.

PURPLE VENUS *Pitar circinata* (Born)   **Pl. 26**
  Range: West Indies.
  Habitat: Shallow water.
  Description: Length 1½ in. Very much like Elegant Venus (below) but without spines. Surface with rather strong concentric lines. Color white, with radiating rays of purplish over the upperparts of shell.

CORDATE VENUS *Pitar cordata* Schwengel   **Pl. 26**
  Range: Off lower Florida Keys and Gulf of Mexico.
  Habitat: Moderately shallow water.
  Description: Length 1½ in. Rather plump, beaks full. Anterior end short and rounded, posterior longer, the tip quite pointed. Ornamentation of concentric lines that are sharp and distinct. Color grayish white.

ELEGANT VENUS *Pitar dione* (Linn.)   **Pls. 3, 26**
  Range: Texas and West Indies.
  Habitat: Moderately shallow water.
  Description: Length about 1½ in. Shell plump, with rather high beaks. Anterior end broadly rounded, posterior gently sloping. Surface bears many deeply cut concentric grooves, and a rather distinct ridge runs in an easy curve from beaks to posterior margin; this ridge has 1 or 2 rows of long spines. Color of whole shell, including spines, delicate lavender; interior white. See Purple Venus (above) for differences.
  Remarks: Considered by many to be the handsomest bivalve shell in w. Atlantic, it very nearly misses our shores. Formerly listed as *Venus dione,* this is the species used by Linnaeus in his original description of the genus *Venus* in 1758. It is a strikingly beautiful shell, not as fragile as it appears, and is generally considered a prize in any collection.

LIGHTNING VENUS *Pitar fulminata* (Menke)   **Pl. 26**
  Range: N. Carolina to Brazil.
  Habitat: Moderately shallow water.
  Description: Length about 1½ in. Roundly oval and plump, with small beaks, the posterior end only slightly longer than the anterior. Sculpture of very fine concentric lines. Color white, with brown or orange spots commonly arranged in a radial pattern. Interior polished white, the margins smooth.

FALSE QUAHOG *Pitar morrhuana* (Linsley) **Pl. 26**
Range: Prince Edward Island to N. Carolina.
Habitat: Shallow water.
Description: Length to 2 in. Shell roundish oval, rather thin, the valves quite convex. Anterior end about half the length of posterior. Margins regularly rounded behind and at base. Heart-shaped lunule (depressed area) in front of moderately elevated beaks. Surface smooth, color rusty gray; interior white.
Remarks: This shell appears something like a small, wave-worn Quahog (p. 59), but it is not as solid, has a smaller tooth structure, and lacks the purple border of the Quahog.

Genus *Callocardia* A. Adams 1864

TEXAS VENUS *Callocardia texasiana* Dall **Pl. 26**
Range: Florida to Texas.
Habitat: Moderately deep water.
Description: Length 1½ to 3 in. Beaks a little closer to anterior end, both ends sloping to rounded tips. Valves inflated and rather thin in substance. Surface smooth, color yellowish white.

Genus *Gouldia* C. B. Adams 1845

SERENE GOULD CLAM **Pl. 24**
Gouldia cerina C. B. Adams
Range: N. Carolina to West Indies.
Habitat: Moderately shallow water.
Description: Length ¼ in. Triangular; high but small beaks centrally located. Both ends rounded. Surface shows tiny concentric and radiating lines, but appears smooth and shiny. Color usually white; sometimes valves are boldly marked with chestnut-brown.

Genus *Macrocallista* Meek 1876

CHECKERBOARD *Macrocallista maculata* (Linn.) **Pl. 3**
Range: N. Carolina to Brazil.
Habitat: Shallow water.
Description: Length 2 to 3 in. long, shell roundish oval. Anterior end rounded, posterior sloping. Surface porcelaneous, color yellowish buff, with squarish spots of violet-brown distributed over whole shell, but with 1 to 2 radiating bands generally present. Interior white and polished, the pattern sometimes showing through. Margins smooth.

Remarks: The well-named Checkerboard is a handsome bivalve, very popular with collectors. It is most plentiful on the Florida west coast.

SUNRAY SHELL *Macrocallista nimbosa* (Sol.) **Pl. 3**
Range: N. Carolina to Florida and west to Texas.
Habitat: Shallow water.
Description: Large and showy, reaches a length of 6 in. Shell smooth, thick, porcelaneous, and elongate-oval, with depressed beaks. Anterior end short and rounded, posterior rounded and elongate. Surface glossy, but with very faint concentric and radiating striations. Inside polished with smooth margins. Color pinkish gray, with radiating lilac bands; in fresh specimens interior is salmon-pink.
Remarks: Half-grown 2- or 3-in. examples are generally more brightly colored than larger shells. The valves, although relatively thick, are quite brittle and break easily.

Genus *Dosinia* Scopoli 1777

WEST INDIAN DISK SHELL **Pl. 25**
Dosinia concentrica (Born)
Range: West Indies to Brazil.
Habitat: Moderately shallow water.
Description: Length 3 in. An orbicular, somewhat compressed shell with sharp beaks almost centrally situated. Sculpture of closely spaced concentric lines, about 9 to a centimeter. Surface shiny, color yellowish white.
Remarks: Very similar to Elegant Disk Shell (below) of Florida shores.

DISK SHELL *Dosinia discus* Reeve **Pl. 25**
Range: Virginia to Florida.
Habitat: Shallow water.
Description: A trim and neat shell, about 3 in. long, resembling Elegant Disk Shell. Circular and considerably compressed, with small but prominent beaks. Surface decorated with numerous and crowded concentric lines. Hinge thick and strong. Color glossy white, with thin yellowish periostracum.
Remarks: The periostracum peels away easily, revealing the shiny white shell.

ELEGANT DISK SHELL *Dosinia elegans* Conrad **Pl. 25**
Range: W. Florida to Texas.
Habitat: Shallow water.
Description: Very similar to Disk Shell, also pure white and up to 3 in. long. Surface bears numerous uniformly spaced

concentric lines, but not so crowded as in *D. discus*.
Remarks: W. J. Clench states (in *Johnsonia,* 1942) that *D. elegans* has from 8 to 10 lines to a centimeter and *D. discus* has 20.

Genus *Gemma* Deshayes 1853

GEM SHELL *Gemma gemma* (Totten) Pl. 24
Range: Nova Scotia to Florida and Texas.
Habitat: Shallow water.
Description: A diminutive shell only ⅕ in. long. Broadly triangular, with beaks about central. Surface shining, with minute, crowded, concentric lines. Inner margins crenulate. Color varies from grayish to lavender, shading to purplish on umbones (beaks).
Remarks: Several names have been applied to this shell, some of which may qualify as subspecific names. *G. gemma* is likely to be lavender over most of the shell; *G. manhattensis* Prime is a form found in w. Long Island Sound that lacks the purplish color, being mostly gray; *G. purpurea* Lea is whitish shading to purplish near the posterior end; *G. fretensis* Rehder is believed to be a synonym of *G. purpurea*.
Sometimes called the Amethyst Gem Shell. Early settlers in Massachusetts sent boxes of them back to England as curiosities. There was a time when it was believed that they were the young of the Quahog.

Genus *Parastarte* Conrad 1862

BROWN GEM CLAM Pl. 24
Parastarte triquetra (Conrad)
Range: Florida to West Indies.
Habitat: Shallow water.
Description: Length generally less than ¼ in. A solid little shell, higher than long, its shape triangular. Beaks elevated, prominent, situated at apex. Margins crenulate. Surface smooth, with minute concentric lines. Color white with a tinge of purple.

Genus *Liocyma* Dall 1870

WAVY CLAM *Liocyma fluctuosa* (Gould) Pl. 24
Range: Greenland to Maine.
Habitat: Moderately deep water.
Description: About ½ in. long. Valves thin, but little inflated. Beaks nearly central. Sculpture of concentric ridges

that fade near the margins. Anterior end shortest and broadest, both ends broadly rounded. Color white beneath a thin yellowish periostracum.

Rock Dwellers: Family Petricolidae

SHELLS elongate, gaping behind, with a weak hinge. They are burrowing mollusks, excavating cavities in clay, coral, limestone, etc. The cavity is gradually enlarged until the clam attains adult size. Distributed in warm, temperate, and even cooler seas around the world.

Genus *Petricola* Lamarck 1801

CORAL PETRICOLA   *Petricola lapicida* (Gmelin)   **Pl. 25**
Range: Bermuda, s. Florida, West Indies.
Habitat: Bores into coral rock.
Description: About ¾ in. long. An oval, well-inflated shell, anterior end short, posterior long and broadly rounded. Surface with weak concentric lines, and short radiating lines that often fork, producing a crosshatched pattern. Color dull chalky white.

FALSE ANGEL WING   **Pl. 25**
Petricola pholadiformis Lam.
Range: Prince Edward Island to Florida and Gulf of Mexico.
Habitat: Intertidal; mudbanks.
Description: Length 2 in. Shell thin, much elongated, and somewhat cylindrical. Anterior end very short and acutely rounded, posterior narrowed, elongate, and slightly gaping. Beaks elevated, with well-defined lunule in front. Surface marked by growth lines and strong radiating ribs. At posterior end ribs are crowded and faint, but at anterior they are large and widely spaced. Color chalky white.
Remarks: Burrowing into peat, mud, or stiff clay, this bivalve bears a striking resemblance to the large and showy Angel Wing, *Cyrtopleura costata* (Linn.); see p. 93.

Genus *Rupellaria* Fleuriau 1802

ATLANTIC RUPELLARIA   **Pl. 25**
Rupellaria typica (Jonas)
Range: N. Carolina to West Indies.

Habitat: Shallow water; bores into coral.
Description: About 1½ in. long. A plump and rough shell, rather elongate. Anterior end well rounded, posterior longer and gaping. Surface bears radiating lines, fine and close near ends of shell, with undulating lines of growth in between. Color grayish white; interior brownish.

Coral Clams: Family Trapeziidae

ELONGATED bivalves, with the beaks almost at the front end. The hinge has 3 cardinal teeth. Distributed in warm seas.

Genus *Coralliophaga* Blainville 1824

CORAL-BORING CLAM   Pl. 26
Coralliophaga coralliophaga (Gmelin)
    Range: Florida to Texas; West Indies.
    Habitat: In burrows, rocks, and coral.
    Description: A thin-shelled bivalve about 1 to 1½ in. long. Shell cylindrical, rounded at both ends, slightly gaping posteriorly. Valves bear faint radiating lines but surface appears fairly smooth. Color yellowish tan; interior white.
    Remarks: Often found in burrows of other mollusks, sometimes in company with the rightful owner.

Surf Clams: Family Mactridae

SHELLS with equal valves, usually gaping slightly at the ends. The hinge has a large spoon-shaped cavity for an inner cartilaginous ligament. Native to all seas, they live at moderate depths, commonly in the surf.

Genus *Mactra* Linnaeus 1767

WINGED SURF CLAM   *Mactra alata* Spengler   Pl. 26
    Range: West Indies.
    Habitat: Shallow water.
    Description: Length 4 in. An oval shell, rather thin and brittle. Beaks prominent. A moderate posterior slope bordered by a distinct ridge extending from beaks to the tip, where it frequently becomes finlike. Interior with spoonlike cavity at hinge. Color white, with thin yellowish periostracum.

FRAGILE SURF CLAM *Mactra fragilis* Gmelin   **Pl. 27**
Range: N. Carolina to Texas; West Indies.
Habitat: Shallow water.
Description: Length averages 2 in. Shell thin and moderately delicate, rather oval in outline. Sculptured with very close concentric lines. Beaks about central. Anterior end rounded, posterior with distinct radiating ridge. Color white, with thin yellowish periostracum.

Genus *Spisula* Gray 1837

STIMPSON'S SURF CLAM   **Pl. 27**
Spisula polynyma (Stimpson)
Range: Greenland to Rhode Island; also n. Pacific south to Puget Sound and Japan.
Habitat: Moderately shallow water.
Description: Closely resembles the Surf Clam. It does not grow quite as large, seldom exceeding 4 in., and it has a less regularly rounded ventral margin. Other points of difference are: lateral teeth (short and plain in this species and long and provided with tiny saw-toothed striations in *S. solidissima*) and the pallial sinus (much larger in *S. solidissima*). Color yellowish white.

SURF CLAM *Spisula solidissima* (Dill.)   **Pl. 27**
Range: Nova Scotia to S. Carolina.
Habitat: Moderately shallow water.
Description: A large and heavy species that resembles Stimpson's Surf Clam and attains a length of 7 in. Shell thick and ponderous in old individuals, roughly triangular. Beaks large and central, with a broad, somewhat flattened area behind them. Hinge very strong, with a large spoon-shaped cavity within, just under beaks. Surface smooth, or slightly wrinkled by growth lines. Color yellowish white, with thin olive-brown periostracum.
Remarks: Also known as the Hen Clam, it is the largest bivalve found on the north Atlantic Coast. Although not as popular as the Quahog, this species is regularly eaten and is considered excellent for clambakes. In the south its place is taken by a smaller subspecies, Southern Surf Clam, *S. s. similis* (Say), shown on Plate 27.

Genus *Mulinia* Gray 1837

LITTLE SURF CLAM *Mulinia lateralis* (Say)   **Pl. 27**
Range: Maine to Florida and Texas.
Habitat: Shallow water.

Description: Length ½ to ¾ in. Shell triangular, smooth, and polished, with beaks nearly central and inclined forward. Areas before and behind beaks are broad, flattened, roughly heart-shaped, and bordered by slightly elevated ridges. Yellowish white, with thin periostracum.

Remarks: An important food item for many of our marine fishes, as well as for our seagoing ducks. It is sometimes known as the Duck Clam.

PUERTO RICO SURF CLAM   Pl. 27
Mulinia portoricensis (Shutt.)
Range: West Indies.
Habitat: Shallow water.
Description: From 1 to 1½ in. long. General shape triangular and swollen, beaks about central. Anterior end sharply rounded, posterior slightly pointed, with distinct slope. Surface smooth, color whitish, with thin brownish periostracum.

Genus *Labiosa* Möller 1832

LINED DUCK *Labiosa lineata* (Say)   Pl. 27
Range: N. Carolina to Florida and Texas.
Habitat: Moderately shallow water.
Description: Some 3 in. long. Shell thin and swollen, with high beaks. Anterior end broadly rounded, posterior gaping and decorated with a cordlike ridge radiating from the beaks. Color white. There is a thin yellowish periostracum.

CHANNELED DUCK *Labiosa plicatella* (Lam.)   Pl. 27
Range: N. Carolina to Texas; West Indies.
Habitat: Moderately shallow water.
Description: About 3 in. long. Shell oval-orbicular, gaping, very thin and fragile. Posterior slope narrowed, beaks high, swollen, and directed backward. Sculpture of evenly spaced, rounded, concentric grooves. Inside, on hinge line, is a spoonlike cavity. Color pure white.
Remarks: Commonly washed ashore as single valves, but one is seldom rewarded by finding a complete specimen. It used to be listed in the genus *Raeta* Gray 1853 and *Anatina* Schumacher 1817, and its specific name was *canaliculata* Say.

Genus *Rangia* Desmoulins 1832

WEDGE RANGIA *Rangia cuneata* (Gray)   Pl. 30
Range: Maryland to Texas.
Habitat: Brackish water.

Description: Length about 1¾ in. A thick and solid shell, somewhat triangular. Beaks high and full, set close to the rounded anterior end. Posterior sloping and bluntly pointed. Color grayish white, with tough gray-brown periostracum; interior white and polished.

Wedge Clams:
Family Mesodesmatidae

OVAL or wedge-shaped shells with very short posterior ends. Hinge with a spoon-shaped cavity in each valve. Lateral teeth with a furrow.

Genus *Mesodesma* Deshayes 1830

ARCTIC WEDGE CLAM   Pl. 27
Mesodesma arctatum (Conrad)
  Range: Gulf of St. Lawrence to New Jersey.
  Habitat: Moderately shallow water.
  Description: Length 1½ to 2 in. Outline wedge-shaped, the very short posterior end forming base of the wedge. Anterior end narrowed and regularly rounded. Shell thick and strong, spoon-shaped cavity at hinge. Color brown, with yellowish periostracum that often reflects a metallic luster. Interior white, muscle impressions strongly delineated.

Genus *Ervilia* Turton 1822

SHINING ERVILIA   *Ervilia nitens* (Montagu)   Pl. 30
  Range: S. Florida to West Indies.
  Habitat: Moderately shallow water.
  Description: Not quite ½ in. long. A plumpish small shell, rather triangular in outline. Beaks close to center. Surface smooth, although there are weak concentric lines. Color white, more or less tinted with pinkish.

Tellins: Family Tellinidae

THE shells of this family are generally equivalve and rather compressed, the anterior end rounded and the posterior more or less pointed. The animals are noted for the length of their siphons. Several hundred species have been described, including many fossil forms. Within this group we find some of the most colorful, highly polished, and graceful of the bivalves. Native to all seas.

Genus *Tellina* Linnaeus 1758

DWARF TELLIN *Tellina agilis* Stimpson   **Pl. 28**
  Range: Gulf of St. Lawrence to N. Carolina.
  Habitat: Shallow water.
  Description: About ½ in. long, moderately elongated and well compressed. Posterior end sloping and unusually short, with a marked ridge running from beaks to that tip. Color white, sometimes yellowish or pink.
  Remarks: Formerly listed as *T. tenera* Say.

LINED TELLIN *Tellina alternata* Say   **Pl. 28**
  Range: N. Carolina to Gulf of Mexico.
  Habitat: Shallow water.
  Description: Length 2 to 3 in. Shell compressed, oblong, narrowed and angulated at posterior end. Anterior gracefully rounded. Surface decorated with numerous parallel, impressed, concentric lines, every other line vestigial on posterior area, which is marked by an angular ridge extending from the beaks to posterior margin. Commonly white, but may be pink or yellow.

ANGULAR TELLIN *Tellina angulosa* Gmelin   **Pl. 28**
  Range: S. Florida to West Indies.
  Habitat: Shallow water.
  Description: About 1½ in. Anterior end rounded, posterior pointed, valves compressed. Sculpture of fine concentric lines, sharp and crowded. Color delicate pink.

WEDGE TELLIN *Tellina candeana* Orb.   **Pl. 28**
  Range: Bermuda, Florida, West Indies.
  Habitat: Moderately shallow water.
  Description: Length ⅜ in. Posterior short and sloping, anterior acutely rounded. Surface polished, but under a lens one can see oblique lines. Color white.

CARIBBEAN TELLIN *Tellina caribaea* Orb.   **Pl. 28**
  Range: West Indies.
  Habitat: Shallow water.
  Description: About 1 in. long. Elongate, thin, and fragile. Posterior end short and rounded, anterior longer and broadly rounded. Surface shiny, with weak concentric lines that are

crossed by slanting cuts, but one needs a hand lens to observe this odd sculpture. Color white or pinkish.

CRYSTAL TELLIN *Tellina cristallina* Spengler **Pl. 28**
Range: S. Carolina to West Indies.
Habitat: Moderately deep water.
Description: About 1 in. long. Anterior end rounded, posterior with straight slope down to a slightly upturned tip that is square. Valves thin and well inflated. Surface shiny, with well-spaced concentric lines. Color transparent white.
Remarks: This species also occurs on the West Coast, ranging from the Gulf of California to Ecuador.

GEORGIA TELLIN *Tellina georgiana* Dall **Pl. 8**
Range: S. Florida to West Indies.
Habitat: Moderately shallow water.
Description: Length 2½ in. Anterior end longer of the two, but the prominent beaks are almost centrally located. Surface with sharp and crowded concentric lines. Color deep pink.

IRIS TELLIN *Tellina iris* Say **Pl. 28**
Range: N. Carolina to Florida.
Habitat: Shallow water.
Description: Length ½ in. Valves compressed. Anterior rounded, posterior rather sharply pointed. Low beaks nearly central. Surface glossy, color white, the valves thin and more or less translucent.

SMOOTH TELLIN *Tellina laevigata* Linn. **Pl. 28**
Range: Florida to West Indies.
Habitat: Shallow water.
Description: About 2 in. long. Outline roundish, anterior end regularly rounded, the slightly shorter posterior end bluntly pointed, with a distinct angle running from the beaks to that tip. Valves moderately inflated. Surface smooth, often glossy. Color white with pale orange rays.

ROSE PETAL TELLIN *Tellina lineata* Turton **Pl. 6**
Range: N. Carolina to West Indies.
Habitat: Shallow water.
Description: About 1 in. long and rosy pink, the umbonal area usually darker in hue. Interior same color as outside. Shell smooth and delicate, nearly as high as long. Anterior end rounded, posterior slightly pointed.
Remarks: A common species, these dainty rose and pink valves — looking like so many rose petals — are washed ashore with every tide. Gathered extensively for the making of shell flowers and other novelties.

SPECKLED TELLIN *Tellina listeri* Röding   **Pl. 3**
Range: N. Carolina to Brazil.
Habitat: Shallow water.
Description: Length 1 to 2 in. Shell rather long and thin, not polished. Anterior end rounded, posterior end rostrate (beaked). Beaks nearly central and low. Surface sculptured with strong, equidistant, concentric lines. Color creamy white, with crowded streaks of brownish purple. Interior polished white.
Remarks: This species has long been listed as *T. interrupta* Wood.

GREAT TELLIN *Tellina magna* Spengler   **Pl. 28**
Range: N. Carolina to West Indies.
Habitat: Moderately shallow water.
Description: Our largest tellin, attains a length of 4½ in. Both ends slope to a rounded ventral margin, with posterior end more pointed. Valves rather thick, hinge strong. Surface marked by growth lines. Color white on left valve, yellowish or orange on the right. Interior frequently pinkish.

MARTINIQUE TELLIN *Tellina martinicensis* Orb.   **Pl. 28**
Range: Florida to West Indies.
Habitat: Moderately shallow water.
Description: Nearly ½ in. long. A moderately inflated shell, posterior end pointed and anterior rounded. Beaks central. Surface shiny, color white.

MERA TELLIN *Tellina mera* Say   **Pl. 28**
Range: Bermuda, Bahamas, s. Florida south through Lesser Antilles and Curaçao.
Habitat: Shallow water.
Description: Nearly 1 in. long. Beaks prominent, somewhat closer to posterior end, which is bluntly pointed. Anterior broadly rounded. Surface smooth and shiny, color white.

SUNRISE TELLIN *Tellina radiata* Linn.   **Pl. 3**
Range: S. Florida to West Indies.
Habitat: Shallow water.
Description: An elongate shell, length to 3 in. Thin but sturdy, with low beaks placed about midway of the shell. Both ends broadly rounded, posterior a little less so than anterior. Surface smooth and very highly polished. Color yellowish white, with bands of pinkish rose radiating from the beaks, and commonly visible on interior.
Remarks: The broad rosy rays, coupled with the pelecypod's high gloss and graceful form, make this a prime favorite with collectors. There is a form of this species that lacks the color-

ful rays; a form consistent in color, generally white or pale yellow, it has been listed as *T. r. unimaculata* Lam.

SAY'S TELLIN   *Tellina sayi* Dall   **Pl. 28**
Range: N. Carolina to Gulf of Mexico.
Habitat: Shallow water.
Description: Length ½ in. Rather well inflated; front end rounded, the posterior slopes gradually to a rounded tip. Beaks small, set rather closer to front end. Color glossy white.
Remarks: Frequently washed ashore. As for many of the tellins, one often finds examples with both valves attached and spread out on the sand like so many mounted butterflies.

CANDY STICK TELLIN   *Tellina similis* Sby.   **Pl. 28**
Range: S. Florida to West Indies.
Habitat: Shallow water.
Description: Moderately elongate, length about 1 in. Shell thin and compressed, beaks rather prominent and situated ⅓ closer to posterior end. Anterior end gracefully rounded, posterior end dropping off rather abruptly to rounded tip. Glossy white, often with short radiating rays of pinkish.

DALL'S DWARF TELLIN   *Tellina sybaritica* Dall   **Pl. 28**
Range: N. Carolina to West Indies.
Habitat: Moderately shallow water.
Description: Less than ½ in. long. Very elongate, with beaks about central. Posterior end sloping, a bit less than anterior. Surface shiny, color generally pinkish but may be deeper red.
Remarks: Probably the smallest member of its genus on our shores.

TAMPA TELLIN   *Tellina tampaensis* Conrad   **Pl. 28**
Range: Florida to Texas; West Indies.
Habitat: Shallow water.
Description: About ¾ in. long. Beaks high and full, anterior end short and regularly rounded, posterior somewhat longer and abruptly pointed. Valves only slightly inflated. Color shiny white. Interior often showing a pinkish tone.

TEXAS TELLIN   *Tellina texana* Dall   **Pl. 28**
Range: Gulf of Mexico.
Habitat: Shallow water.
Description: Thin and compressed, about ½ in. long. Slightly inflated. Angular slope on posterior. Sculpture of concentric lines but surface appears smooth. Color white, sometimes with a tinge of pinkish, the valves often having an iridescent luster.

DEKAY'S DWARF TELLIN   Pl. 28
Tellina versicolor DeKay
   Range: Cape Cod to Florida and Texas; West Indies to
Trinidad.
   Habitat: Shallow water.
   Description: About ½ in. long. Beaks fairly prominent, ante-
rior end broadly rounded, posterior end short, sloping, bluntly
pointed. Surface polished and iridescent. Color may be white,
pink, reddish, or weakly rayed with rosy pink.

Genus *Arcopagia* Brown 1827

FAUST TELLIN   *Arcopagia fausta* (Pulteney)   Pl. 28
   Range: N. Carolina to West Indies.
   Habitat: Moderately shallow water.
   Description: A large and moderately heavy shell, length as
much as 4 in. Both ends slope to a rounded ventral margin,
the posterior end more pointed. Valves rather thick, hinge
strong. Surface marked by coarse growth lines. Color whitish; interior often showing tinges of yellow.

Genus *Macoma* Leach 1819

HANLEY'S MACOMA   *Macoma aurora* (Hanley)   Pl. 29
   Range: S. Florida to Brazil.
   Habitat: Moderately shallow water.
   Description: Length ¾ in. An oval shell, anterior end longer
than posterior. Both ends rounded, posterior more acutely.
Beaks fairly prominent. Surface smooth, color white, occa-
sionally pinkish. Interior glossy, tinged with yellow.

BALTIC MACOMA   *Macoma balthica* (Linn.)   Pl. 29
   Range: Arctic to Georgia.
   Habitat: Shallow water.
   Description: Length to 1½ in. Shell moderately thin, with
rounded outline, posterior end somewhat constricted. Beaks
rather prominent and nearly central. Surface bears many
very fine concentric lines of growth. Color pinkish white with
a dull finish and there is a thin olive-brown periostracum,
usually lacking on the upperparts of shell.
   Remarks: This bivalve is abundant in muddy bays and
coves, and commonly travels partway up many creeks and
rivers. It occurs on our West Coast and in Europe.

SHORT MACOMA   *Macoma brevifrons* (Say)   Pl. 29
   Range: New Jersey to S. America.
   Habitat: Moderately shallow water.

Description: About 1 in. long. A small dull white shell, the only sculpture being indistinct lines of growth. Commonly a yellowish tint over the umbones, and interior may be buffy yellow. Anterior end long and broadly rounded, posterior end sloping rather sharply to a bluntly pointed tip. Thin brownish periostracum.

CHALKY MACOMA *Macoma calcarea* (Gmelin) **Pl. 29**
Range: Greenland to Long Island.
Habitat: Moderately shallow water.
Description: Length 2 in. Elongate-oval in shape, posterior end narrowed and slightly twisted. Dull chalky white, inside and outside, with dull grayish periostracum generally present only toward base of shell.

CONSTRICTED MACOMA **Pl. 29**
Macoma constricta (Brug.)
Range: Florida to West Indies; Texas.
Habitat: Shallow water.
Description: About 2½ in. long. Shell moderately inflated, with broadly rounded margins. Posterior end partially truncate, and notched below, with a feeble fold extending from beaks to ventral margin. Color white, with a thin yellowish periostracum.

SLENDER MACOMA *Macoma extenuata* Dall **Pl. 29**
Range: Gulf of Mexico.
Habitat: Moderately deep water.
Description: About ½ in. long. An elongate shell, anterior end broadly rounded, posterior more sharply rounded. Beaks set about ⅓ closer to back end. Color white, surface polished and somewhat iridescent.

HENDERSON'S MACOMA **Pl. 29**
Macoma hendersoni Rehder
Range: S. Florida to West Indies.
Habitat: Shallow water.
Description: Nearly ½ in. long. Shell rather thin and fragile, anterior end rounded, posterior sloping to a square tip, somewhat upturned. Beaks central and prominent. Surface sculptured with wavy concentric lines. Color dull white.
Remarks: Some authorities consider this a subspecies of *M. orientalis* Dall.

NARROWED MACOMA *Macoma tenta* (Say) **Pl. 29**
Range: Prince Edward Island to Florida and Gulf of Mexico; West Indies.
Habitat: Shallow water.

Description: A small, thin, delicate bivalve, the length slightly less than ½ in. as a rule, but it may be as much as ¾ in. Anterior end long and broadly rounded, posterior short and abruptly sloping. Surface bears tiny but sharp lines of growth. Color pinkish white, though surface is iridescent and reflects a whole rainbow of hues.

Genus *Strigilla* Turton 1822

ROSY STRIGILLA *Strigilla carnaria* Linn. **Pl. 29**
Range: N. Carolina to West Indies.
Habitat: Shallow water.
Description: Length slightly under 1 in. Shell moderately solid, rather circular in outline, and fairly well inflated. Surface appears smooth but there is a sculpture of extremely fine radial lines that become oblique and wavy over posterior area. Pale rose color, deepest over umbones (beaks); interior rosy pink.
Remarks: The rosy color makes this common species an attractive shell for decorative purposes.

WHITE STRIGILLA *Strigilla mirabilis* Phil. **Pl. 29**
Range: N. Carolina to Texas; West Indies.
Habitat: Shallow water.
Description: Nearly ½ in. long, oval, and well inflated. Beaks situated slightly closer to front end. Surface bears the same peculiar sculpture as in Rosy Strigilla. Color white.

PEA STRIGILLA *Strigilla pisiformis* (Linn.) **Pl. 29**
Range: S. Florida to West Indies; Bahamas.
Habitat: Shallow water.
Description: Length ⅓ in. Moderately inflated, this is an oval, slightly oblique shell, with much the same sculpture as others of its genus. Color pink, darker over the umbonal region. On inside the color is darkest centrally, often white at the margins.

Genus *Phylloda* Schumacher 1817

CRENULATE TELLIN *Phylloda squamifera* (Desh.) **Pl. 29**
Range: N. Carolina to Florida.
Habitat: Moderately shallow water.
Description: Length about 1 in. This is a rather thin shell, the front end broadly rounded and the posterior sloping to a somewhat truncate tip. Sculpture of fine but sharp concentric lines. Noticeable thornlike crenulations on the dorsal margins. Color white, occasionally yellowish or orange.

Genus *Quadrans* Bertin 1878

LINEN TELLIN *Quadrans lintea* (Conrad) **Pl. 29**
Range: Cape Hatteras to Gulf of Mexico; West Indies.
Habitat: Moderately deep water.
Description: Not quite 1 in. long. Shell thin and rather delicate, beaks small and pointed. Anterior end rounded, posterior sharply sloping, with a slightly truncate tip. Surface sculptured with finely chiseled concentric lines, upturned on posterior slope to form a filelike edge. Color white; interior polished.

Genus *Tellidora* H. & A. Adams 1856

CRESTED TELLIN *Tellidora cristata* (Récluz) **Pl. 29**
Range: N. Carolina to Florida; Texas.
Habitat: Shallow water.
Description: Length about 1 in. Shell compressed, left valve flatter than right, outline somewhat triangular. Beaks central and prominent. Ventral margin broadly rounded. Surface bears faint concentric ridges that form teeth on the lateral margins, giving the shell a saw-toothed appearance. Color pure white.

Genus *Apolymetis* Salisbury 1929

GROOVED MACOMA *Apolymetis intastriata* (Say) **Pl. 28**
Range: Florida to West Indies.
Habitat: Shallow water.
Description: A large but thin shell about 3 in. long. Posterior end long and broadly rounded, anterior sloping and profoundly folded, so the shell has an oddly twisted appearance at its front end. Valves well inflated. Color pure white, sometimes with a tinge of yellow.
Remarks: Some modern malacologists place this pelecypod in the genus *Psammotreta* Dall 1900.

Bean Clams: Family Donacidae

GENERALLY small, wedge-shaped clams, the posterior end elongated and acutely rounded and the anterior sharply sloping and short. Distributed in all seas that are warm, where they live in the sand close to shore.

Genus *Donax* Linnaeus 1758

CARIBBEAN COQUINA *Donax denticulata* Linn. **Pl. 29**
Range: West Indies.
Habitat: Shallow water.
Description: A sturdy shell, slightly more than 1 in. long. Elongate-oval, with fine radiating lines on surface. Inner margins crenulate. Color may be brown, violet, or yellowish, often with darker rays.

FOSSOR COQUINA *Donax fossor* Say **Pl. 29**
Range: Long Island to New Jersey.
Habitat: Shallow water.
Description: About ½ in. long. Shell rather thick and solid, inner margins crenulate. Shaped much like the Coquina (below), of which it may turn out to be a variety. Surface decorated with fine radiating lines. Color white or bluish white, sometimes with faint rays.

ROEMER'S COQUINA *Donax roemeri* (Phil.) **Pl. 29**
Range: Texas.
Habitat: Shallow water.
Description: This is a slightly smaller form than Fossor Coquina, and occurs on the Texas coast; it may be a subspecies of the common Coquina (below). It is not as elongate, and the posterior end is blunter. Color tends to be more yellowish.

STRIATE COQUINA *Donax striatus* Linn. **Pl. 29**
Range: West Indies.
Habitat: Shallow water.
Description: Length 1 in. Anterior end long and bluntly rounded, posterior sloping, abruptly rounded. Basal margin slightly concave. Surface smooth, a few radiating lines on posterior slope. Color bluish white.

COQUINA *Donax variabilis* Say **Pl. 29**
Range: Virginia to Florida and west to Texas.
Habitat: Shallow water.
Description: About ¾ in. long. Shell wedge-shaped, posterior end prolonged and acutely rounded, anterior short and obliquely truncated. Surface bears numerous fine radiating lines, inner margins crenulate. Displays a bewildering variety of colors, ranging from pure white to yellowish, rose, lavender, pale blue, and deep purple. It is usually decorated with radiating reddish-brown bands and sometimes with concentric colored lines. Now and then a shell has both, producing a

plaid pattern. Inner surface smooth, and also shows varied coloration, often deep purple.

Remarks: Known by several other popular names, among them Butterfly Shell, Wedge Shell, and Pompano. It burrows in loose sand at the midwater line, where in favorable situations individuals may be gathered by the handful with scarcely any intermixture of sand; despite their diminutive size they are often so gathered and made into a delicious chowder. Dead shells usually remain in pairs, connected at the hinge and spread out like so many butterflies. The color patterns are almost countless; out of 50 shells it is sometimes difficult to find 2 exactly alike. Suites of this little pelecypod are often used in biology classes to demonstrate an extreme in color variation within a single species.

See Fossor Coquina (above) for a similar shell that may be only *variabilis,* which occasionally is established north of Virginia, then is subsequently killed off during winter.

Genus *Iphigenia* Schumacher 1817

GREAT FALSE COQUINA   **Pl. 29**
Iphigenia brasiliensis (Lam.)
   Range: Florida to West Indies.
   Habitat: Shallow water.
   Description: Length to 3 in. A fairly large and sturdy clam, broadly triangular. Beaks prominent and almost central. Anterior end sharply rounded, posterior with angled slope. Margins not crenulate. Color buffy white, but in life there is a rather substantial tan periostracum. Interior shiny white, often pinkish on teeth.

Gari Shells:
Family Sanguinolariidae

SHELLS somewhat like the tellins, with which they were once grouped. The animals have very long siphons. Distributed chiefly in warm seas.

Genus *Sanguinolaria* Lamarck 1799

ATLANTIC SANGUIN   *Sanguinolaria cruenta* (Sol.)   **Pl. 30**
   Range: S. Florida, Gulf states; West Indies.
   Habitat: Moderately shallow water.
   Description: About 1½ in. long. Shape oval, with a flattish

top. Both ends and basal margin regularly rounded. Beaks very low. Surface fairly smooth. Color rosy pink over the beak area, fading to whitish toward the margins.
Remarks: Formerly listed as *S. sanguinolaria* (Gmelin).

Genus *Asaphis* Modeer 1793

GAUDY ASAPHIS *Asaphis deflorata* (Linn.)   **Pl. 30**
    Range: S. Florida to West Indies.
    Habitat: Shallow water.
    Description: Length averages 2 in. but sometimes is longer. Shell thin but strong, rather well inflated. Surface bears numerous radiating lines, most pronounced on posterior slope. Lines at both ends crossed by wavy growth lines. Color variable, ranging from white to yellow, orange, and purple, with majority of specimens tending toward latter. Interior may be yellow, orange, or purple.

Genus *Tagelus* Gray 1847

PURPLISH TAGELUS *Tagelus divisus* (Spengler)   **Pl. 31**
    Range: Cape Cod to Florida and Gulf states; West Indies.
    Habitat: Shallow water.
    Description: Length 1½ in. Shell thin and fragile, elongate with parallel margins, both ends rounded. Beaks nearly central. Surface smooth, generally shiny, with a thin yellowish-brown periostracum. Color of shell purplish gray, faintly rayed with purple. Interior often deep purple.

STOUT TAGELUS *Tagelus plebeius* (Lightf.)   **Pl. 31**
    Range: Massachusetts to Florida and Gulf states.
    Habitat: Shallow water.
    Description: Length 3 to 4 in. Shell elongate, stout, gaping, and abruptly rounded at both ends. Beaks blunt and but little elevated; situated just off middle of shell. Surface coarsely wrinkled concentrically. Color white or yellowish, with a thin yellowish-brown periostracum. Most of the shells that have been empty for any length of time are a dull chalky white.
    Remarks: Formerly listed as *T. gibbus* (Spengler).

Genus *Heterodonax* Mörch 1853

FALSE COQUINA *Heterodonax bimaculata* (Linn.)   **Pl. 29**
    Range: Florida to West Indies.
    Habitat: Shallow water.

Description: About ¾ in. long. Triangular-oval in shape, rather solid. Posterior end short and squarish, anterior more rounded. Common shade is bluish white, and most valves bear 2 red or purple spots, and in addition there may be rays of purplish. Interior commonly spotted with brown or purple.

Semeles: Family Semelidae

SHELLS roundish oval and but little inflated, with more or less obscure folds on the posterior ends. There is a deep pallial sinus. Chiefly confined to warm seas.

Genus *Semele* Schumacher 1817

CANCELLATE SEMELE   Pl. 30
Semele bellastriata Conrad
  Range: N. Carolina to West Indies.
  Habitat: Moderately shallow water.
  Description: Length to 1 in. Round-oval, with anterior end somewhat shorter and more sharply rounded. Surface has both radiating and concentric lines, so that with some specimens it appears slightly beaded. Color may be solid purplish gray, but often yellowish with streaks of brown. Interior polished and frequently brightly colored with yellow or lavender.

TINY SEMELE   *Semele nuculoides* Conrad   Pl. 30
  Range: Florida to West Indies.
  Habitat: Shallow water.
  Description: Length ¼ in. An oval shell, front end short and sloping, posterior longer and rounded. Surface shining, with minute concentric lines. Color white or yellowish.

WHITE SEMELE   *Semele proficua* (Pulteney)   Pl. 30
  Range: N. Carolina to West Indies.
  Habitat: Shallow water.
  Description: Length to 1½ in. Shell rather thin and compressed, rounded oval. Beaks small and turned forward, with small lunule in front. Sculpture of extremely fine but sharp concentric lines. Color creamy white, occasionally variegated with pinkish rays. Very thin periostracum. Interior yellowish and polished, occasionally lightly speckled with purplish or pink.

PURPLE SEMELE   *Semele purpurascens* (Gmelin)   Pl. 30
  Range: N. Carolina to West Indies.
  Habitat: Shallow water.

Description: An oval shell with both ends rounded, about 2 in. long. Posterior end twice the length of anterior. Surface bears fine concentric lines, but appears quite smooth. Color pale yellow, rather blotched with purple or brown or orange. Some individuals may be uniformly deep yellow.

Genus *Cumingia* Sowerby 1833

SOUTHERN CUMINGIA *Cumingia antillarum* Orb. **Pl. 29**
  Range: S. Florida to West Indies.
  Habitat: Shallow water.
  Description: Length ⅜ in. A roughly oval shell, beaks moderately elevated. Interior with spoonlike pit beneath beaks. Surface bears concentric lines, with minute radiating lines in between. Color white.

COMMON CUMINGIA   **Pl. 29**
Cumingia tellinoides (Conrad)
  Range: Nova Scotia to Florida.
  Habitat: Shallow water.
  Description: About ¾ in. long. Shell oval-triangular, quite thin, the anterior end broadly rounded, posterior considerably pointed and gaping. Surface covered by numerous sharp, elevated, concentric lines. Color white.

Genus *Abra* Lamarck 1818

COMMON ATLANTIC ABRA *Abra aequalis* (Say) **Pl. 30**
  Range: N. Carolina to Texas; West Indies.
  Habitat: Moderately deep water.
  Description: A small, rather plump shell just under ½ in. long. Shell rounded and slightly oblique, decorated with minute concentric wrinkles near the margins, the rest of surface relatively smooth. Color brownish, often tinged with buff.

DALL'S LITTLE ABRA *Abra lioica* Dall **Pl. 30**
  Range: Massachusetts to West Indies.
  Habitat: Moderately deep water.
  Description: Length about ¼ in. Shell thin and but little inflated. Beaks slightly closer to anterior end. Surface rather shiny, the color buff.

ELONGATED ABRA   **Pl. 30**
Abra longicallis americana Verrill & Bush
  Range: Arctic Ocean to West Indies.
  Habitat: Deep water.

Description: About ½ in. long. Somewhat elongate, beaks full. Both ends rounded, the posterior more sharply. Surface smooth and polished, the color pale brown.
Remarks: *A. longicallis* (Scacchi) is European, the type having been described from Norway.

Razor Clams: Family Solenidae

MEMBERS of this family are the true razor clams, and they are so called wherever they occur. The shells have equal valves, are usually greatly elongated, and gape at both ends as a rule. They are distributed in the sandy bottoms of coastal waters in nearly all seas, and all are edible.

Genus *Solen* Linnaeus 1758

WEST INDIAN RAZOR CLAM   Pl. 30
Solen obliquus Spengler
    Range: West Indies.
    Habitat: Shallow water.
    Description: Length 5 to 6 in. Remarkably elongate, margins parallel. Low beaks at extremely short anterior end. Both ends squarish. Growth lines the only sculpture. Color whitish, with brown periostracum.

LITTLE GREEN RAZOR CLAM   *Solen viridis* Say   Pl. 30
    Range: Rhode Island to n. Florida and n. Gulf of Mexico.
    Habitat: Intertidal.
    Description: Length 2 to 3 in. Shell thin, elongate, somewhat compressed. Hinge line nearly straight. Gaping at both ends. Surface smooth and rather glossy, with very faint concentric wrinkles. Shiny green periostracum.
    Remarks: Specimens may be distinguished from young examples of the Common Razor Clam, *Ensis directus* (below), by their straighter shells and slightness of curve.

Genus *Ensis* Schumacher 1817

COMMON RAZOR CLAM   *Ensis directus* (Conrad)   Pl. 30
    Range: Labrador to Florida.
    Habitat: Intertidal.
    Description: Most specimens 5 to 7 in. long, but the species may reach 10 in. Shell thin, greatly elongated, gaping, and noticeably curved. Sides parallel, ends squarish. Surface has

glossy greenish periostracum and a long triangular space marked by concentric lines of growth. Shell itself whitish.

Remarks: Burrows vertically in sandbars, and the speed with which it can sink down out of sight in wet sand is astonishing. Also a successful if somewhat erratic swimmer. The long foot is extended and folded back against the shell, then suddenly straightened out as if it were a steel spring, and the clam is propelled like an arrow for 3 or 4 feet. Razor clams are tender if not too large and have an excellent flavor. See Remarks under Little Green Razor Clam, *Solen viridis* (above).

DWARF RAZOR CLAM *Ensis minor* Dall **Pl. 30**
Range: Florida to Texas.
Habitat: Intertidal.
Description: Proportionally narrower and more fragile than Common Razor Clam, seldom exceeding 4 in. Many regard this as a subspecies of the Common Razor Clam, *E. directus*.

Genus *Solecurtus* Blainville 1825

CORRUGATED RAZOR CLAM **Pl. 31**
Solecurtus cumingianus Dunker
Range: N. Carolina to Texas.
Habitat: Moderately deep water.
Description: About 1½ in. long. Shape much like Stout Tagelus, *Tagelus plebeius* (p. 83), with beaks closer to front and with both ends rounded. An odd sculpture, consisting of a number of prominent concentric rings, on which is superimposed a series of distinct lines that are oblique on the front portion of the valves and sharply vertical on the rear. White, with yellowish-gray periostracum.

ST. MARTHA'S RAZOR CLAM **Pl. 31**
Solecurtus sanctaemarthae (Orb.)
Range: S. Florida to West Indies.
Habitat: Shallow water.
Description: Length to about 1 in. An oblong shell, squarish and gaping at both ends. Beaks low, set closer to front end. Surface similar to that of Corrugated Razor Clam. Color white, with thin yellowish periostracum.

Genus *Siliqua* Mühlfeld 1811

RIBBED POD *Siliqua costata* (Say) **Pl. 30**
Range: Gulf of St. Lawrence to N. Carolina.

Habitat: Shallow water.
Description: About 2 in. long. Oval-elliptical, thin and rather fragile, moderately elongate. Broadly rounded at each end. Beaks very low, situated closer to front. Interior strengthened by a prominent vertical rib that extends from the beaks, bending slightly backward, and, expanding, loses itself about halfway across the valve. Color pinkish white, with thin yellowish-green periostracum.

Rock Borers: Family Hiatellidae

SHELLS usually elongate, valves unequal. Some very large. Members of this family commonly bore into sponge, coral, and limestone, but some burrow deeply in mud. The surface is generally irregular and rough, and lacking in colors. Distributed from the Arctic to the tropics.

Genus *Hiatella* Daudin 1801

ARCTIC ROCK BORER   *Hiatella arctica* (Linn.)   **Pl. 31**
Range: Arctic Ocean to West Indies.
Habitat: Moderately shallow water.
Description: A rough and unattractive bivalve about 1½ in. long and dingy white. Shell oblong-oval, coarse and irregular in shape. Beaks rather prominent; from them run 2 faint ridges to rear margin. Both ends rounded, the posterior nearly 3 times as long as the anterior. Surface coarsely marked with lines of growth and irregularly undulated.
Remarks: It occasionally does some damage by excavating its burrows in the cement work of breakwaters or embankments. Each individual, after boring its home, attaches itself to the walls by a byssal cord and remains fixed for life. Young specimens are often found attached to rocks and pebbles close to shore. This shell is quite common as a fossil in Pleistocene rocks throughout the Northeast.

Genus *Cyrtodaria* Daudin 1799

PROPELLER CLAM   *Cyrtodaria siliqua* (Spengler)   **Pl. 30**
Range: Labrador to Rhode Island.
Habitat: Moderately deep water.
Description: A thick and heavy shell about 3 in. long. Elongate-oval, beaks closer to front end. Valves widely gaping and slightly twisted like a propeller. Surface bears con-

centric grooves. A thick and horny, glossy black periostracum frequently projects beyond the valves. Beaks very low, usually eroded.

Remarks: The animal is very large, and the valves never completely cover the fleshy parts. Periostracum easily flakes off dry shells.

Genus *Panomya* Gray 1857

ROUGH SOFT-SHELLED CLAM   **Pl. 31**
Panomya arctica (Lam.)

   Range: Circumpolar; Arctic Ocean to Georges Bank (off Cape Cod).

   Habitat: Intertidal; gravelly muds.

   Description: Rough and sturdy, rather squarish in outline, attains a length of about 3 in. Posterior end squarely cut off, gapes widely; front end slopes to a rounded point and then to the ventral margin, which is straight. Color chalky white, with a dark, almost black periostracum that peels easily from the shell, so even living examples are rarely completely covered.

   Remarks: The shell usually fails to encase the animal completely.

Genus *Panope* Ménard 1807

ATLANTIC GEODUCK   *Panope bitruncata* Conrad   **Pl. 31**

   Range: N. Carolina to Florida.

   Habitat: Moderately shallow water.

   Description: Length 5 to 6 in. Well inflated, with prominent beaks, regularly rounded before and rather sharply truncated behind, where it gapes widely. Surface marked with growth lines, color dull grayish white.

   Remarks: A rare shell, once believed to be extinct, but a few live specimens have been collected in recent years. It might be considered the Atlantic cousin of the well-known North Pacific Gweduc, or "Gooey-duck."

Soft-shelled Clams:
Family Myacidae

VALVES usually unequal and gaping. Left valve contains a spoonlike structure called a chondrophore, to which the resilium (internal cartilage in hinge) is attached. Distributed in all seas.

Genus *Mya* Linnaeus 1758

SOFT-SHELLED CLAM   *Mya arenaria* Linn.   **Pl. 31**
  Range: Labrador to N. Carolina.
  Habitat: Intertidal; gravelly muds.
  Description: Length 3 to 6 in. Shell gapes at both ends. Anterior rounded, posterior somewhat pointed. Surface roughened and wrinkled by lines of growth. An erect spoon-like tooth (chondrophore) is located under beak in left valve. Dull gray or chalky white, with thin grayish periostracum.
  Remarks: Known by other popular names such as Long-necked Clam, Long Clam, and Steamer Clam, this species lives in the muds and gravels between tides, where it is exposed to air twice each day. It lives buried with only the tips of its siphons at the surface; as one walks over its territory the mollusk's position is revealed by a vertical spurt of water, ejected as the alarmed clam suddenly withdraws its siphons. Although generally regarded as inferior to the Quahog, this is a very important food mollusk, and it enjoys a steady popularity in the markets. Persistent clamming has made specimens as large as 6 in. rare and difficult to find.

TRUNCATE SOFT-SHELLED CLAM   **Pl. 31**
Mya truncata Linn.
  Range: Greenland to Massachusetts.
  Habitat: Intertidal.
  Description: Length 2 to 3 in. Shell rather thick and solid, oblong. Anterior end rounded, posterior end abruptly truncated and widely gaping, the edges slightly flaring. Beaks moderately prominent, chondrophore in left valve. Surface roughly wrinkled. Color dingy white, with a tough yellowish-brown periostracum.
  Remarks: A cold-water species, seldom found south of Maine. It is known on our West Coast as far south as Puget Sound, and also in Japan and Norway. It occurs commonly as a Pleistocene fossil throughout the northern New England shoreline.

Basket Clams: Family Corbulidae

SHELLS small but solid, valves very unequal, one valve generally overlapping the other. Slightly gaping at the anterior end. An upright, conical tooth is present in each valve. White, and usually concentrically ribbed. Distributed in nearly all temperate seas.

Genus *Notocorbula* Iredale 1930

FAT BASKET CLAM   **Pl. 31**
Notocorbula operculata (Phil.)
    Range: N. Carolina and Gulf of Mexico; West Indies.
    Habitat: Shallow water.
    Description: Thin-shelled and glossy, length about ⅜ in. Beaks high and curved inward. Valves so inflated the shell is almost globular; one valve slightly larger than other. Color white, more or less tinged with pink, and with a yellowish periostracum.
    Remarks: Formerly listed as *Corbula disparilis* Orb.

Genus *Corbula* Bruguière 1792

EQUAL-VALVED BASKET CLAM   **Pl. 31**
Corbula aequivalvis Phil.
    Range: West Indies.
    Habitat: Moderately shallow water.
    Description: Nearly ½ in. long. A solid and stubby shell, with a sharply flattened posterior slope. Beaks full and central. Valves nearly equal. Sculpture of distinct concentric lines. Color white.

CARIBBEAN BASKET CLAM   **Pl. 31**
Corbula caribaea Orb.
    Range: S. Florida to West Indies.
    Habitat: Shallow water.
    Description: Length ¼ in. Valves moderately inflated, the right one projecting beyond and partly enclosing the left. Front end rounded, posterior pinched out and decidedly rostrate (beaked). Surface bears concentric lines that are obscure on anterior end. Color white.

COMMON BASKET CLAM   *Corbula contracta* Say   **Pl. 31**
    Range: Massachusetts to West Indies.
    Habitat: Shallow water.
    Description: About ½ in. long. Shell solid and inflated, anterior end rounded, posterior somewhat pointed. An angular ridge runs from beaks to rear ventral margin, giving that portion of the shell a distinct slope. Right valve overlaps left at rear. Surface sculptured with concentric ribs. Dull white, with thin brownish periostracum.

SNUB-NOSED BASKET CLAM   **Pl. 31**
Corbula nasuta Say
    Range: N. Carolina to West Indies.

Habitat: Shallow water.
Description: Shell rugged and strong, about ¼ in. long. Anterior rounded, posterior drawn out to a blunt point. Right valve overlaps left. Sculpture of distinct concentric lines. White or yellowish white, often with a rim of brownish periostracum at margins.

SWIFT'S BASKET CLAM   Pl. 31
Corbula swiftiana C. B. Adams
Range: Massachusetts to West Indies.
Habitat: Shallow water.
Description: Less than ½ in. long. Shell somewhat triangular, the posterior ridge quite prominent. Concentric lines feeble or lacking in some specimens. Right valve overlaps left. Color white, with thin yellowish periostracum.

Genus *Paramya* Conrad 1860

OVATE PARAMYA   *Paramya subovata* Conrad   Pl. 30
Range: N. Carolina to Florida.
Habitat: Moderately shallow water.
Description: About ½ in. long. Anterior end rounded, posterior enlarged and bluntly rounded, rendering the shell rather square in outline. Beaks moderately prominent. Color dull white or yellowish white.

Gaping Clams:
Family Gastrochaenidae

BURROWING or boring mollusks living in coral, limestone, or dead shells of other mollusks. The valves are equal and gape considerably. Chiefly dwellers in warm seas.

Genus *Rocellaria* Blainville 1828

ATLANTIC ROCELLARIA   Pl. 32
Rocellaria hians (Gmelin)
Range: N. Carolina to West Indies.
Habitat: Soft coral rocks.
Description: Length to about 1 in. Beaks close to front end, which slopes sharply to ventral margin. Posterior gapes widely and is somewhat twisted. Sculpture of very fine concentric lines. Color yellowish gray.
Remarks: Sometimes excavates burrows in other shells, especially large examples of *Spondylus*. Formerly in the genus *Gastrochaena* Spengler 1783.

Genus *Spengleria* Turton 1861

ATLANTIC SPENGLER CLAM   **Pl. 32**
Spengleria rostrata (Spengler)
  Range: Florida to West Indies.
  Habitat: Bores into coral and limestone.
  Description: A small yellowish-white shell about 1 in. long.
  Valves sturdy, elongate, squarish, slightly twisted, and widely
  gaping. Anterior end rounded, posterior end squarely trun-
  cate. There is an elevated, transversely ribbed, radiating area
  extending from beaks to posterior margin.

Piddocks: Family Pholadidae

THESE are boring clams, capable of penetrating wood, coral,
and moderately hard rocks, although some live in clay and
mud and a few in wood. The shells are white, thin, and brittle, generally elongate, and narrowed toward the posterior end.
They gape at both ends. The front end bears a sharp abrading
sculpture, and they have accessory shelly plates on the dorsal
surface. Distributed in all seas.

Genus *Pholas* Linnaeus 1758

WING SHELL   *Pholas campechiensis* Gmelin   **Pl. 32**
  Range: N. Carolina to Gulf of Mexico; West Indies to Brazil.
  Habitat: Shallow water.
  Description: About 4 in. long. Shell greatly elongated, thin,
  and brittle, gaping at both ends. Rayed all over with rather
  distinct ribs, those on anterior end sharp and rasplike. Hinge
  plate reflected over the umbones (beaks) and supported by
  several vertical shelly plates. There are 2 accessory plates on
  dorsal surface. Apophyses (shelly braces) long and narrow.
  Color white.
  Remarks: This graceful shell closely resembles the Angel
  Wing, *Cyrtopleura costata* (below), but it is smaller and slimmer, with a sculpture that is much less coarse.

Genus *Cyrtopleura* Tryon 1862

ANGEL WING   *Cyrtopleura costata* (Linn.)   **Pl. 32**
  Range: Cape Cod to Gulf of Mexico; West Indies to Brazil.
  Habitat: Shallow water.
  Description: Length 5 to 7 in. Shell quite thin and brittle.

Rounded at front, narrowed and prolonged at rear. Sculptured with strong imbricate radiating ribs. Coarse lines of growth rise over the ribs in an undulating manner. Ribs at front end (boring end) sharp and scaly. Shelly brace (called the apophysis) under each beak projects into shell's interior and is for the attachment of the foot muscles. Valves gape widely, touching only at a point near the top. Color pure white, occasionally with pinkish margins. Thin grayish periostracum during life. See Wing Shell, *Pholas campechiensis* (above), for similarity, and also the False Angel Wing, *Petricola pholadiformis* (p. 68).

Remarks: Formerly listed as *Barnea costata* Linn. A single glance at the snowy-white, graceful shell is enough to convince one that the name Angel Wing is a good one. It has long been a great favorite with collectors. A staple article of food in parts of the West Indies, it is not uncommon in Florida but is rare north of Virginia.

Genus *Barnea* Risso 1826

TRUNCATE BORER   *Barnea truncata* (Say)   Pl. 32
Range: Massachusetts to Florida.
Habitat: Intertidal; mud and peat banks.
Description: About 2 in. long, sometimes slightly more. Valves thin and fragile, somewhat elongated, posterior end broadly truncate at tip. Surface transversely and longitudinally wrinkled, and studded on front end with small erect scales. Valves gape widely, and there is a small elongate accessory shelly plate situated between the valves just in front of the beaks. Color white, with thin grayish periostracum.
Remarks: The shell is so fragile that it is difficult to dig a specimen free without crushing it. The best way to obtain perfect specimens is to dig out a large block of mud and put it in the nearest tide pool, where it may disintegrate slowly; thus the earth is washed from the clam instead of the clam's being pried out of the earth.

Genus *Zirfaea* Gray 1842

GREAT PIDDOCK   *Zirfaea crispata* (Linn.)   Pl. 32
Range: Labrador to New Jersey.
Habitat: Bores into clay and soft rocks.
Description: Length 2 to 3 in. Shell sturdy, rather oblong, rounded posteriorly and somewhat pointed at front. Widely gaping at both ends. Surface bears numerous coarse, wrinkle-like ridges that become lamellar on anterior half of shell. Valves divided into nearly equal portions by a broad channel running from beaks to the middle of ventral margin. In live

shells, a membranous expansion covers upperpart of shell. Strong apophysis (shelly brace) in each beak cavity. Color pure white. Accessory plate transverse, posterior to umbos.

Genus *Martesia* Blainville 1824

WOOD PIDDOCK *Martesia cuneiformis* (Say)   **Pl. 32**
Range: N. Carolina to Florida and Gulf of Mexico; south to Brazil.
Habitat: Bores into wood.
Description: Length ½ to ¾ in. Shell pear-shaped and rather chubby. Anterior gapes widely in young but pedal gape (opening for foot) closed by a callum in adult. Valves divided into sections by a narrow sulcus (groove) from beaks to ventral margin. Anterior sculptured with closely packed concentric ridges, posterior with smooth, rounded, concentric ridges. Long and thin apophyses. There is a triangular accessory plate over the beaks that shows a deep crease down its center. Color grayish.

STRIATE WOOD PIDDOCK   **Pl. 32**
Martesia striata (Linn.)
Range: N. Carolina to Florida; West Indies to Brazil.
Habitat: Bores into wood.
Description: Length about 2 in. Shell wedge-shaped, anterior gapes widely in young but is closed by a callum in the adult. Anterior slope sculptured with filelike ridges, posterior margin rounded. Surface transversely wrinkled and striated with elevated, minutely crenulated lines. Umbonal-ventral groove prominent. Apophyses long and thin. Accessory plate circular, cushionlike in adult. Color grayish white.
Remarks: Wood piddocks may be collected by searching in old waterlogged wood. The boring is done by the abrading surface of the valves; the clam generally works against the grain of the wood.

Genus *Xylophaga* Turton 1822

ATLANTIC WOOD BORER   **Pl. 32**
Xylophaga atlantica Richards
Range: Quebec to Virginia.
Habitat: Waterlogged wood.
Description: Less than ¼ in. long, color white. Shell gapes widely. Valves reinforced by sturdy central ribs. Apophyses lacking. Siphons short, combined, the excurrent slightly shorter than the incurrent. Dorsal accessory plate is of 2 triangular parts.

Remarks: This bivalve has all the appearances of belonging to the next group, Teredinidae, the shipworms, but lack of pallets (a pair of calcareous structures) and presence of small accessory plates on dorsal surface place it among the Pholadidae. Until fairly recently this borer was listed as *X. dorsalis* Turton, which is a European species. *X. atlantica* was named by Richards in 1942.

Shipworms: Family Teredinidae

SHIPWORMS, or pileworms, are wood borers. The front part of shipworms is covered by a tiny bivalve shell, but the posterior part is enclosed in a long shelly tube. The shell is white, very much inflated, and gaping, and is decorated with closely set lines. A pair of calcareous structures called pallets at the extreme rear end is used to close the tube; the individual characters of these pallets are determining factors in identifying the different genera and species. It is next to impossible to tell some of them apart with any degree of certainty by their shells alone.

Genus *Teredo* Linnaeus 1758

COMMON SHIPWORM   *Teredo navalis* Linn.   **Pl. 32**
    Range: Whole Atlantic Coast; Europe.
    Habitat: Bores into wood.
    Description: A small, globular, bivalve shell about ¼ in. long so far as the vestigial shell is concerned, but most of the animal lives in a shelly tube that may be several inches long. Anterior part of valves provided with sharp scaly ribs for boring, posterior smooth and widely gaping. Interior with stout umbonal-ventral rib. Color white. Pallet simple, white. See Fimbriate Shipworm (below), which this resembles.
    Remarks: The Common Shipworm lives in wood and is a great destroyer of timbers, both in ships and in wharves. It usually bores with the grain of the wood, and never intersects its neighbor's tunnel. Infested timbers are so honeycombed by the elongated galleries that they eventually disintegrate and crumble away. Painting or soaking the wood with creosote appears to discourage the mollusk to some extent. Although this tiny clam costs the shipping industry vast sums of money yearly, its record is not entirely bad, for it must be said that it performs a valuable service as a scavenger, ridding our harbors of wooden derelicts and other hazards to navigation.

Genus *Teredora* Bartsch 1921

HAMMER SHIPWORM   *Teredora malleola* (Turton)   **Pl. 32**
Range: Bermuda; West Indies.
Habitat: Burrows usually in floating wood.
Description: Length ⅜ in. Vestigial shell globular, widely gaping. Valves 3-lobed, with median furrow. Color white. Pallets with thumbnail-like depression.
Remarks: This species has been called *Teredo thompsoni* Tryon.

Genus *Bankia* Gray 1842

FIMBRIATE SHIPWORM   **Pl. 32**
Bankia fimbriatula Moll & Roch
Range: Florida to West Indies.
Habitat: Bores into wood.
Description: Nearly ¼ in. long, this shell is similar to Common Shipworm, *Teredo navalis* (above), since white, with an abrading anterior and a gaping posterior; it has the same long, curving, toothlike structure (apophysis) descending from each beak cavity. Pallets elongate, composed of numerous segments.
Remarks: There are other species of shipworms that can be found in drifting timbers along the Atlantic Coast, all having the same habit of burrowing with the grain of the wood; their specific identification is generally a task for the specialists. Since these small mollusks live in drifting timbers, their exact range is somewhat unpredictable.

Paper Shells: Family Lyonsiidae

SHELLS small and fragile, valves unequal. The hinge bears a narrow ledge to which the ligament is attached. Interior pearly. These mollusks are represented in both the Atlantic and Pacific Oceans, living in shallow water on sandy bottoms as a rule.

Genus *Lyonsia* Turton 1822

SAND LYONSIA   *Lyonsia arenosa* (Möller)   **Pl. 33**
Range: Greenland to Maine.
Habitat: Shallow water.
Description: A blocky shell about ½ in. long. Full beaks ⅓

closer to front end, with rear dorsal margin in straight line to the tip. Color silvery white, shell thin and translucent, quite pearly within.

PEARLY LYONSIA *Lyonsia beana* (Orb.)   **Pl. 33**
Range: N. Carolina to West Indies.
Habitat: Shallow water; often attaches to sponges.
Description: About 1 in. long. Very unequal valves and rather unpredictable in shape. Posterior prolonged and gaping; one valve usually quite concave, the other more flat. Color shiny white to brownish, often polished.

FLORIDA LYONSIA *Lyonsia floridana* (Conrad)   **Pl. 33**
Range: W. Florida to Texas.
Habitat: Shallow water.
Description: A trim but fragile bivalve about ¾ in. long. Shell very inequivalve, thin and translucent, and somewhat pearly. Surface bears widely spaced radiating lines and a paper-thin grayish periostracum.
Remarks: Some authorities regard this as a subspecies of the Glassy Lyonsia, *L. hyalina.*

GLASSY LYONSIA *Lyonsia hyalina* (Conrad)   **Pl. 33**
Range: Nova Scotia to S. Carolina.
Habitat: Shallow water.
Description: Length about ¾ in. Shell thin and fragile, translucent and pearly. Anterior end rounded, posterior elongated, narrowed, and slightly truncate. Beaks prominent.
Remarks: The surface is covered with radiating wrinkles, minutely fringed, that entangle grains of sand, with which the margins of the shell are often coated; an enlarged view on Plate 33 illustrates this.

Genus *Lyonsiella* Sars 1872

GRANULATE LYONSIELLA   **Pl. 33**
Lyonsiella granulifera Verrill
Range: Off Chesapeake Bay.
Habitat: Deep water.
Description: Length ¾ in. Shell higher than long, somewhat triangular and well inflated. Beaks prominent, centrally situated, and turned forward. Dorsal margin convex behind the beaks, concave in front. Surface smooth, color white; interior pearly.

Pandoras: Family Pandoridae

SHELLS small, very thin and flat, with inconspicuous beaks; valves unequal. The exterior is white, the interior very pearly. These clams are distributed in all seas and generally prefer a sandy or gravelly bottom.

Genus *Pandora* Chemnitz 1795

SAND PANDORA   *Pandora arenosa* Conrad   **Pl. 33**
  Range: N. Carolina to Florida.
  Habitat: Shallow water.
  Description: About ½ in. long. Very thin and flat; anterior rounded and posterior with square tip. Weak posterior ridge. One valve perfectly flat, the other slightly inflated. Color white; interior pearly.
  Remarks: Sometimes listed as *P. carolinensis* Bush, but Bush described it in 1885, whereas Conrad's name goes back to 1848.

GOULD'S PANDORA   *Pandora gouldiana* Dall   **Pl. 33**
  Range: Gulf of St. Lawrence to New Jersey.
  Habitat: Shallow water.
  Description: About 1 in. long. Shell irregularly wedge-shaped, rounded before, with a recurved truncate tip behind. The valves are asymmetrical, the right one flat and the left slightly convex. Exceedingly thin and flat. White or grayish white; interior pearly.

THREE-LINED PANDORA   *Pandora trilineata* Say   **Pl. 33**
  Range: N. Carolina to Florida; Texas.
  Habitat: Shallow water.
  Description: Length about 1 in. More than twice as long as high, with a marked concave area between beaks and up-turned posterior tip. 3 distinct lines along dorsal concavity. Color white; interior very pearly.

Thracias: Family Thraciidae

THIS group contains some rather large pelecypods, and some that are quite small. Valves unequal, and more or less gaping at each end. Beaks prominent, one commonly perforated.

Genus *Thracia* Blainville 1824

CONRAD'S THRACIA *Thracia conradi* Couthouy   **Pl. 33**
Range: Nova Scotia to Long Island.
Habitat: Moderately deep water.
Description: Shell oval and inflated, length up to 4 in.
Beaks about central and slightly turned backward; beak of
right valve perforated to receive point of left beak. Right
valve larger and more convex than left and projects somewhat beyond it. Hinge toothless. Anterior end rounded, posterior bluntly pointed. Surface coarsely wrinkled by growth
lines. Color dingy white, with thin brownish periostracum;
interior chalky white.

SHINING THRACIA *Thracia nitida* Verrill   **Pl. 33**
Range: Off Chesapeake Bay.
Habitat: Deep water.
Description: Length 1 in. Broadly oval, gaping considerably
at back and slightly in front. Beaks full and prominent,
slightly incurved. Shell thin and fragile and well inflated.
Color whitish, with a thin greenish periostracum; interior
white.

NORTHERN THRACIA   **Pl. 33**
Thracia septentrionalis Jeffreys
Range: Greenland to Rhode Island.
Habitat: Moderately deep water.
Description: Length usually under 1 in. An oval shell with
both ends rounded. Small beaks set closer to front end.
Valves somewhat compressed. Brownish, with thin peri-
ostracum; interior white.

Genus *Cyathodonta* Conrad 1849

WAVY THRACIA   *Cyathodonta semirugosa* (Reeve)   **Pl. 34**
Range: West Indies.
Habitat: Shallow water.
Description: Length about 1 in. A thin and delicate shell,
beaks almost central. Anterior end broadly rounded, posterior
truncate with flaring edges. Surface with broad undulating
concentric folds. Color white.

Spoon Shells:
Family Periplomatidae

THE shells are white, small, and usually fragile. Slightly gaping, they are called spoon shells from a low, spoon-shaped tooth (chondrophore) in each valve. A small triangular prominence lying next to this structure is generally lost when the animal is removed from its shell.

Genus *Periploma* Schumacher 1817

FRAGILE SPOON SHELL   **Pl. 33**
Periploma fragile (Totten)
  Range: Labrador to New Jersey.
  Habitat: Moderately shallow water.
  Description: Length about ½ in. Right valve rather more concave than and projecting a little beyond the left. Somewhat oval; anterior gradually rounded, posterior narrowed and slightly truncate. An elevated angular ridge extends from beaks to posterior margin. Surface minutely wrinkled, color pearly white.

UNEQUAL SPOON SHELL   **Pl. 33**
Periploma inequale C. B. Adams
  Range: S. Carolina to Texas.
  Habitat: Moderately deep water.
  Description: Length about ½ in. Front end short and abruptly rounded, posterior long and broadly rounded. Surface smooth and valves moderately inflated. Chondrophore relatively large and points backward. Color white.

LEA'S SPOON SHELL *Periploma leanum* (Conrad) **Pl. 33**
  Range: Gulf of St. Lawrence to N. Carolina.
  Habitat: Moderately shallow water.
  Description: About 1½ in. long. Left valve almost flat, right convex and somewhat truncate at posterior tip; a faint ridge proceeds from the beaks to this tip. Spoon-shaped tooth (chondrophore) nearly horizontal. Pearly white, with thin yellowish periostracum.

PAPER SPOON SHELL *Periploma papyratium* Say **Pl. 33**
  Range: Labrador to Rhode Island.
  Habitat: Moderately deep water.
  Description: About 1 in. long. A delicate shell with a short and angled anterior end and a long and rounded posterior

end. Valves moderately inflated. Spoonlike chondrophore points downward. White, with a thin yellowish periostracum.

WAVY SPOON SHELL   *Periploma undulata* Verrill   **Pl. 33**
Range: New Jersey to N. Carolina.
Habitat: Deep water.
Description: Length ½ in. Anterior end rounded, posterior bluntly pointed. Beaks nearly central and somewhat prominent. A slight undulation runs from beak to anterior ventral margin. A somewhat roughened and slightly elevated ridge runs from beak to lower angle of posterior end and fine radiating lines cover the posterior surface above it. Chondrophore small and more oblique than in Paper Spoon Shell. Periostracum thin and tinged with rusty brown toward margins.

Deep-water Clams:
Family Pholadomyidae

THESE deep-water pelecypods are well represented as fossils, particularly in the Jurassic rocks, but now are nearly extinct. The valves are equal, gaping behind, and are thin and white. Sculpture consists of strong radiating ribs. The hinge shows 1 obscure tooth in each valve.

Genus *Panacea* Dall 1905

FURROWED PANACEA   **Pl. 34**
Panacea arata (Verrill & Smith)
Range: South of Martha's Vineyard.
Habitat: Deep water.
Description: Length about 1½ in. Triangular, thin-shelled; posterior end long and bluntly pointed, anterior very short and abruptly truncated. Beaks prominent. Surface bears deep, rather wide concave radiating furrows, separated by elevated sharp-edged ribs. At extreme posterior end the ribs are small and crowded. Color white.

Deep-water Clams:
Family Poromyidae

SHELLS small, with valves unequal. Thin, and somewhat pearly within. Surface commonly granulated. Widely distributed, chiefly in deep water.

Genus *Poromya* Forbes 1844

GRANULATE SEA MUSSEL   Pl. 34
Poromya granulata (Nyst & Westen.)
   Range: Maine to Florida.
   Habitat: Deep water.
   Description: Less than ½ in. long. Shell oval, moderately
   inflated, beaks full and centrally situated. Both ends
   rounded. Surface bears tiny granules but appears smooth.
   Color yellowish brown.

SMOOTHISH SEA MUSSEL   Pl. 34
Poromya sublevis Verrill
   Range: Off Delaware and Chesapeake Bays.
   Habitat: Deep water.
   Description: Short and high, about ¾ in. long. Well inflated,
   beaks large and curved forward. Both ends short and bluntly
   rounded, the posterior rather oblique. Surface smooth, minute radiating lines. Color white, with thin yellowish periostracum; interior polished white.

Dipper Shells:
Family Cuspidariidae

SMALL pear-shaped bivalves, mostly confined to deep water.
They are called dipper shells from their elongated handlelike
posterior end (rostrum).

Genus *Cuspidaria* Nardo 1848

LITTLE DIPPER SHELL   Pl. 34
Cuspidaria fraterna Verrill & Bush
   Range: Massachusetts to Virginia.
   Habitat: Moderately deep water.
   Description: Short but sturdy, about ½ in. long. Beaks full,
   valves well inflated. Posterior "handle" (rostrum) short and
   squarely truncate. Surface smooth, color yellowish brown.

NORTHERN DIPPER SHELL   Pl. 34
Cuspidaria glacialis (Sars)
   Range: Nova Scotia to Maryland.
   Habitat: Moderately deep water.
   Description: About 1 in. long. Beaks high, anterior regularly
   rounded, posterior narrowed to form a definite "handle," but

since it is not as long or as pronounced as in some of this group the shell appears to be rather squat. Color cream to white and with grayish-white periostracum. See similar Stout Dipper Shell for differences.

Remarks: Probably the commonest dipper shell along our coast, but not likely to be found close to shore.

STOUT DIPPER SHELL   Pl. 34
Cuspidaria media Verrill & Bush

Range: South of Martha's Vineyard.

Habitat: Deep water.

Description: About ½ in. long, this species resembles the Northern Dipper Shell, but it is smaller, and decidedly more swollen, with the posterior end more narrow and proportionally longer. Front end evenly rounded, beaks rather prominent. Surface smooth, color yellowish white.

LONG-HANDLED DIPPER SHELL   Pl. 34
Cuspidaria microrhina Dall

Range: Off Florida.

Habitat: Deep water.

Description: Length 1¼ in. Valves moderately inflated, anterior end regularly rounded, posterior drawn out to form a rigid "handle." Beaks elevated. Surface smooth, posterior with transverse ridges. Color grayish.

ROSTRATE DIPPER SHELL   Pl. 34
Cuspidaria rostrata (Spengler)

Range: Arctic Ocean to West Indies; also Europe.

Habitat: Deep water.

Description: Nearly 1 in. long. Valves well inflated, anterior end sharply rounded, posterior end abruptly narrowed and drawn out to form a relatively long rostrum ("handle"). Surface bears delicate lines of growth which form strong wrinkles on rostrum. Color yellowish white; interior shiny white.

Genus *Cardiomya* A. Adams 1864

COSTELLATE DIPPER SHELL   Pl. 34
Cardiomya costellata (Desh.)

Range: Off N. Carolina to West Indies.

Habitat: Deep water.

Description: Length ¼ in. A small shell, beaks nearly central. Front end regularly rounded, posterior end produced into a narrow rostrum. A series of strong radiating ribs decorate the front half of shell, which is scalloped at the margin. Surface shiny, color bluish white.

SCULPTURED DIPPER SHELL Pl. 34
Cardiomya glypta Bush
   Range: Cape Hatteras to West Indies.
   Habitat: Deep water.
   Description: An ornate minute shell scarcely more than ⅛ in. long. Anterior rounded, posterior rostrate (beaked). Surface strongly sculptured with broad radiating folds. Color pale brown.

MANY-RIBBED DIPPER SHELL Pl. 34
Cardiomya multicostata (Verrill & Smith)
   Range: Off Martha's Vineyard to N. Carolina.
   Habitat: Deep water.
   Description: Length ½ in. Anterior rounded and marked with numerous crowded radiating lines, posterior relatively short and narrowed. Color bluish white.
   Remarks: Since it is known that the ribs on the Costellate Dipper Shell (above) increase in number with age, it has been suggested that this pelecypod is simply an old example of *C. costellata,* and that may well be the case.

WEST INDIAN DIPPER SHELL Pl. 34
Cardiomya perrostrata (Dall)
   Range: West Indies.
   Habitat: Moderately shallow water.
   Description: Length ½ in. Beaks full, anterior end broadly rounded, posterior abruptly narrowed to moderately long rostrum. Sculpture of strong radiating ribs, rostrum smooth. Color white.

Genus *Myonera* Dall & Smith 1886

GIANT SEA MUSSEL *Myonera gigantea* (Verrill) Pl. 34
   Range: Off Virginia.
   Habitat: Deep water.
   Description: Length 1 in. Valves thin and delicate, considerably inflated. Surface bears crowded growth lines. Anterior end rounded, posterior pointed and truncate. Color white, with grayish periostracum.
   Remarks: This rare shell has been dredged from nearly 12,000 ft.

LAMELLATED SEA MUSSEL Pl. 34
Myonera lamellifera (Dall)
   Range: Off Florida.
   Habitat: Deep water.

Description: Length ½ in. Regularly rounded anterior end, sharply pointed posterior. Beaks elevated. Surface bears broad concentric folds. Color white.

Verticords:
Family Verticordiidae

THESE are mostly small mollusks, very pearly inside. The valves are equal or nearly so, well inflated, and quite solid in substance. Chiefly dwellers in deep water.

Genus *Verticordia* Sowerby 1844

ELEGANT VERTICORD   Pl. 34
Verticordia elegantissima Dall
  Range: Florida to West Indies.
  Habitat: Deep water.
  Description: Length ½ in. Beaks full, nearly central. Front end rounded, rear slopes to a rounded point. Surface decorated with fine radiating lines studded with sharp points. Color white; interior very pearly.

ORNATE VERTICORD *Verticordia ornata* Orb.   Pl. 34
  Range: Massachusetts to West Indies.
  Habitat: Deep water.
  Description: Length ¼ in. Beaks hooked forward, front end short and acutely rounded. Front half of shell sculptured with strong radiating folds that make that margin crenulate, longer posterior portion smooth. Valves considerably compressed. Color dull yellowish gray; interior very pearly.

Genus *Halicardia* Dall 1894

FLEXED SEA-HEART CLAM   Pl. 34
Halicardia flexuosa (Verrill & Smith)
  Range: Off Martha's Vineyard.
  Habitat: Deep water.
  Description: An obliquely swollen and somewhat triangular shell about 1¼ in. high and nearly the same in length. Beaks considerably incurved and turned forward owing to slopes that divide surface into 3 distinct areas. Front end shorter than posterior. Color grayish white; interior rather pearly.

II

Gastropods

Abalones: Family Haliotidae

SHELLS spiral, depressed, spire small. Body whorl constitutes most of the shell. There is a row of round or oval holes along the left margin, those near the edge open. Interior pearly and often multicolored. Animals live attached to rocks like limpets. All are edible. Common in Pacific and Indian Oceans, also European waters; only 1 species, exceedingly rare, known to live on our Atlantic Coast.

Genus *Haliotis* Linnaeus 1758

POURTALÈS ABALONE *Haliotis pourtalesi* Dall **Pl. 36**
Range: Off Florida Keys and n. Cuba.
Habitat: Deep water.
Description: Length about ¾ in. About 3 whorls, spire low. Margin shows 5 open holes. Sculpture consists of rather strong wavy ridges that follow the curvature of the volutions. Color yellowish brown, with patches of orange on last (body) whorl; interior pearly white.
Remarks: A single specimen, less than 1 in. long, was dredged (alive) by Count Louis François de Pourtalès in 1869. It was taken in 1200 ft. off s. Florida. This shell, along with much other dredged material, was sent to Dr. William Stimpson in Chicago, who was at that time engaged in a monumental work on East Coast mollusks; the whole collection, including Stimpson's notes and manuscript, was destroyed in the great Chicago fire of 1871. The unique abalone was described by William H. Dall about ten years later, working entirely from memory. Forty-two years later a second specimen was collected off Sand Key, Florida. Several imperfect shells or fragments have been taken since, including 1 nearly perfect example from off Key Largo in 1944. This would certainly rate as one of our rarest marine shells, and for a complete discussion of its interesting history the reader is referred to *Johnsonia,* Vol. 2, No. 21 (1946), from which this brief account has been condensed.

Keyhole Limpets:
Family Fissurellidae

SHELLS conical, oval at base. The apex is perforated, or there is a slit, or notch, in the margin of the shell. Surface usually strongly ribbed. The slit ("keyhole") distinguishes these snails from the true limpets. When very young, they have a spiral shell with a marginal slit. Shelly material is added slowly until the margin below the slit is united, and then the spiral is absorbed as the hole enlarges. Distributed in warm and temperate seas as a rule.

Genus *Fissurella* Bruguière 1798

POINTED KEYHOLE LIMPET Pl. 35
Fissurella angusta Gmelin
    Range: Florida Keys to British Guiana.
    Habitat: Intertidal.
    Description: About ¾ in. long. Rather flattish, somewhat pointed in front. Sculpture of about 10 rather nodulous ribs. Color whitish, with reddish-brown blotches; interior greenish, pale brown around orifice.

BARBADOS KEYHOLE LIMPET Pl. 35
Fissurella barbadensis Gmelin
    Range: S. Florida to n. S. America.
    Habitat: Intertidal.
    Description: Length about 1 in. Solid, conical, highly elevated, with round or oval orifice at top. Surface rough, with a heavy assortment of radiating ribs that project at the margins and give the shell a scalloped edge. Variable in color, generally some shade of gray, green, or pink, with brown blotches; interior usually has alternating rings of dull green and white.

ROCKING-CHAIR KEYHOLE LIMPET Pl. 35
Fissurella fascicularis Lam.
    Range: N. Carolina to West Indies.
    Habitat: Intertidal.
    Description: Length ¾ in. Rather flat. Base oval, front and back ends raised noticeably. Apex opening somewhat long and narrow, with a small notch on each side at center which forms a cross. Sculpture of numerous radiating ribs. Color superficially a faded magenta, with a few reddish rays near the perforation; interior white, pinkish around orifice.
    Remarks: Formerly listed as *F. pustula* Lam. and *F. punc-*

tata Fischer. When placed on a level surface the shell rests only upon the middle of each side, like rockers.

KNOBBY KEYHOLE LIMPET — Pl. 35
Fissurella nodosa Born
  Range: S. Florida to West Indies.
  Habitat: Intertidal.
  Description: About 1 in. long. Moderately elevated, with elongated opening at summit. Sculptured with 20 or so radiating ribs that bear nodules concentrically arranged. Base broadly oval, margins deeply scalloped. Color may be white, brown, gray, or pinkish; interior white.

ROSY KEYHOLE LIMPET — Pl. 35
Fissurella rosea Gmelin
  Range: S. Florida to West Indies.
  Habitat: Intertidal.
  Description: Length about 1 in. Shell conical but rather flat, the summit bearing a small, usually round opening. Marginal outline egg-shaped, the front end narrower. Sculpture of close, unequal radiating lines. Color variable — commonly delicate pink with paler rays, but some individuals are deep pink or purple; interior greenish at margins, white at center, often with pinkish around orifice.

Genus *Diodora* Gray 1821

ARCUATE KEYHOLE LIMPET — Pl. 35
Diodora arcuata (Sby.)
  Range: S. Florida to Trinidad.
  Habitat: Shallow water.
  Description: Length ½ in. A well-arched species, the small orifice close to the front end. Sculpture of weak radiating lines. Color whitish, faintly rayed with pale brown.

LITTLE KEYHOLE LIMPET — Pl. 35
Diodora cayenensis (Lam.)
  Range: Maryland and south; Gulf of Mexico, West Indies, and south to Brazil.
  Habitat: Intertidal.
  Description: Length 1 to 1½ in. Shell moderately solid and highly elevated, with elongated hole at summit. Surface sculptured with numerous sharp and distinct radiating lines (every 4th one larger) and wrinkled by growth lines. Margins finely crenulate. Usually white but may be buff, pink, or dark gray; interior white or bluish gray.
  Remarks: Formerly listed as *D. alternata* (Say). In the genus *Diodora* the orifice is bordered on the inside by a deep notch toward the front.

DYSON'S KEYHOLE LIMPET Pl. 36
Diodora dysoni (Reeve)
  Range: Florida, West Indies, and south to Brazil.
  Habitat: Shallow water.
  Description: Length about ½ in. Base oval, the ends somewhat pointed. Not highly elevated; perforation small, triangular as a rule and not far from middle of shell. Small knob
  behind orifice. Surface with strong radiating ribs, usually 3
  small ones between each larger rib. Color white, with broad
  bands of black.

LISTER'S KEYHOLE LIMPET Pl. 35
Diodora listeri (Orb.)
  Range: Bermuda; Florida to West Indies.
  Habitat: Intertidal.
  Description: Length about 1½ in. Shell solid, oval, highly
  elevated. Apex slightly in front of middle of shell, with slitlike opening. Surface has alternating large and small ribs
  crossed by strong cordlike lines. Margins scalloped. Color
  white or buff, sometimes with darker rays; interior white.

MINUTE KEYHOLE LIMPET Pl. 35
Diodora minuta (Lam.)
  Range: S. Florida to West Indies.
  Habitat: Intertidal.
  Description: Length about ½ in. Shell well arched, elongated, apex closer to front end. Sides slightly raised, so shell
  rests on its ends. Surface ornamented with beaded ridges that
  radiate from the summit. Color white, blotched or rayed with
  brown or black; interior white, with black ring around orifice.

TANNER'S KEYHOLE LIMPET Pl. 35
Diodora tanneri (Verrill)
  Range: Delaware Bay to Cape Hatteras.
  Habitat: Deep water.
  Description: Length nearly 2 in. Shape oval, well elevated.
  Anterior end slightly smaller than posterior. Orifice round
  and closer to front end, the slope of which is concave and
  steep; posterior slope is gradual and convex. Surface with fine
  and regular radiating lines, margins smooth. Color grayish
  white.
  Remarks: Our largest and perhaps most beautiful member of
  the keyhole tribe, but not likely to be picked up on the
  beach. Verrill's type specimen came from more than 600 ft.
  and was described as *Fissurella tanneri* in 1882.

VARIEGATED KEYHOLE LIMPET Pl. 35
Diodora variegata (Sby.)
  Range: Bahamas to West Indies.

Habitat: Shallow water.
Description: Length ½ in. Rather thin and elongate, moderately arched. Orifice near front end. Fine radiating ribs and concentric lines produce a somewhat reticulated pattern. Color white, with brownish rays.
Remarks: Some authorities regard this as a subspecies of the Minute Keyhole Limpet, *D. minuta* (Lam.), above.

GREEN KEYHOLE LIMPET Pl. 36
Diodora viridula (Lam.)
Range: S. Florida; Caribbean islands to Trinidad.
Habitat: Intertidal.
Description: Length about 1 in. Moderately arched, orifice elongated, situated closer to front end. Strong radiating ribs that are white, with smaller greenish ribs between. Interior white. Black stain around perforation, inside and outside.

Genus *Lucapina* Sowerby 1835

SOWERBY'S KEYHOLE LIMPET Pl. 36
Lucapina sowerbii (Sby.)
Range: S. Florida, West Indies, and south to Brazil.
Habitat: Intertidal.
Description: Length ¾ in. An elongated shell with rounded ends and straight sides. Apex but little raised, shell moderately elevated. Perforation large and rounded, set close to anterior end. Surface bears alternating large and small ribs that are crossed by numerous concentric lines. Color white to buff, blotched with brown; interior white, no ring around orifice.
Remarks: Formerly listed as *L. adspersa* (Phil.)

CANCELLATED KEYHOLE LIMPET Pl. 36
Lucapina suffusa (Reeve)
Range: Florida to West Indies.
Habitat: Intertidal.
Description: Length about 1 in. Shell conical, somewhat depressed, oval. Summit with round orifice, slightly ahead of middle of shell. Surface ornamented by strong radiating ribs, alternating in size and cancellated by regular, concentric ridges. Color buffy or white; interior white, with bluish ring around orifice.
Remarks: Formerly listed as *L. cancellata* (Sby.).

Genus *Lucapinella* Pilsbry 1890

FILE KEYHOLE LIMPET Pl. 36
Lucapinella limatula (Reeve)
Range: N. Carolina to West Indies.

Habitat: Moderately deep water.
Description: Length ½ to ¾ in. Shell oval, only moderately elevated; orifice near center and large, often somewhat triangular. Sculpture of alternating larger and smaller radiating ribs, made scaly by concentric wrinkles. Color brownish, with spotted whitish rays; interior white.

Genus *Emarginula* Lamarck 1801

RUFFLED EMARGINULA   **Pl. 36**
Emarginula phrixodes Dall
  Range: N. Carolina to West Indies.
  Habitat: Deep water.
  Description: About ¼ in. long. An oval shell, moderately elevated and arched. No summit perforation, but there is a slit at margin. Surface bears radiating and concentric ribs; the radiating ribs make the margins somewhat scalloped. Color translucent white.

PYGMY EMARGINULA   **Pl. 36**
Emarginula pumila A. Adams
  Range: S. Florida to Brazil.
  Habitat: Shallow water.
  Description: Length about ½ in. Base oval, shell well arched, apex pointing backward. A very small slit at anterior margin. Sculpture of radiating ribs and rather strong concentric lines. Color yellowish to whitish, darker between ribs; interior with greenish rays.

Genus *Hemitoma* Swainson 1840

EMARGINATE LIMPET   **Pl. 36**
Hemitoma emarginata (Blain.)
  Range: Florida Keys to West Indies.
  Habitat: Moderately deep water.
  Description: Length ¾ to 1 in. A highly elevated, caplike shell, the apex strongly curved backward. Small slit at anterior margin when young. Sculpture of concentric threads and radiating lines. Base oval, front end truncated and provided with a sharp notch. Color whitish; interior pale green.

EIGHT-RIBBED LIMPET   **Pl. 36**
Hemitoma octoradiata (Gmelin)
  Range: Florida Keys to West Indies.
  Habitat: Shallow water.
  Description: Length 1 in. Shell solid, flattened to moderately arched. Apex closer to anterior, and but slightly curved backward. Ornamentation of 8 nodular ribs, with smaller

lines between. Very small notch at front end. Color grayish white; interior grayish green, commonly with a darker zone near the rim.

Genus *Rimula* Defrance 1827

BRIDLE RIMULA *Rimula frenulata* Dall **Pl. 36**
Range: N. Carolina to West Indies.
Habitat: Moderately deep water.
Description: Less than ¼ in. long. Elongate and canoe-shaped, with a sharp apex that overhangs the short front end. Surface bears fine radiating and concentric lines, and midway of the rear slope there is an elongated slit that fills up in front as the shell increases in size. Translucent white; interior glossy.

Genus *Puncturella* Lowe 1827

NOAH'S PUNCTURED SHELL **Pl. 36**
Puncturella noachina (Linn.)
Range: Circumpolar; Arctic Ocean to Massachusetts.
Habitat: Shallow to deep water.
Description: About ½ in. long. Steeply conical, with oval base, laterally compressed. Sharply pointed apex; just in front of tip is a narrow slit. Sculpture of rather sharp radiating ribs. Color dull white.
Remarks: Sometimes listed as *P. princeps* (Mighels & Adams).

Limpets: Family Acmaeidae

SHELLS conical, oval, open at base. There is no opening at the summit. Not spiral at any stage of growth. Mollusks of the shore region, adhering to rocks and grasses.

Genus *Acmaea* Eschscholtz 1830

SOUTHERN LIMPET *Acmaea antillarum* (Sby.) **Pl. 36**
Range: S. Florida to West Indies.
Habitat: Intertidal.
Description: About ¾ in. long. Rather flat, base oval, apex closer to front end. Gray or buff, with narrow black lines radiating from apex which sometimes combine to form broad rays; interior polished, bluish gray, with broadly checked border.
Remarks: Formerly listed as *A. candeana* (Orb.).

JAMAICA LIMPET *Acmaea jamaicensis* (Gmelin) **Pl. 37**
Range: Florida Keys to West Indies.
Habitat: Shallow water.
Description: Slightly more than ½ in. long. Apex nearly central, and well elevated. Sculpture of fine radiating lines, usually pale yellowish or white on a darker background. Interior white, with light brown callus, and often with brown-spotted rim.
Remarks: Sometimes listed under a later name, *A. albicosta* C. B. Adams.

CUBAN LIMPET *Acmaea leucopleura* (Gmelin) **Pl. 37**
Range: S. Florida to West Indies.
Habitat: Intertidal; commonly on large snail shells.
Description: Length ½ to ¾ in. Roundish, well elevated at summit. Shell marked with radiating black lines on white background. Apex generally eroded. Interior white with a checkered rim, and the callus presents an arrow-shaped design in dark brown.
Remarks: Sometimes listed as *A. cubensis* (Reeve).

SPOTTED LIMPET *Acmaea pustulata* (Helb.) **Pl. 37**
Range: S. Florida to West Indies.
Habitat: Intertidal.
Description: About ¾ in. long. A heavy, oval, thick shell with parallel sides; only moderately arched. Apex nearly central. Surface decorated with coarse radiating ribs (few) and sometimes by weak concentric lines. Color yellowish white, usually with reddish specks; interior polished white, callus yellowish.
Remarks: Formerly listed as *A. punctulata* (Gmelin).

DWARF LIMPET *Acmaea rubella* (Fabr.) **Pl. 36**
Range: Arctic Ocean to Maine.
Habitat: Moderately deep water.
Description: A small oval shell about ¼ in. long. Apex well elevated and closer to front end, which slopes rather steeply. Shell smooth, thin, and fragile. Color whitish or rusty brown.

TORTOISESHELL LIMPET **Pl. 36**
Acmaea testudinalis (Müller)
Range: Arctic Ocean to New York.
Habitat: Shallow water; on rocks.
Description: Averages about 1 in. long, but some specimens may be close to 2 in. Shell conical, oblong-oval, moderately arched. Surface relatively smooth, color bluish white, checkered with dark brown marks radiating from the summit. Interior dark glossy brown, with a checkered gray and brown

border; central area and border separated by a paler band.
Remarks: The largest specimens are found in the vicinity of Eastport, Maine. The Eelgrass Limpet, *A. t. alveus* (Conrad), shown on Plate 36, is a form that lives on the narrow leaves of eelgrass instead of on rocks. It is smaller — seldom more than ½ in. long — well arched, and has parallel sides; often is boldly checkered with yellowish or whitish dots, which are sometimes plainer on inside of shell than on outside. This form may be found from Maine to Connecticut.

Blind Limpets: Family Lepetidae

SHELLS small, oval, somewhat depressed. Apex nearly central. Distributed chiefly in cold seas and moderately deep water.

Genus *Lepeta* Gray 1842

NORTHERN BLIND LIMPET   Pl. 37
Lepeta caeca (Müller)
  Range: Greenland to Massachusetts.
  Habitat: Moderately deep water.
  Description: About ¼ in. long. Oval, conical, but not highly arched. Numerous minute radiating lines, crossed by equally fine concentric lines. The resulting sculpture, when viewed under a lens, is a distinct network. Color white or grayish white, inside and outside.

Addison's Limpets: Family Addisoniidae

THIS family was erected by W. H. Dall because the following species, although outwardly like other limpets, is quite different inside, particularly in respect to its gills and dentition. It was named in honor of Addison E. Verrill, distinguished Yale zoologist, who was a contemporary of Dall.

Genus *Addisonia* Dall 1882

PARADOX LIMPET   *Addisonia paradoxa* Dall   Pl. 37
  Range: Rhode Island to Virginia.
  Habitat: Moderately deep water.

Description: A thin oval shell, ½ in. long. Apex very close to front end and slightly hooked. Surface has extremely fine radial grooves and a somewhat shiny appearance. Color white; interior well polished.

Pearly Top Shells:
Family Trochidae

HERBIVOROUS snails, widely distributed in warm and cold seas, living among seaweeds in shallow water, and upon wave-washed rocks. The shells are composed largely of iridescent nacre, although the pearly luster may be concealed by pigmentation of various colors on the outer surface. They are varied in shape but are commonly pyramidal. The operculum is thin and horny.

Genus *Margarites* Gray 1847

RIDGED TOP SHELL　　　　　　　　　　　　　　　　Pl. 37
Margarites cinereus (Couthouy)
　　Range: Greenland to Massachusetts.
　　Habitat: Moderately shallow water.
　　Description: About ½ in. high. Shell thin and of depressed conical shape, with from 5 to 7 whorls, made angular by prominent revolving ridges. Umbilicus broad and deep. Aperture circular, with sharp outer lip. Color ashy white, sometimes tinged with green.

GREENLAND TOP SHELL　　　　　　　　　　　　　Pl. 37
Margarites groenlandicus (Gmelin)
　　Range: Greenland to Massachusetts.
　　Habitat: Moderately shallow water.
　　Description: Height ⅓ in., diam. about ½ in. 4 or 5 whorls, somewhat flattened above and undulated near sutures by short folds or wrinkles. Shell thin but strong, sculptured by numerous revolving lines. Umbilicus broad and funnel-shaped, operculum thin and horny. Color dull reddish brown; aperture pearly.

SMOOTH TOP SHELL　　　　　　　　　　　　　　　Pl. 37
Margarites helicinus (Phipps)
　　Range: Greenland to Massachusetts.
　　Habitat: Moderately shallow water.
　　Description: Diam. about ⅓ in. 4 whorls, body whorl rela-

tively large. Sutures well impressed. Surface smooth and shining, color pale yellowish horn color, often with iridescent sheen.

Remarks: This little snail lives on eelgrass and other marine vegetation, well below the low-water level. After storms they may be found clinging to plants that have been torn from their moorings and cast up on the beach.

Genus *Solariella* Wood 1842

CHANNELED TOP SHELL Pl. 37
Solariella lacunella Dall
    Range: Virginia to Florida.
    Habitat: Moderately deep water.
    Description: Height about ¼ in. Some 5 whorls, with fine revolving lines, those on the shoulders beaded. Sutures rather deeply channeled. Aperture circular, umbilicus small. Color white or pale pearly gray.

LAMELLOSE TOP SHELL Pl. 37
Solariella lamellosa Verrill & Smith
    Range: Massachusetts to West Indies.
    Habitat: Moderately deep water.
    Description: About ⅛ in. high. 4 or 5 whorls, moderately tall spire, apex rather blunt. Sutures channeled, umbilicus deep. Sculpture of 3 prominent revolving ridges and numerous weak vertical lines. Color yellowish white.

OBSCURE TOP SHELL Pl. 37
Solariella obscura (Couthouy)
    Range: Labrador to Virginia.
    Habitat: Moderately deep water.
    Description: Height ¼ in., diam. about the same. Whorls (5) slightly angular because of single beaded line at shoulders. Umbilicus open but small, aperture circular. Color gray, often with pinkish tint.

LITTLE RIDGED TOP SHELL Pl. 37
Solariella vericosa (Mighels & Adams)
    Range: Labrador to Maine.
    Habitat: Moderately deep water.
    Description: Height ¼ in. 4 or 5 whorls, fairly well elevated spire. Aperture round, lip thin and sharp. Small but deep umbilicus. Volutions decorated with distinct vertical folds. Color deep gray.

Genus *Lischkeia* Fischer 1879

BEADED TOP SHELL   **Pl. 37**
Lischkeia infundibulum (Watson)
   Range: Nova Scotia to N. Carolina.
   Habitat: Deep water.
   Description: About 1 in. high, with 6 or 7 rounded whorls
that produce a tall, pointed spire, the sutures rather indistinct. Each volution bears 2 revolving ridges studded with
tiny spikelike beads. 5 sharp revolving ridges on base of
shell. Umbilicus deep. Color yellowish gray.

OTTO'S TOP SHELL   *Lischkeia ottoi* (Phil.)   **Pl. 37**
   Range: Newfoundland to N. Carolina.
   Habitat: Moderately deep water.
   Description: Height about ¾ in. Stoutly top-shaped, with 5
rounded whorls tapering to a sharp apex. Aperture large and
nearly round, lip thin and sharp. Deep umbilicus. Each volution with sharp revolving lines on lower portion and strongly
beaded lines on upperpart. Color creamy white, with pearly
iridescence.
   Remarks: A handsome shell, seldom seen in collections. It
used to be known as *Margarites regalis* Verrill & Smith.

Genus *Synaptocochlea* Pilsbry 1890

FALSE STOMATELLA   *Synaptocochlea picta* (Orb.)   **Pl. 37**
   Range: Bermuda, Florida, West Indies.
   Habitat: Moderately shallow water.
   Description: About ¼ in. long. Only 2 or 3 whorls and
mostly all body whorl. Apex short, aperture large and flaring. Sculpture of weak revolving lines. Color white, blotched
with reddish brown. Columella white.

Genus *Calliostoma* Swainson 1840

ADELE'S TOP SHELL   **Pl. 37**
Calliostoma adelae Schwengel
   Range: S. Florida.
   Habitat: Moderately shallow water.
   Description: About ¾ in. high. Top-shaped, solid; 8 to 10
flat-sided whorls. Apex pointed. Volutions with beaded revolving lines, 2 larger ridgelike lines at suture, giving the shell a
keeled appearance. Narrow umbilicus. Color pale brown, mottled with white.

BAIRD'S TOP SHELL Pl. 37
Calliostoma bairdi Verrill & Smith
Range: Massachusetts to Florida.
Habitat: Moderately deep water.
Description: About 1¼ in. high. Some 6 whorls, tapering regularly to a sharply pointed apex. No shoulders, sutures indistinct. Each volution decorated with revolving rows of small beads, with 1 row near the suture larger than others. Color yellowish brown, more or less spotted with squarish red dots.
Remarks: Like the others in this group, the shell has a pearly layer under its pebbly exterior.

FLORIDA TOP SHELL Pl. 37
Calliostoma euglyptum (A. Adams)
Range: Cape Hatteras to Mexico.
Habitat: Shallow water.
Description: Strong and solid, height about 1 in. Top-shaped, apex well elevated, base flat. 5 or 6 whorls, sutures rather indistinct. Ornamentation consists of revolving rows of beads. Columella oblique, thickened at base. Color dingy white, mottled with reds and browns.

JAVA TOP SHELL *Calliostoma javanicum* (Lam.) Pl. 37
Range: S. Florida to West Indies.
Habitat: Moderately shallow water.
Description: About 1¼ in. high. Solid, top-shaped. 9 or 10 flat-sided whorls, with sharp peripheral keel. Sutures indistinct. Sculpture of fine beaded ridges. Columella arched and twisted, umbilicus funnel-shaped. Color pale brownish, mottled with reds and orange.

MOTTLED TOP SHELL Pl. 37
Calliostoma jujubinum (Gmelin)
Range: Cape Hatteras to West Indies.
Habitat: Shallow water.
Description: Height 1 in. Robust and conical, apex acutely pointed. 8 to 10 whorls, sutures indistinct. Surface sculptured with revolving cords, those on the shoulders broken into beads. Umbilicus narrow and funnel-like. Color brownish with gray and white streaks.

PEARLY TOP SHELL Pl. 37
Calliostoma occidentale (Mighels & Adams)
Range: Nova Scotia to Massachusetts.
Habitat: Moderately shallow water.
Description: About ½ in. high, diam. about the same. Shell thin, acutely pointed, with flattened base. 5 or 6 whorls, en-

circled by strong revolving ridges, the uppermost broken into a series of disjointed dots. Outer lip thin and made wavy by terminating ridges. No umbilicus. Creamy white to pearly white; aperture iridescent.

PSYCHE'S TOP SHELL   *Calliostoma psyche* Dall   **Pl. 37**
Range: Cape Hatteras to Florida.
Habitat: Moderately deep water.
Description: About ¾ in. high. Shell thin but strong, with 8 flat-sided whorls. Keel at periphery. Sculpture of beaded lines, sutures indistinct. No umbilicus. Color yellowish, marked with patches of brownish red.

BEAUTIFUL TOP SHELL   **Pl. 37**
Calliostoma pulchrum (C. B. Adams)
Range: N. Carolina to West Indies.
Habitat: Moderately shallow water.
Description: About ¾ in. high. Top-shaped, with flat sides. 8 to 10 whorls, with angled keel at periphery. Sutures indistinct. Sculpture of finely beaded revolving lines. Color mottled with small reddish-brown spots, rather evenly spaced, and with wavy vertical streaks of grayish.

YUCATAN TOP SHELL   **Pl. 37**
Calliostoma yucatecanum Dall
Range: N. Carolina to Yucatan.
Habitat: Moderately shallow water.
Description: About ½ in. high. 8 whorls, producing a slightly convex spire, apex sharp. Sutures distinct. Sculpture of numerous revolving cords. Aperture moderately large, umbilicus deep. Color pinkish white, with scattered short brown dashes.

Genus *Cittarium* Philippi 1847

MAGPIE SHELL   *Cittarium pica* (Linn.)   **Pl. 37**
Range: West Indies.
Habitat: Shallow water.
Description: Height about 4 in. Large, solid, top-shaded. 5 or 6 whorls and moderately sharp apex. Surface irregular, umbilicus deep, aperture roundish. Operculum leathery, with numerous whorls. Color dark green, heavily splashed with white zigzag markings.
Remarks: This large and showy species is found as a dead shell in s. Florida. It is frequently found in old Indian shell heaps along the Florida coast, suggesting that the species may have been abundant on our shores in the past. It occurs

quite commonly as a Pleistocene fossil in Florida and Bermuda. The shell is very pearly, and when the outer layer has been removed it can be given a high polish. Until recently it was listed as *Livona pica* (Linn.).

Genus *Tegula* Lesson 1832

GREEN-BASE TOP SHELL   Pl. 38
Tegula excavata (Lam.)
   Range: S. Florida to West Indies.
   Habitat: Shallow water.
   Description: About ¾ in. high. 4 or 5 flattish whorls, with no shoulders, but with sharply delineated sutures. Base concave. Aperture rather small, with umbilicus distinct. Color purplish brown or black, but most specimens are eroded in places to show the pearly layer. The excavated base and aperture often are greenish.

COLORFUL TOP SHELL   *Tegula fasciata* (Born)   Pl. 38
   Range: S. Florida to West Indies.
   Habitat: Shallow water.
   Description: About ½ in. high. Shell roundly pyramidal, consisting of 4 or 5 whorls, with smooth, sometimes shiny surface. Aperture oval, feebly toothed within, and there may be 2 teeth at base of columella. White callus may extend partially over umbilicus. Operculum thin and horny. Color grayish pink, rather heavily mottled and marked with brownish or black. There is commonly a paler band on body whorl.

WEST INDIAN TOP SHELL   Pl. 38
Tegula lividomaculata C. B. Adams
   Range: Florida Keys to West Indies.
   Habitat: Shallow water.
   Description: Height ¾ in., with 5 or 6 rather rounded whorls decorated with distinct revolving lines that are weakly beaded. Sutures well impressed. Deep umbilicus bordered by a pair of strong lines, or ridges. Aperture oval, operculum horny. Color gray, somewhat mottled with reds and blacks.
   Remarks: Formerly listed as *T. scalaris* Anton.

Seguenzias: Family Seguenziidae

THESE are tiny but beautifully sculptured shells occurring in deep water. The shell is trochiform (top-shaped), usually translucent and pearly, and is boldly decorated with revolving ridges and sharp vertical lines.

Genus *Seguenzia* Jeffreys 1876

PRECIOUS SEGUENZIA   *Seguenzia eritima* Verrill   **Pl. 38**
   Range: South of Martha's Vineyard.
   Habitat: Deep water.
   Description: Height ⅛ in. A stoutly conical shell with about 7 whorls, strongly angulated and carinated (ridged) in the middle. Sutures distinct. Sculpture of sharp revolving lines and strong vertical ridges. Aperture angulated, with notch at its upper angle and a pair of notches at base. Color tannish gray.
   Remarks: This delicately ornate shell has been dredged from a depth of more than 6000 ft.

GIRDLE SEGUENZIA   **Pl. 38**
Seguenzia monocingulata (Seguenza)
   Range: Maine to Gulf of Mexico.
   Habitat: Deep water.
   Description: Height ⅛ in. A delicate little shell of 6 or 7 whorls. Body whorl bears 3 sharp and prominent revolving ridges, the spaces between them concave and crossed by numerous regularly spaced thin and curving vertical riblets. Color translucent pearly white.
   Remarks: Formerly listed as *S. formosa* Jeffreys.

Turbans: Family Turbinidae

SHELLS generally heavy and solid, turbinate (top-shaped), and commonly brightly colored with pearly underlayers. The surface may be smooth, wrinkled, or spiny. There is a heavy calcareous operculum, and no umbilicus as a rule. These are herbivorous gastropods, native to tropic seas throughout the world. Many are used for ornamental purposes. The famous cat's-eye of s. Pacific island jewelry is the colorful operculum of a member of this family (*Turbo petholatus* Linn.).

Genus *Arene* H. & A. Adams 1854

STAR ARENE   *Arene cruentata* (Mühl.)   **Pl. 38**
   Range: S. Florida to West Indies.
   Habitat: Shallow water.
   Description: About ¼ in. high, diam. ½ in. 4 or 5 sharply angled whorls, each with a series of prominent triangular

spines at the periphery. Spire low, aperture round, umbilicus small but deep. Color white, variously marked with red and brown.

THREE-RIDGED ARENE   **Pl. 38**
Arene tricarinata (Stearns)
   Range: N. Carolina to West Indies.
   Habitat: Shallow water.
   Description: Height ⅛ in. 3 or 4 whorls, low spire, distinct sutures. Sculpture of 3 revolving ridges, beaded at the shoulders. Aperture round, umbilicus deep. Color whitish, blotched with reddish.
   Remarks: Formerly listed as *Liotia gemma* Toumey & Holmes.

Genus *Cyclostrema* Marryat 1818

TORTUGAS CYCLOSTREME   **Pl. 38**
Cyclostrema tortuganum Dall
   Range: Florida Keys.
   Habitat: Deep water.
   Description: Diam. ⅜ in. A flatly coiled shell of 3 or 4 whorls. Deep umbilicus on lower side. Decorated with a double row of pronounced knobs at periphery, surface in between with sharp crosslines. Color dull yellowish white.

Genus *Turbo* Linnaeus 1758

CHANNELED TURBAN   **Pl. 38**
Turbo canaliculatus Hermann
   Range: S. Florida to West Indies.
   Habitat: Shallow water.
   Description: Height 2 to 3 in. About 5 well-rounded whorls, each sculptured on its upper half by deeply incised revolving lines. Aperture round, outer lip sharp, no umbilicus. Thick, round, calcareous operculum. A deep, smooth channel at suture. Surface smooth, often polished. Color greenish yellow, usually mottled and checked with green and brown; aperture pearly.
   Remarks: Formerly listed as *T. spenglerianus* Gmelin.

KNOBBY TURBAN   *Turbo castaneus* Gmelin   **Pl. 38**
   Range: N. Carolina to Texas; West Indies.
   Habitat: Shallow water.
   Description: Height about 1½ in. Shell solid and roundly top-shaped, apex rather sharply pointed. 5 or 6 whorls, deco-

rated with revolving ridges of beads, those on the shoulders more pronounced. Aperture large and round, reflecting somewhat on columella. Operculum thick and calcareous. Color buffy brown, more or less blotched with darker brown; occasional specimens may be dull greenish.

Genus *Astraea* Röding 1798

AMERICAN STAR SHELL   Pl. 38
Astraea americana (Gmelin)
>**Range:** Florida to West Indies.
>**Habitat:** Shallow water.
>**Description:** About 1 in. high. 7 or 8 whorls, apex bluntly pointed. Solid and stony, with well-elevated spire. Volutions decorated with vertical folds that terminate in knobs on the shoulders. Outer lip crenulate, operculum calcareous. Color greenish or grayish white.

CARVED STAR SHELL   *Astraea caelata* (Gmelin)   **Pl. 38**
>**Range:** S. Florida to West Indies.
>**Habitat:** Shallow water.
>**Description:** About 3 in. high. A large species, rugged and conical, sculptured with a series of strong, oblique ribs that become tubular at the periphery, and with equally prominent revolving ribs. 6 or 7 whorls. Aperture large and oblique, columella somewhat curved. Heavy calcareous operculum, dull white and finely pustulate. Color greenish white, mottled with reds and brown.

LONG-SPINED STAR SHELL   Pl. 38
Astraea phoebia Röding
>**Range:** S. Florida to West Indies.
>**Habitat:** Shallow water.
>**Description:** Height about 1 in., diam. 2 in. or more. Shell strong and solid, spire very low, base flat. 6 or 7 whorls, the margins sharply keeled. Ornamentation consists of a series of triangular sawlike spines that project beyond the periphery of shell. Color iridescent silvery white.
>**Remarks:** Formerly listed as *A. longispina* (Lam.). The silvery color and ornate appearance make it a great favorite with collectors and manufacturers of shell jewelry, but empty shells left on the beach soon lose their luster. There is considerable variation in the length of the spines, and those with short spines are often labeled *A. brevispina* (Lam.), but that name belongs to a West Indian snail (Short-spined Star Shell) characterized by a small patch of orange at the umbilicus; see Plate 38.

IMBRICATED STAR SHELL **Pl. 38**
Astraea tecta (Lightf.)
> **Range:** S. Florida to West Indies.
> **Habitat:** Shallow water.
> **Description:** About 1½ in. high. Shell strong and rough, steeply pyramidal with sharp apex. About 7 whorls, sutures distinct. Vertical folds on volutions which form squarish tube-like ribs, so surface is unusually rough and nodular. Color dull greenish brown.
> **Remarks:** Formerly listed as *A. imbricata* (Gmelin). Some authorities consider it a subspecies of the American Star Shell, *A. americana* (above).

GREEN STAR SHELL *Astraea tuber* (Linn.) **Pl. 4**
> **Range:** S. Florida to West Indies.
> **Habitat:** Shallow water.
> **Description:** About 2 in. high. A rugged and solid shell of about 5 whorls. Sculptured with strong obliquely vertical ribs swollen at the base and shoulders to present a double row of nodes on each volution. No umbilicus, operculum calcareous. Color green, with the pearly layer generally showing through in several places, especially at apex. Aperture very pearly, and there is a broad pearly area on the inner lip.

Genus *Homalopoma* Carpenter 1864

WHITE DWARF TURBAN **Pl. 38**
Homalopoma albida (Dall)
> **Range:** S. Florida to Texas; West Indies.
> **Habitat:** Moderately deep water.
> **Description:** Height ¼ in. A small top-shaped shell of about 5 whorls, sutures rather indistinct. Sculpture of sharp revolving lines. Commonly a small tooth at base of columella. Aperture round, operculum calcareous. No umbilicus. Color grayish white.

Pheasant Shells:
Family Phasianellidae

THE shell is fairly high spired, graceful in shape, porcelaneous, and shining. There is no periostracum. Tropical species attain a height of 2 or 3 inches and are among our most showy shells, usually brightly colored with pinks and browns. Only a few members of this family occur on our shores, and they all are small.

Genus *Tricolia* Risso 1826

CHECKERED PHEASANT SHELL   **Pl. 38**
Tricolia affinis (C. B. Adams)
  Range: S. Florida to West Indies.
  Habitat: Shallow water.
  Description: About ¼ in. high. 4 or 5 whorls, the spire
  tapering gradually to a blunt apex. Sutures distinct. Surface
  polished. Color pale buff or gray, variously checkered and
  spotted with pink and reddish brown and often with a narrow
  encircling band of yellow or orange on body whorl.

UMBILICATE PHEASANT SHELL   **Pl. 38**
Tricolia thalassicola Rob.
  Range: Off N. Carolina and south through West Indies to
  Brazil.
  Habitat: Shallow water.
  Description: Height ¼ in. 5 whorls, sutures well impressed.
  Apex blunt. Tiny chinklike umbilicus. Shell polished, color
  creamy white, checkered and flecked with red and gray.
  Remarks: Formerly listed as *Phasianella umbilicata* Orb.

Nerites: Family Neritidae

SMALL, usually bright-colored shells, mostly globular in shape,
and commonly with toothed apertures. They inhabit warm
countries, where they are often found abundantly in shallow
seas, in brackish waters, in freshwater, and in some cases even
on dry land.

Genus *Nerita* Linnaeus 1758

ANTILLEAN NERITE   *Nerita fulgurans* Gmelin   **Pl. 39**
  Range: Bermuda; s. Florida to West Indies.
  Habitat: Shallow, brackish water.
  Description: Height about ¾ in., diam. about 1 in. 4 or 5
  whorls, little or no spire. Sculpture of close-set revolving
  lines. Aperture wide; broad shelf on inner lip, with feeble
  teeth. 2 nublike teeth at top of outer lip. Color yellowish
  gray and mottled with brown.
  Remarks: This gastropod shows a marked preference for
  brackish water.

BLEEDING TOOTH *Nerita peloronta* Linn.   **Pls. 4, 39**
Range: Bermuda; s. Florida to West Indies.
Habitat: Shallow water.
Description: Height 1 to 1½ in. Shell thick and heavy, spire low. Surface bears several broadly rounded revolving ribs. Outer lip feebly toothed within, inner lip (columellar margin) sports 1 or 2 glistening white, strong central teeth surrounded by a rich red stain. Operculum shelly. Color yellowish white, marked with zigzag bars of red and black, the black predominating.
Remarks: Popular names are sometimes misleading or meaningless, but one has only to look into the aperture to see how perfectly the name Bleeding Tooth fits this snail. It is most active at night, when it roams about the rocks between tides feeding upon algae.

CHECKERED NERITE *Nerita tessellata* Gmelin   **Pl. 39**
Range: Bermuda; Florida to Texas; West Indies.
Habitat: Shallow water.
Description: Diam. ½ to ¾ in., height about same. 4 or 5 whorls separated by indistinct sutures. Surface sculptured with some 12 rounded spiral ribs on each volution; deep narrow grooves in between. Numerous small teeth on columellar margin. Color checkered black and white.

VARIEGATED NERITE *Nerita versicolor* Gmelin   **Pl. 39**
Range: Florida to West Indies.
Habitat: Shallow water.
Description: About 1 in. high. Shell thick and porcelaneous, semiglobose, with little if any spire. About 4 whorls, decorated with broad and rounded revolving ribs; narrow grooves in between. Outer lip toothed within; several robust teeth are present on inner lip. Color whitish, with zigzag bars of black and red.

Genus *Neritina* Lamarck 1816

CLENCH'S NERITE *Neritina clenchi* Russell   **Pl. 39**
Range: S. Florida to West Indies; Cen. America.
Habitat: Fresh or brackish water.
Description: Height about 1 in., glossy. 3 to 4 whorls, apex moderately pointed. Less globular than many of the genus. Color and patterns quite variable. An area at the parietal wall is stained orange-yellow.

SCALY NERITE *Neritina meleagris* Lam.   **Pl. 39**
Range: West Indies; Cen. to S. America.

Habitat: Intertidal, brackish water.
Description: Height about ½ in. 3 whorls, body whorl large, spire short. Minute teeth on columella edge. Aperture wide, no umbilicus. Surface smooth and polished. Color bluish gray or olive, with darker spots rimmed with white.

NETTED NERITE *Neritina piratica* Russell   **Pl. 39**
Range: Cuba and south through West Indies to Brazil.
Habitat: Brackish water, swamps.
Description: About ¾ in. high. Nearly globular, with short spire. 3 or 4 whorls. Surface smooth, color yellowish or olive-green. A network of close black lines produces a pattern that appears dotted.

SPOTTED NERITE *Neritina punctulata* Lam.   **Pl. 39**
Range: West Indies; Cen. America.
Habitat: Freshwater.
Description: Height 1 in. About 3 whorls. Surface smooth but dull. Color purplish or greenish, with many paler spots. Heavy black smudge present on body whorl near the aperture.
Remarks: This is a freshwater snail but is included here since empty shells are frequently to be found on marine beaches.

GREEN NERITE *Neritina reclivata* (Say)   **Pl. 39**
Range: Florida to Texas; West Indies.
Habitat: Brackish water.
Description: Height ½ in., diam. about the same. Shell semiglobular, with 3 or 4 whorls; body whorl constitutes most of shell. Surface smooth, color greenish, commonly with tiny black lines; many examples are pure dark green. Columellar margin glistening white.
Remarks: This snail should be looked for in the tidal areas of streams.

VIRGIN NERITE *Neritina virginea* (Linn.)   **Pl. 39**
Range: Bermuda; Florida to Texas; West Indies.
Habitat: Brackish water.
Description: Diam. usually less than ½ in. Shell semiglobular, consists of 3 or 4 whorls, and body whorl makes up most of the shell. Aperture oval, lip thin. Whole shell highly polished. Color extremely variable — usually some shade of gray-green, tan, or yellow, scrawled all over with lines, circles, and dots of black; often shows banding.
Remarks: This colorful snail travels well up rivers and streams. Very similar species are found in tropic lands all around the world.

Genus *Puperita* Gray 1857

ZEBRA NERITE   *Puperita pupa* (Linn.)   **Pl. 39**
  Range: S. Florida to West Indies.
  Habitat: Intertidal.
  Description: Diam. about ½ in. Shell thin but sturdy, globular, no spire to speak of, and consists of 2 or 3 whorls. Sutures fairly distinct. Outer lip thin and sharp, and there is a broad, flat, polished area at base of columella. Operculum shelly, with flexible border. Color creamy white, spirally striped with fine, irregular black lines. Aperture varies from yellow to bright orange.
  Remarks: Formerly listed as *Neritina pupa* (Linn.) Commonly found in splash pools just above the high-tide mark.

Genus *Smaragdia* Issel 1869

EMERALD NERITE   **Pl. 39**
Smaragdia viridis viridemaris Maury
  Range: Bermuda; s. Florida to West Indies.
  Habitat: Shallow water.
  Description: Diam. about ¼ in. 2 or 3 whorls, practically no spire. Aperture large, lip thin and sharp, broad polished area on inner lip. Surface highly polished. Color pale to brighter green, often with a few whitish streaks on the shoulders.
  Remarks: Typical *S. viridis* (Linn.) is a Mediterranean snail.

Chink Shells:
Family Lacunidae

THESE snails are conical, stout, and thin-shelled. The aperture is half-moon-shaped, and the distinguishing character is a lengthened groove, or chink, alongside the columella.

Genus *Lacuna* Turton 1827

CHINK SHELL   *Lacuna vincta* Turton   **Pl. 40**
  Range: Labrador to New Jersey.
  Habitat: Moderately shallow water.
  Description: About ⅓ in. high. Shell thin and conical, about 5 rather stout whorls separated by moderately deep sutures. Apex pointed. Surface bears minute lines of growth. Aper-

ture semilunar, outer lip thin and sharp; inner lip flattened and excavated by a smooth, elongated groove that terminates in a tiny umbilicus and forms a prominent chink beside the columella. Brownish white, commonly banded with purplish.

Periwinkles:
Family Littorinidae

A LARGE family of shore-dwelling snails found clinging to rocks and plants between the tides, sometimes well up beyond the high-tide limits. The shell is usually sturdy, has few volutions, and is without an umbilicus. Distribution worldwide.

Genus *Littorina* Férussac 1827

SOUTHERN PERIWINKLE   **Pl. 40**
Littorina angulifera (Lam.)
    Range: Bermuda; s. Florida to West Indies.
    Habitat: Mangrove roots and leaves.
    Description: About 1¼ in. high. 6 or 7 whorls, sharp apex. Shell thin but strong. Sutures slightly channeled. Sculpture of minute spiral lines. Outer lip with central groove in lower part. Operculum horny. Color variable — may be gray, reddish, yellow, or purplish, with dark oblique markings.
    Remarks: Sometimes listed as *L. scabra* (Linn.), which is a closely allied species living in s. Pacific.

GULF PERIWINKLE *Littorina irrorata* (Say)   **Pl. 40**
    Range: New Jersey to Florida and Texas.
    Habitat: Intertidal.
    Description: Height about ¾ in. Heavy and robust, sharp apex. 5 whorls, decorated with fine spiral ridges. Outer lip stout, tapering rapidly to a thin edge. Aperture oval, operculum horny. Color soiled white, with spiral rows of chestnut-brown dots. Aperture commonly yellowish.
    Remarks: In the past this periwinkle existed in the north, and fossil specimens (post-Pleistocene) are commonly washed ashore from peat beds on the Connecticut shore of Long Island Sound.

COMMON PERIWINKLE *Littorina littorea* (Linn.)   **Pl. 40**
    Range: Labrador to New Jersey; Europe.

Habitat: Intertidal.
Description: Height ½ to 1¼ in. Shell solid and heavy, with 6 or 7 whorls. Outer lip thick, and black on inside, base of columella white. No umbilicus. Apex sharp, but shell appears rather squat. Surface fairly smooth, brownish olive to nearly black, usually spirally banded with dark brown.
Remarks: This snail, so common in Europe (where it is sold, roasted in its shell, from pushcarts in the city streets), was long believed to be a fairly recently introduced species. Apparently none of the early conchologists ever saw a specimen. The first one reported taken was at Halifax, Nova Scotia, in 1857. From there the species spread down the coast as far as New Jersey, where southward migration was halted, probably by water temperature and possibly by the sandy Jersey beaches, for this is a rock lover. Recent investigations of ancient Indian mounds in the vicinity of Halifax have brought to light unquestioned examples of this species, which may be more than 1000 years old. It is likely that currents off Nova Scotia prevented the free-swimming larvae from reaching the New England shores until eventually a ship, possibly sailing from Halifax to Maine, inadvertently transported some eggs — perhaps in seaweeds used in packing. Once established in northern New England this hardy snail had nothing to prevent its colonizing all of our Northeast.

SPOTTED PERIWINKLE   **Pl. 40**
Littorina meleagris Potiez & Michaud
    Range: S. Florida to West Indies.
    Habitat: Intertidal.
    Description: About ¼ in. high. 3 or 4 whorls, the body whorl making up most of the shell. There is a respectable spire, however, with a pointed apex. Surface smooth and shining, color gray, with revolving lines of darker hue, the lines often broken into short dashes.
    Remarks: Formerly listed as *L. guttata* Phil.

DWARF PERIWINKLE   *Littorina mespillum* Mühl.   **Pl. 40**
    Range: S. Florida to West Indies.
    Habitat: Splash pools at high-water line.
    Description: Height ¼ in. About 4 whorls, spire low. Surface smooth, white or pale brown, sometimes with rows of small round black dots. Aperture glossy brown. Tiny umbilicus. In life there is a thin brownish periostracum.

CLOUDY PERIWINKLE   *Littorina nebulosa* Lam.   **Pl. 40**
    Range: West Indies.
    Habitat: Intertidal.

Description: Height about ¾ in. Fairly tall, with 5 whorls and a pointed apex. Sutures distinct, no umbilicus. Shell rather solid, smooth, and with a dull surface. Color bluish white to yellowish, often with reddish-brown spots arranged in axial lines; aperture purplish.

SMOOTH PERIWINKLE *Littorina obtusata* (Linn.)   **Pl. 40**
Range: Labrador to New Jersey; Europe.
Habitat: Intertidal; commonly on seaweeds.
Description: Height ½ in. Shell stout and globular, smooth and shining, with very faint revolving lines. About 4 whorls, the last large and the others scarcely rising above it. Aperture nearly circular, outer lip sharp. Operculum horny, umbilicus lacking. Color variable, usually yellow, but may be orange, whitish, or reddish brown. Some, especially juveniles, may be banded.
Remarks: Formerly listed as *L. palliata* (Say).

ROUGH PERIWINKLE *Littorina saxatilis* (Olivi)   **Pl. 40**
Range: Arctic Ocean to New Jersey; Europe.
Habitat: Intertidal.
Description: About ½ in. high. Shell ovate, strong, and coarse. 4 or 5 convex whorls, well-defined sutures. Moderately elevated spire, pointed apex. Sculpture of revolving grooves. Yellowish or ash-colored. Some young shells are smooth and variously mottled or spotted with yellow and black.
Remarks: Formerly listed as *L. rudis* (Maton).

ZEBRA PERIWINKLE *Littorina ziczac* (Gmelin)   **Pl. 40**
Range: S. Florida to West Indies.
Habitat: Intertidal.
Description: About ½ in. high. Shell sturdy, sharply pointed, with well-defined keel near base of body whorl. Sculpture of very fine and widely spaced spiral grooves. Whorl count 5 or 6. Aperture oval and small, operculum horny. Color whitish, with many wavy stripes of dark brown or black.

Genus *Nodilittorina* Martens 1897

PRICKLY PERIWINKLE   **Pl. 40**
Nodilittorina tuberculata (Menke)
Range: Bermuda; s. Florida to West Indies.
Habitat: Intertidal.
Description: About ¾ in. high. About 5 whorls, sutures not

very distinct. Aperture large, inner lip somewhat flattened at base. Sculpture of several revolving rows of pointed knobs. Color dark brown.

Genus *Echininus* Clench & Abbott 1942

FALSE PRICKLY PERIWINKLE   **Pl. 40**
Echininus nodulosus (Pfr.)
  Range: S. Florida to West Indies.
  Habitat: Intertidal.
  Description: Just under 1 in. high. About 8 whorls, decorated with revolving rows of pointed knobs, larger and more pronounced along the sutures, which are generally lighter-colored than the rest of the shell. Aperture nearly round, lip moderately thick, no umbilicus. Color mottled gray and black.

Genus *Tectarius* Valenciennes 1833

KNOBBY PERIWINKLE   **Pl. 40**
Tectarius muricatus (Linn.)
  Range: Bermuda; s. Florida to West Indies.
  Habitat: Intertidal.
  Description: About ¾ in. high. Shell sturdy and top-shaped, with about 7 well-rounded whorls, slightly shouldered above, and with a sharp apex. Surface decorated with revolving rows of beads, about 5 rows to a volution. Aperture moderately small and oval. Outer lip somewhat thickened, and there is a small groove down the columella toward the base. Color yellowish gray.
  Remarks: This gastropod is often found on rocks well above the high-tide limits. It is able to survive for long periods out of water; indeed, Henry A. Pilsbry recorded a specimen that revived after being isolated in a cabinet for a year.

Swamp Snails:
Family Hydrobiidae

SHELLS small, with several rounded whorls and a moderately tall spire. They inhabit both marine and nonmarine environments, and are usually abundant in marshes and ditches near the sea.

Genus *Hydrobia* Hartman 1840

SWAMP HYDROBIA *Hydrobia minuta* (Totten)   **Pl. 39**
  Range: Labrador to New Jersey.
  Habitat: Salt-marsh pools.
  Description: About ⅛ in. high. Shell thin and semitransparent, with about 5 rounded whorls, surface faintly marked and wrinkled by growth lines. Aperture oval, operculum horny. Surface shiny, color yellowish brown.
  Remarks: Formerly in the genus *Paludestrina* Orb. 1840.

Genus *Truncatella* Risso 1826

CARIBBEAN TRUNCATELLA   **Pl. 39**
Truncatella caribaeensis Reeve
  Range: Bermuda and West Indies.
  Habitat: Under stones above high tide.
  Description: About ¼ in. high. Tall and slender, with 7 or 8 whorls. Sutures moderately impressed. Sculpture of numerous fine vertical lines, about 25 on each volution. Color white to orange-yellow.
  Remarks: In this genus most shells lose the earlier whorls and appear to be truncated, but occasional specimens retain them and taper to a fine point.

BEAUTIFUL TRUNCATELLA   **Pl. 39**
Truncatella pulchella Pfr.
  Range: Florida to West Indies.
  Habitat: Under stones above high tide.
  Description: About ¼ in. high. Approximately 6 whorls, each with 15 to 18 rather weak vertical lines. Sutures well impressed. Outer lip thin and sharp. Color white to pale yellow.
  Remarks: See Remarks under Caribbean Truncatella.

SHOULDERED TRUNCATELLA   **Pl. 39**
Truncatella scalaris Michaud
  Range: Florida, Bahamas, West Indies.
  Habitat: Under stones above high tide.
  Description: Height about ³⁄₁₆ in. 6 or 7 whorls; indented sutures suggest shoulders. 7 to 10 rather coarse vertical lines, with minute revolving lines between. Outer lip thickened. Color yellowish orange.
  Remarks: See Remarks under Caribbean Truncatella (above).

Rissos: Family Rissoidae

VERY small shells, living for the most part on stones, shells, and sponges. They are variable in structure and occur in nearly all seas. Hundreds of species have been described.

Genus *Rissoina* Orbigny 1840

CARIBBEAN RISSO *Rissoina bryerea* (Montagu) **Pl. 39**
Range: Florida Keys to West Indies.
Habitat: Moderately shallow water.
Description: About ¼ in. high. A rather slender shell of 7 or 8 whorls, the sutures somewhat indistinct. Sculpture of weak vertical grooves. Aperture small, operculum horny. Surface shiny, color white.

CANCELLATE RISSO *Rissoina cancellata* (Phil.) **Pl. 39**
Range: S. Florida to West Indies.
Habitat: Moderately shallow water.
Description: Height ¼ to nearly ½ in. Elongate, with 8 whorls, well rounded and decorated with both vertical and revolving lines, so surface has a cancellate appearance. Aperture oval, lip rather thick. Color yellowish white.

CHESNEL'S RISSO *Rissoina chesneli* (Michaud) **Pl. 39**
Range: N. Carolina to West Indies.
Habitat: Moderately shallow water.
Description: Height ¼ in. About 6 whorls, sutures distinct and volutions moderately well rounded. Aperture round, lip thickened. Sculpture of vertical grooves, sharp and well defined. Surface shiny, color white or yellowish white.

DECUSSATE RISSO *Rissoina decussata* Montagu **Pl. 39**
Range: N. Carolina to West Indies.
Habitat: Moderately shallow water.
Description: About ¼ in. high. 6 or 7 but slightly rounded whorls, sutures indistinct. Aperture semilunar, lip thickened. Surface marked with very fine vertical striations. Color white.

FISCHER'S RISSO *Rissoina fischeri* Desj. **Pl. 39**
Range: West Indies.
Habitat: Shallow water.
Description: Usually less than ⅛ in. high. 6 or 7 rounded whorls, with deep sutures. Strong vertical ridges, about 12 on last volution. Aperture small, lip thin and sharp. Color white.

MANY-RIBBED RISSO Pl. 39
Rissoina multicostata C. B. Adams
 Range: S. Florida to West Indies.
 Habitat: Shallow water.
 Description: Slightly more than ⅛ in. high. 7 or 8 whorls, sutures moderate. Sculpture of numerous fine vertical lines. Color yellowish white.

STRIATE RISSO *Rissoina striosa* C. B. Adams Pl. 39
 Range: S. Florida to West Indies.
 Habitat: Shallow water.
 Description: Nearly ½ in. high. About 8 sloping whorls, sutures weakly impressed. Apex sharp. Sculpture of crowded vertical lines, commonly obscure or lacking on later volutions. Color yellowish white.

Genus *Cingula* Fleming 1828

POINTED CINGULA *Cingula aculeus* (Gould) Pl. 39
 Range: Nova Scotia to Maryland.
 Habitat: Shallow water.
 Description: Height ⅛ in. A tiny shell of 5 or 6 rounded whorls. Spire well elevated, sutures distinct. Surface smooth, aperture oval, no umbilicus. Color brownish.

Genus *Alvania* Risso 1826

WEST INDIAN ALVANIA *Alvania auberiana* Orb. Pl. 38
 Range: West Indies.
 Habitat: Moderately shallow water.
 Description: About ⅛ in. high. A sturdy little shell of 4 or 5 whorls, sculptured with both vertical and revolving lines. Sutures rather deep, whorls slightly shouldered. Color yellowish white.

CARINATE ALVANIA Pl. 38
Alvania carinata (Mighels & Adams)
 Range: Gulf of St. Lawrence to Maine.
 Habitat: Moderately deep water.
 Description: A tiny snail only ¹⁄₁₀ in. high. Moderately conical and quite sturdy. 5 rather convex whorls, sculptured on upperparts with strong vertical lines, the lower half with revolving lines. Aperture nearly circular. Color brown.

NORTHERN ALVANIA *Alvania janmayeni* Friele Pl. 38
 Range: Labrador to N. Carolina.
 Habitat: Moderately deep water.

Description: About ⅛ in. high. 4 or 5 whorls, sutures distinct. Sculpture of prominent revolving lines, the one on the shoulder broken into beads or knobs. Spire rather strongly turreted, the early volutions showing strong vertical ribs. Color brown.

Genus *Benthonella* Dall 1889

TREASURE BENTHONELLA Pl. 40
Benthonella gaza Dall
    Range: Georgia to West Indies.
    Habitat: Moderately deep water.
    Description: Height about ¼ in. 6 or so well-rounded whorls, sutures sharply defined. Apex pointed, aperture round, tiny umbilicus. Surface smooth and shiny. Color white, apex sometimes brown.

Genus *Zebina* H. & A. Adams 1854

SMOOTH RISSO *Zebina browniana* (Orb.) Pl. 39
    Range: S. Carolina to West Indies.
    Habitat: Shallow water.
    Description: About ⅛ in. high. 7 or 8 sloping whorls, sutures indistinct. Apex sharp. Surface smooth and polished, color white, occasionally with pale brownish encircling bands.
    Remarks: Formerly in the genus *Rissoina*.

Vitreous Snails:
Family Vitrinellidae

SMALL, rather flatly coiled snails, living chiefly in deep water. Shells are translucent, and often polished. Small umbilicus, round aperture.

Genus *Vitrinella* C. B. Adams 1850

DALL'S VITRINELLA *Vitrinella dalli* (Bush) Pl. 40
    Range: Off Cape Hatteras.
    Habitat: Moderately deep water.
    Description: Diam. ⅛ in. 4 whorls, flatly coiled. Surface bears minute revolving lines, with just a suggestion of wrinkles near the sutures. Often well polished. Color grayish white.

THREADED VITRINELLA Pl. 40
Vitrinella multistriata (Verrill)
  Range: N. Carolina to West Indies.
  Habitat: Moderately shallow water.
  Description: Diam. about ⅛ in. A rather flatly coiled shell of about 4 whorls. Sculpture of many fine, threadlike revolving lines, but surface appears polished. Umbilicus large. Color white.

Genus *Cyclostremiscus*
Pilsbry & Olsson 1945

BEAU'S CIRCLE-MOUTH SHELL Pl. 40
Cyclostremiscus beaui (Fischer)
  Range: Carolinas to West Indies.
  Habitat: Moderately shallow water.
  Description: Diam. about ½ in. Flatly coiled, with 4 rounded whorls. Sculpture of prominent revolving lines, with relatively deep grooves in between. Umbilicus large. Color whitish or pale brown.
  Remarks: Formerly in the genus *Vitrinella*.

WOVEN CIRCLE-MOUTH SHELL Pl. 40
Cyclostremiscus trilix (Bush)
  Range: N. Carolina to Florida.
  Habitat: Deep water.
  Description: Diam. ⅛ in. About 4 flatly coiled whorls, no spire, and rather deep umbilicus. Last whorl relatively large, aperture round. Distinct line on upper center of each volution. Surface polished, color grayish white.
  Remarks: Formerly in the genus *Vitrinella*.

Caecum: Family Caecidae

VERY minute tubular shells; at first spiral-shaped but soon becoming cylindrical. The spiral portion is nearly always lost. The family contains but a single genus, with a great many species. Distributed in warm and temperate seas.

Genus *Caecum* Fleming 1813

COOPER'S ATLANTIC CAECUM Pl. 41
Caecum cooperi S. Smith
  Range: Massachusetts to Florida.
  Habitat: Moderately shallow water.
  Description: About ⅛ in. long. Tubular and slightly curved,

marked with faint longitudinal lines. Sometimes there are a few rings at larger end, giving that portion a somewhat cancellate look. As with nearly all of this group, the spiral nuclear part is lost and the open tip is plugged with shelly material; in this species the tip is often distinctly prong-shaped. Color white or yellowish white.

BEAUTIFUL CAECUM   Pl. 41
Caecum pulchellum Stimpson
   Range: Massachusetts to Florida.
   Habitat: Moderately shallow water.
   Description: About $\frac{1}{10}$ in. long. Tubular, composed of a series of regular, rounded rings. Nuclear (initial) part generally lost, open tip plugged with shelly material. Color yellowish white.
   Remarks: This mollusk would be passed up by most searchers at the seashore, for it is so tiny that it could very easily be missed. Under a lens, however, it proves to be an attractive and unusual shell.

Flat Snails: Family Skeneopsidae

VERY tiny flattish shells with few volutions, found under stones in moderately shallow to deep water.

Genus *Skeneopsis* Iredale 1915

FLAT-COILED SKENEOPSIS   Pl. 40
Skeneopsis planorbis (Fabr.)
   Range: Greenland to Florida.
   Habitat: Moderately shallow water.
   Description: A minute species, diam. no more than $\frac{1}{8}$ in., with very low spire. Rather flatly coiled, 4 whorls, sutures well defined. Umbilicus relatively wide and deep. Surface smooth, color dull brown.
   Remarks: These diminutive snails may be found living on oysters, sponges, corals, and all sorts of objects offshore, but they are so tiny that they are usually overlooked.

Genus *Separatista* Gray 1847

BELTED SEPARATISTA   Pl. 39
Separatista cingulata (Verrill)
   Range: Nova Scotia to Delaware.
   Habitat: Deep water.
   Description: Diam. less than $\frac{1}{4}$ in. Flatly coiled, with low

spire. 3 or 4 whorls, the last dipping down at aperture, which is large and somewhat flaring, the lip thin and sharp. Small umbilicus. A series of weak revolving lines on volutions. Color grayish white.

Turret Shells: Family Turritellidae

GREATLY elongated, many-whorled shells, generally turreted. A large family, living chiefly in Pacific waters. Relatively few representatives are present on our Atlantic coast. Many of the s. Pacific species are very colorful and much sought after by collectors.

Genus *Turritella* Lamarck 1799

BORING TURRET SHELL   **Pl. 41**
Turritella acropora Dall
   Range: N. Carolina to West Indies.
   Habitat: Moderately shallow water.
   Description: About 1 in. high. 10 to 12 rather flat-sided whorls, the sutures not very distinct. Sculpture of fine revolving threadlike lines. Aperture small, squarish, operculum horny. Color yellowish or brownish red.

COMMON TURRET SHELL   **Pl. 41**
Turritella exoleta (Linn.)
   Range: S. Florida to West Indies.
   Habitat: Moderately shallow to deep water.
   Description: Height to 3 in., with 16 to 18 whorls. Shell slender and tall, with sharp apex. Sutures distinct. Upper few volutions, which are not occupied by the animal, divided by a septum at each half turn. Volutions decidedly concave at center, sculptured with fine revolving lines. Aperture squarish, lip thin. Color creamy white, usually splashed with chocolate-brown.

VARIEGATED TURRET SHELL   **Pl. 41**
Turritella variegata (Linn.)
   Range: Texas, West Indies.
   Habitat: Moderately shallow water.
   Description: Height 2 to 4 in. A gracefully elongated, spikelike species. About 12 rather flattish whorls, sutures well impressed. Sculpture of revolving lines. Aperture squarish, operculum horny. Color whitish or purplish, with reddish vertical streaks.

Genus *Tachyrhynchus* Mörch 1868

ERODED TURRET SHELL   Pl. 41
Tachyrhynchus erosum (Couthouy)
  Range: Labrador to Massachusetts; Alaska to Br. Columbia.
  Habitat: Moderately deep water.
  Description: About ¾ in. high. An elongate, high-spired shell of 9 or 10 whorls, each grooved with 5 blunt furrows that give the surface a spiral ornamentation. Aperture nearly round, operculum horny. Color chalky white, with a brown periostracum.
  Remarks: Apex commonly eroded or broken in majority of adults.

RETICULATE TURRET SHELL   Pl. 41
Tachyrhynchus reticulatum (Mighels & Adams)
  Range: Greenland to Maine.
  Habitat: Moderatcly deep water.
  Description: Somewhat less than ¾ in. high. A slim shell of 8 or 9 nicely rounded whorls, sutures well impressed. Sculpture consists of both revolving and vertical grooves, so surface is slightly reticulate. Color brown.

Genus *Vermicularia* Lamarck 1799

FARGO'S WORM SHELL   Pl. 41
Vermicularia fargoi Olsson
  Range: W. Florida to Texas.
  Habitat: Shallow water.
  Description: Length 3 to 5 in. In its youthful stages the shell is tightly coiled and looks like a small *Turritella,* but as it grows older it becomes free and wanders off in an irregular and seemingly aimless fashion. Surface with longitudinal keels, rendering the aperture somewhat square. Color reddish brown.
  Remarks: This genus used to be placed in the family Vermetidae. See next species for differences.

KNORR'S WORM SHELL   Pl. 41
Vermicularia knorri Desh.
  Range: N. Carolina to West Indies.
  Habitat: Chiefly in sponges.
  Description: Length 2 to 3 in. Very much like Fargo's Worm Shell, but the tip and first few volutions are pure white, the rest of the shell buffy brown.

Remarks: This genus used to be placed in the family Vermetidae.

COMMON WORM SHELL   Pl. 41
Vermicularia spirata Phil.
   Range: Massachusetts to West Indies.
   Habitat: Shallow water.
   Description: Length to 6 in., possibly more. First few whorls tightly coiled, most of shell tubular and unpredictable; several individuals often found growing together in an intricate, tangled mass. Prominent longitudinal ridges. Color yellowish brown.
   Remarks: This curious creature appears more like a worm than a mollusk, but it is a true gastropod. The body is greatly elongated, and the head bears tentacles, eyes, and a toothed tongue (radula) that is thoroughly snail-like. The shell has been likened to a petrified angleworm. Some authorities believe that this species does not occur north of Florida, and that the usually smaller specimens found off Massachusetts should be called *V. radicula* Stimpson. This genus used to be placed in the family Vermetidae.

Sundials:
Family Architectonicidae

SHELL solid, circular, and but little elevated. The umbilicus is broad and deep, commonly bordered by a knobby keel. Confined chiefly to warm seas.

Genus *Architectonica* Röding 1798

SUNDIAL   *Architectonica nobilis* Röding   Pls. 5, 42
   Range: N. Carolina to Texas; West Indies.
   Habitat: Shallow water.
   Description: Diam. about 1½ in., height about ¾ in. Shell circular and somewhat flattened, with 6 or 7 whorls and rather distinct sutures. Base flat. Surface finely checked by crossing spiral lines and radiating ridges that produce a pattern of raised granules. Umbilicus wide and deep and strongly crenulate. Aperture round, lip thin and sharp. Color white or gray, spotted and marbled with brown and purple.
   Remarks: Formerly in the genus *Solarium* Lam., it has been listed as *S. verrucosum* (Phil.) and *S. granulatum* (Lam.).

Related species, some of them much larger but otherwise almost identical, are found in the Pacific and Indian Oceans.

Genus *Heliacus* Orbigny 1842

BISULCATE SUNDIAL *Heliacus bisulcatus* Orb. **Pl. 42**
Range: N. Carolina to West Indies.
Habitat: Moderately deep water.
Description: Diam. about ½ in. 4 or 5 whorls, low spire. Sculptured with prominent beaded cords. Umbilicus wide and deep, crenulate at edge. Color yellowish brown.

CYLINDER SUNDIAL **Pl. 42**
Heliacus cylindricus (Gmelin)
Range: S. Florida to West Indies.
Habitat: Moderately shallow water.
Description: Diam. about ½ in., height approximately the same. 4 or 5 whorls, spire moderately elevated. Sculpture of revolving lines, cut by sharp vertical beads. Umbilicus narrow but deep, the edges crenulate. Color brownish to black, commonly dotted with white.
Remarks: This is believed to be the same as *Torinia cyclostoma* (Menke).

Genus *Philippia* Gray 1847

KREBS' SUNDIAL *Philippia krebsi* Mörch **Pl. 42**
Range: N. Carolina to West Indies.
Habitat: Moderately deep water.
Description: Diam. about ½ in. A rather flatly coiled shell of 4 or 5 whorls. Spire low. Surface smooth and polished. Umbilicus moderate-sized, bordered by 2 beaded lines. Color yellowish white, with brownish spots at periphery.

Genus *Pseudomalaxis* Fischer 1885

NOBLE FALSE DIAL **Pl. 42**
Pseudomalaxis nobilis (Verrill)
Range: Virginia to West Indies.
Habitat: Deep water.
Description: Diam. about ⅜ in. A flatly coiled snail, with 4 or 5 whorls, each volution rendered squarish by ridges at the shoulders, and the aperture is squarish too. No spire at all. Color white, the shell not polished.
Remarks: A most attractive little shell but unfortunately quite rare.

Worm Shells: Family Vermetidae

WORMLIKE mollusks usually attached to some firm substrate, such as rocks or other shells and sometimes on sponges. The early whorls are similar to those of other young gastropods but the later whorls are nonattached to one another and sometimes become considerably elongated. Worldwide, generally in warm seas.

Genus *Petaloconchus* Lea 1843

ERECT WORM SHELL *Petaloconchus erectus* Dall **Pl. 41**
Range: S. Florida to West Indies.
Habitat: Moderately shallow water; on dead shells.
Description: Diam. of each tube ⅛ in. Spreads out over the surface of corals or some dead shell, occasionally one tube growing over another; the last half inch or so usually stands straight up. Color whitish.

FLORIDA WORM SHELL **Pl. 41**
Petaloconchus floridanus Olsson & Harb.
Range: Florida to West Indies.
Habitat: Shallow water.
Description: Diam. of each tube about ¼ in. Sometimes living solitarily but usually in small colonies, attached to a stone or dead shell. Surface reticulate, color yellowish brown.

IRREGULAR WORM SHELL **Pl. 41**
Petaloconchus irregularis (Orb.)
Range: S. Florida to West Indies.
Habitat: Moderately shallow water; on stones.
Description: Diam. of each tube nearly ¼ in. A tangled mass of reddish-brown, twisted and coiled tubelike shells. Surface roughened by coarse revolving ridges.

BLACK WORM SHELL **Pl. 41**
Petaloconchus nigricans (Dall)
Range: W. Florida.
Habitat: Shallow water.
Description: Diam. of each tube ⅛ in. Individually a loosely coiled, twisted, elongated tubelike shell growing in almost any contorted manner, but it is usually intertwined with others of its kind to form an intricate mass. Shell ribbed horizontally and roughened by wrinkles and lines of growth. Color reddish brown to black.
Remarks: This wormlike gastropod often forms reefs just off shore with the accumulation of its twisted shells.

Genus *Serpulorbis* Sasso 1827

DECUSSATE WORM SHELL Pl. 41
Serpulorbis decussata (Gmelin)
   Range: N. Carolina to West Indies.
   Habitat: Shallow water.
   Description: Diam. of tube about ¼ in., sometimes a little more. A wormlike gastropod living singly and attached to a stone or dead shell; adhering surface is flat, generally in a coiled position. Surface decorated with longitudinal lines or ridges. Color yellowish brown.
   Remarks: See next species for comparison.

RIISE'S WORM SHELL *Serpulorbis riisei* (Mörch) Pl. 41
   Range: West Indies.
   Habitat: Shallow water.
   Description: Very much like Decussate Worm Shell, but slightly larger and with a somewhat pebbly surface. Color pale brown.

Slit Worm Shells:
Family Siliquariidae

SHELL tubular, irregular, often spiral at first. Tube with continuous longitudinal slit. Attached to stones and dead shells.

Genus *Siliquaria* Bruguière 1789

SLIT WORM SHELL *Siliquaria anguillae* (Mörch) Pl. 41
   Range: West Indies.
   Habitat: Shallow water.
   Description: Diam. about ¼ in., length of shell may be as much as 6 in. A loosely coiled snail living unattached. Noticeable slit along upper surface of shell. Ornamentation of weak longitudinal ridges. Color brownish.

Planaxis: Family Planaxidae

SMALL brownish snails living under stones in shallow water and characterized by a spirally grooved ornamentation. The aperture is oval, well notched below. Common in warm seas.

Plate 1

NEW ENGLAND SHELLS
All shells approximately ⅔ natural size

1. **FILE YOLDIA,** *Yoldia limatula*   p. 7
   Thin-shelled; posterior pointed.

2. **RIBBED MUSSEL,** *Modiolus demissus*   p. 17
   Strong radiating ribs; interior silvery white.

3. **DOGWINKLE,** *Thais lapillus*   p. 199
   Thick-shelled; color variable.

4. **TEN-RIDGED WHELK,** *Neptunea decemcostata*   p. 210
   10 red-brown encircling ridges.

5. **DUCKFOOT,** *Aporrhais occidentalis*   p. 166
   Lip greatly expanded; rounded vertical folds.

6. **QUAHOG,** *Mercenaria mercenaria*   p. 59
   Roundish oval, sturdy.

Plate 2

Plate 3

FLORIDA PELECYPODS
All shells approximately ⅔ natural size

Plate 4

FLORIDA GASTROPODS
All shells approximately ⅔ natural size

1. **GREEN STAR SHELL,** *Astraea tuber* — p. 125
   Heavy and solid; pearly within.

2. **APPLE MUREX,** *Murex pomum* — p. 192
   Rough and spiny; 3 varices.

3. **VIOLET SNAIL,** *Janthina janthina* — p. 160
   Thin-shelled and fragile; darker below.

4. **BLEEDING TOOTH,** *Nerita peloronta* — p. 127
   Globular; red stain around white teeth.

5. **JUNONIA,** *Scaphella junonia* — p. 229
   Graceful, with spiral rows of dots.

6. **COLORFUL MOON SHELL,** *Natica canrena* — p. 173
   Smooth; ridged shelly operculum.

7. **SOZON'S CONE,** *Conus sozoni* — p. 238
   Sharp apex.

Plate 5

FLORIDA GASTROPODS
All shells approximately ⅞ natural size

1. **BANDED TULIP SHELL,** *Fasciolaria hunteria*   p. 218
   Smooth; threadlike encircling lines.

2. **CROWN CONCH,** *Melongena corona*   p. 214
   Shoulders with strong curved spines.

3. **SUNDIAL,** *Architectonica nobilis*   p. 142
   Flatly coiled; crenulate umbilicus.

4. **MEASLED COWRY,** *Cypraea zebra*   p. 176
   Polished, heavily spotted.

5. **LIGHTNING WHELK,** *Busycon contrarium*   p. 215
   Left-handed; young streaked, old commonly dull grayish white.

Plate 6

VARIOUS FLORIDA AND GULF OF MEXICO SHELLS

All shells approximately ⅔ natural size

1. **LETTERED OLIVE,** *Oliva sayana*   p. 222
   Cylindrical, highly polished.

2. **GOLDEN OLIVE,** *Oliva sayana citrina*
   See Remarks under Lettered Olive (p. 222).

3. **BABY BONNET,** *Cypraecassis testiculus*   p. 181
   Fine vertical lines, inner lip with shield.

4. **BROWN-LINED LATIRUS,** *Latirus infundibulum*   p. 218
   Spiral ridges, vertical folds.

5. **NUTMEG,** *Cancellaria reticulata*   p. 230
   Cancellate surface, folds on columella.

6. **ROSE PETAL TELLIN,** *Tellina lineata*   p. 74
   Smooth; anterior rounded; not highly polished.

7. **BUTTERCUP,** *Anodontia alba*   p. 48
   Orbicular; bright yellow to orange inside.

8. **TURNIP WHELK,** *Busycon coarctatum*
   Bulbous body whorl, long canal.

Plate 7

CARIBBEAN GASTROPODS
All shells approximately natural size

Plate 8

CARIBBEAN SHELLS
All shells approximately natural size

Plate 9

VEILED CLAMS AND NUT SHELLS

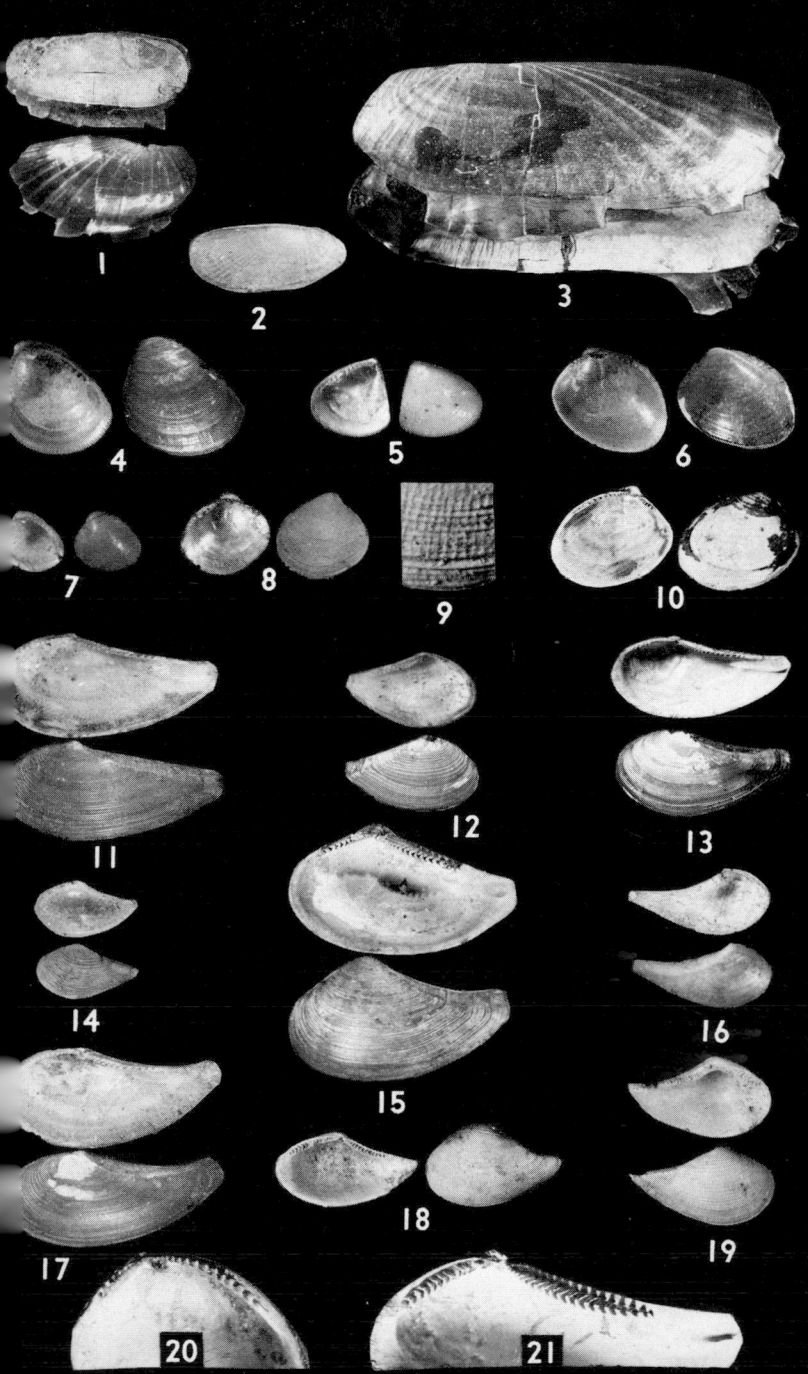

Plate 10

YOLDIAS AND ARK SHELLS

1. **MESSANEAN NUT SHELL,** *Ledella messanensis* ×3   p. 6
   Tiny, oval; ends rounded; pale brown.

2. **SHORT YOLDIA,** *Yoldia sapotilla* ×1   p. 7
   Oval; anterior rounded, posterior narrowed; green.

3. **SHINING YOLDIA,** *Yoldia lucida* ×5   p. 7
   Tiny; prismatic colors; see text.

4. **IRIS YOLDIA,** *Yoldia iris* ×5   p. 6
   Tiny; prismatic colors; see text.

5. **AX YOLDIA,** *Yoldia thraciaeformis* ×1   p. 8
   Squarish; yellowish brown; strong periostracum.

6. **ARCTIC YOLDIA,** *Yoldia arctica* ×1   p. 6
   Squarish, well inflated; lead-gray.

7. **OVAL YOLDIA,** *Yoldia myalis* ×1   p. 7
   Elongate-oval; yellowish olive.

8. **LOVABLE NUT SHELL,** *Tindaria amabilis* ×2   p. 9
   Plump; ends rounded; concentric grooves; dull yellowish.

9. **BLUNT NUT SHELL,** *Malletia obtusa* ×2   p. 8
   Oblong; beaks small; greenish yellow.

10. **DILATED NUT SHELL,** *Malletia dilatata* ×2   p. 8
    Stout, squarish; narrow concentric lines; white.

11. **OVATE NUT SHELL,** *Malletia subovata* ×3   p. 8
    Full beaks, weak teeth; yellowish green.

12. **HAIRY ARK,** *Barbatia cancellaria* ×1   p. 10
    Purplish brown; hairy periostracum.

13. **RETICULATE ARK,** *Barbatia domingensis* ×1   p. 10
    Network pattern; yellowish white.

14. **BRIGHT ARK,** *Barbatia candida* ×1   p. 10
    Shell compressed; fine beaded lines; white.

15. **MOSSY ARK,** *Arca imbricata* ×1   p. 9
    Sturdy; broad area between beaks; shaggy periostracum;
    purplish white.

Plate 11

ARK SHELLS

1. **INCONGRUOUS ARK,** curving basal line
2. **CHEMNITZ'S ARK,** straight basal line
3. **INCONGRUOUS ARK,** *Anadara brasiliana* ×1   p. 11
   Solid; ventral line curved; white.
4. **TRANSVERSE ARK,** *Anadara transversa* ×1   p. 13
   Oblong; white; dark brown periostracum.
5. **CHEMNITZ'S ARK,** *Anadara chemnitzi* ×1   p. 11
   Solid; basal or ventral line straight; white.
6. **BLOOD ARK,** *Anadara ovalis* ×1   p. 12
   Strongly ribbed; white; animal has red blood.
7. **CUT-RIBBED ARK,** *Anadara lienosa floridana* ×1   p. 12
   Grooved ribs; white; brownish periostracum.
8. **BALL-SHAPED ARK,** *Bathyarca glomerula* ×3   p. 13
   Slightly higher than long; white; brownish periostracum.
9. **SCALLOPLIKE ARK,** *Bathyarca pectunculoides* ×3   p. 13
   Anterior short, beaks full; yellowish brown.
10. **ADAMS' ARK,** *Arcopsis adamsi* ×2   p. 11
    Small, solid; ends rounded; yellowish white.
11. **STOUT ARK,** *Barbatia tenera* ×1   p. 11
    Full beaks, fine radiating lines; white.
12. **DEEP-SEA ARK,** *Bentharca profundicola* ×2   p. 14
    Anterior short, beaks small and full; yellowish brown.
13. **PONDEROUS ARK,** *Noetia ponderosa* ×1   p. 13
    Solid; posterior slope; white; thick black periostracum.
14. **EARED ARK,** *Anadara notabilis* ×1   p. 12
    Sturdy, oblong; thin dorsal edge at posterior; white.

Plate 12

BITTERSWEETS, MUSSELS, AND OTHERS

1. **EARED LIMOPSIS**, *Limopsis aurita* ×3   p. 14
   Concentric lines; pale brown hairy periostracum.

2. **SULCATE LIMOPSIS**, *Limopsis sulcata* ×2   p. 15
   Concentric grooves; brownish furry periostracum.

3. **FLAT LIMOPSIS**, *Limopsis plana* ×3   p. 15
   Compressed, oblique; pale brown hairy periostracum.

4. **GREGARIOUS LIMOPSIS**, *Limopsis affinis* ×2   p. 14
   Oblique, inflated; soft, hairy, yellowish-brown periostracum.

5. **MINUTE LIMOPSIS**, *Limopsis minuta* ×3   p. 14
   Weakly cancellate; brownish furry periostracum.

6. **LINED BITTERSWEET**, *Glycymeris undata* ×1   p. 16
   Solid, moderately inflated; blotched with brown; see text.

7. **AMERICAN BITTERSWEET**, *Glycymeris americana* ×1   p. 15
   Compressed; grayish tan, with yellowish-brown mottlings.

8. **GROOVED BITTERSWEET**, *Glycymeris decussata* ×1   p. 16
   Solid, moderately inflated; see text.

9. **COMB BITTERSWEET**, *Glycymeris pectinata* ×1   p. 16
   Radiating ribs; spotted bands of yellowish brown.

10. **HORSE MUSSEL**, *Modiolus modiolus* ×½   p. 18
    Large, broad; brownish to bluish black.

11. **GLASSY TEARDROP**, *Dacrydium vitreum* ×5   p. 22
    Tiny, triangular, smooth; white.

12. **BLUE MUSSEL**, *Mytilus edulis* ×1   p. 17
    Elongate-triangular; blue-black; see text.

13. **YELLOW MUSSEL**, *Brachidontes citrinus* ×1   p. 18
    Fan-shaped on one side; yellowish brown; interior purplish.

14. **SCORCHED MUSSEL**, *Brachidontes exustus* ×1   p. 18
    Fan-shaped on one side; yellowish-brown periostracum.

15. **BENT MUSSEL**, *Brachidontes recurvus* ×1   p. 18
    Obliquely curved; strong radiating ribs; bluish black.

16. **TULIP MUSSEL**, *Modiolus americanus* ×1   p. 17
    Thin-shelled, plump; yellowish brown.

17. Beak of *Modiolus*

18. Beak of *Congeria*

19. Beak of *Brachidontes*

20. Beak of *Mytilus*

Plate 13

MUSSELS, DATE MUSSELS, AND TREE OYSTERS

1. **LATERAL MUSSEL,** *Musculus lateralis* ×1   p. 20
   Fine lines at both ends; pale brown.
2. **LITTLE BLACK MUSSEL,** *Musculus niger* ×1   p. 20
   Radiating lines at both ends, smooth in between; brownish black.
3. **DISCORDANT MUSSEL,** *Musculus discors* ×1   p. 20
   Well inflated; surface smooth; brownish-black periostracum.
4. **SILVER CLAM,** *Idasola argenteus* ×5   p. 22
   Tiny, oblong, smooth; pale brown.
5. **GLANDULAR BEAN MUSSEL,** *Crenella glandula* ×2   p. 21
   Small, oval; radiating and concentric lines; yellowish brown.
6. **LITTLE BEAN MUSSEL,** *Crenella faba* ×1   p. 20
   Small; radiating lines; yellowish brown.
7. **DUSKY MUSSEL,** *Botula fusca* ×1   p. 21
   Beaks hooked; smooth; dark chestnut-brown.
8. **CHESTNUT MUSSEL,** *Lioberus castaneus* ×1   p. 21
   Cylindrical; light brown; grayish periostracum.
9. **RIDGED MUSSEL,** *Gregariella coralliophaga* ×2   p. 19
   Strong posterior ridge; brownish to blackish.
10. **PAPER MUSSEL,** *Amygdalum dendriticum* ×1   p. 19
    Elongate, fragile; delicate pale green periostracum.
11. **ARTIST'S MUSSEL,** *Gregariella opifex* ×3   p. 19
    Small; posterior tips drawn out and frayed; reddish brown.
12. **TWO-FURROWED DATE MUSSEL**   p. 22
    Lithophaga bisulcata ×1
    Radiating furrows; pale brown; calcareous encrustations.
13. **PLATFORM MUSSEL,** *Congeria leucopheata* ×1   p. 23
    Elongate; platform under beaks; grayish brown.
14. **SCISSOR DATE MUSSEL,** *Lithophaga aristata* ×1   p. 22
    Cylindrical; posterior tips cross; pale brown.
15. **BLACK DATE MUSSEL,** *Lithophaga nigra* ×1   p. 22
    Radiating lines both ends, smooth in between; brown or black.
16. **GIANT DATE MUSSEL,** *Lithophaga antillarum* ×1   p. 21
    Cylindrical; thin brownish periostracum.
17. **FLAT TREE OYSTER,** *Isognomon alatus* ×1   p. 24
    Vertical grooves in hinge; brown, black, or purplish; interior pearly.
18. **LISTER'S TREE OYSTER,** *Isognomon radiatus* ×1   p. 24
    Vertical grooves in hinge; greenish brown; interior pearly.

Plate 14

PEN SHELLS

Plate 15

PEARL OYSTERS AND SCALLOPS

1. **SCALY PEARL OYSTER,** *Pinctada radiata* ×1   p. 25
   Hinge line straight; color variable, usually brown or green; interior pearly.

2. **WINGED PEARL OYSTER,** *Pteria colymbus* ×1   p. 24
   Posterior wing; brownish purple; interior pearly.

3. **RAVENEL'S SCALLOP,** *Pecten raveneli* ×1   p. 27
   Deeply cupped lower valve, mottled flat upper; pinkish to purplish, sometimes golden yellow.

4. **ZIGZAG SCALLOP,** *Pecten ziczac* ×½   p. 27
   Deeply cupped lower valve; zigzag black lines on upper.

5. **PAPER SCALLOP,** *Amusium papyraceus* ×1   p. 31
   Smooth, polished; ribbed interior; upper valve reddish brown; lower valve white.

Plate 16

SCALLOPS

Plate 17

SCALLOPS AND FILE SHELLS

1. **NUCLEUS SCALLOP,** *Aequipecten nucleus* ×1   p. 30
   See text.

2. **BAY SCALLOP,** *Aequipecten irradians irradians* ×1   p. 30
   Plump; wings nearly equal; color variable (see text).

3. **GULF SCALLOP**   p. 30
   Aequipecten irradians amplicostatus ×1
   Plump; fewer ribs; lower valve white (see text).

4. **SOUTHERN SCALLOP**   p. 30
   Aequipecten irradians concentricus ×1
   Plump; upper valve mottled, lower pale; see text.

5. **SPATHATE SCALLOP,** *Aequipecten phrygium* ×1   p. 31
   Small, flattish; greenish gray, with bands of dull pink.

6. **ANTILLEAN SCALLOP,** *Lyropecten antillarum* ×1   p. 29
   Flattish; valves thin; color variable; see text.

7. **FRAGILE SCALLOP,** *Pseudamussium fragilis* ×1   p. 31
   Thin and delicate; broad concentric ridges; silvery white.

8. **SMALL-EARED FILE SHELL**   p. 33
   Limatula subauriculata ×2
   Elongate-oval; 2 stronger ribs at center; white.

9. **OVATE FILE SHELL,** *Limea subovata* ×4   p. 34
   Tiny, elongate-oval; white.

10. **TRANSPARENT SCALLOP,** *Pseudamussium vitreus* ×1   p. 32
    Thin, translucent; silvery gray.

11. **DALL'S SCALLOP,** *Propeamussium dalli* ×5   p. 32
    Thin and fragile; ribs on interior; silvery white.

Plate 18

FILE AND JINGLE SHELLS, SPINY OYSTERS

Plate 19

CATS' PAWS, OYSTERS, AND OTHERS

1. **SPINY CAT'S PAW,** *Plicatula spondyloidea* ×1   p. 35
   See text.

2. **CAT'S PAW,** *Plicatula gibbosa* ×1   p. 35
   Trigonal, thick; wavy folds; white, with gray or reddish lines.

3. **CARIBBEAN OYSTER,** *Crassostrea rhizophorae* ×½   p. 36
   See text.

4. **COON OYSTER,** *Ostrea frons* ×1   p. 36
   Elongate; fluted margins; brown.

5. **COON OYSTER** ×1 on twig

6. **CRESTED OYSTER,** *Ostrea equestris* ×1   p. 36
   Oval; fluted margins; yellowish gray.

7. **SPONGE OYSTER,** *Ostrea permollis* ×1   p. 36
   Thick but soft periostracum; golden brown.

8. **SPONGE OYSTER** ×1 on sponge

9. **COMMON OYSTER,** *Crassostrea virginica* ×½   p. 37
   Irregular; lead-colored; muscle scar deep violet.

10. **FLORIDA MARSH CLAM,** *Pseudocyrena floridana* ×1   p. 40
    Thin, sturdy; posterior prolonged; purplish white.

11. **CAROLINA MARSH CLAM,** *Polymesoda caroliniana* ×1   p. 39
    Oval, swollen; greenish periostracum.

12. **BLACK CLAM,** *Arctica islandica* ×½   p. 39
    Roughly circular; shiny black or deep brown periostracum.

Plate 20

ASTARTES, CARDITAS, AND OTHERS

1. **WAVED ASTARTE,** *Astarte undata* ×1 — p. 41
   Solid; strong concentric ridges; reddish brown.
2. **GIBB'S CLAM,** *Eucrassatella speciosa* ×1 — p. 42
   Solid; hinge thick and sturdy; concentric lines; yellowish brown.
3. **CHESTNUT ASTARTE,** *Astarte castanea* ×1 — p. 41
   Somewhat kidney-shaped; ridges weak; chestnut-brown.
4. **STRIATE ASTARTE,** *Astarte striata* ×1 — p. 41
   Small, solid; concentric wrinkles; dark brown.
5. **DOMINGO CARDITA,** *Cardita dominguensis* ×4 — p. 43
   Moderately inflated; radiating ribs; pinkish white.
6. **NORTHERN ASTARTE,** *Astarte borealis* ×1 — p. 40
   Solid, oval; concentric furrows on upperpart; deep brown.
7. **LENTIL ASTARTE,** *Astarte subaequilatera* ×1 — p. 41
   Solid; strong concentric ridges; yellowish brown.
8. **BROAD-RIBBED CARDITA,** *Cardita floridana* ×1 — p. 43
   Heavy, solid; scaly ribs; yellowish white, blotched.
9. **NORTHERN CARDITA,** *Venericardia borealis* ×1 — p. 43
   Solid; strong radiating ribs; shaggy brownish periostracum.
10. **THREE-TOOTHED CARDITA** — p. 44
    Venericardia tridentata ×2
    Trigonal, inflated; beaded ribs; grayish brown.
11. **FLAT CARDITA,** *Venericardia perplana* ×3 — p. 44
    Trigonal, not inflated; pinkish brown, sometimes mottled.
12. **WEST INDIAN CARDITA,** *Cardita gracilis* ×1 — p. 43
    Elongate; coarse ribs on posterior; grayish brown.
13. **THICKENED CRASSINELLA,** *Crassinella mactracea* ×1 — p. 42
    Triangular, solid; hinge sturdy; yellowish green.
14. **LUNATE CRASSINELLA,** *Crassinella lunulata* ×3 — p. 42
    Flattish; concentric furrows; white or pinkish.
15. **FLAT LEPTON,** *Mysella planulata* ×4 — p. 52
    Tiny, oval, polished; brownish periostracum.
16. **RED KELLIA,** *Kellia rubra* ×4 — p. 52
    Roundly oval, well inflated; pinkish white.
17. **SHINING KELLIELLA,** *Kelliella nitida* ×4 — p. 52
    Plump; beaks high; white.
18. **FALSE MUSSEL,** *Mytilopsis domingensis* ×1 — p. 23
    Mussel-like in shape; grayish; brown periostracum.
19. **PIMPLED DIPLODON,** *Diplodonta semiaspera* ×1 — p. 45
    Small, inflated; pimpled surface; chalky white.
20. **FLAT DIPLODON,** *Diplodonta notata* ×1 — p. 44
    Compressed; surface pitted; white.
21. **ATLANTIC DIPLODON,** *Diplodonta punctata* ×1 — p. 45
    Inflated, roundish; pure white.

(contd. on legend page for Plate 21)

Plate 21

CLEFT CLAMS AND LUCINES

1. **GOULD'S CLEFT CLAM,** *Thyasira gouldi* ×2   p. 45
   Small, roundish; weak cleft; thin yellowish periostracum.

2. **CLEFT CLAM,** *Thyasira insignis* ×1   p. 45
   Beaks pointed; white, thin yellowish periostracum.

3. **ATLANTIC CLEFT CLAM,** *Thyasira trisinuata* ×2   p. 46
   Posterior slope with weak cleft; grayish white.

4. **LENTICULAR MYRTEID,** *Myrtea lens* ×2   p. 48
   Roundish; low beaks, posterior flattened; dull white.

5. **CROSSHATCHED LUCINE**   p. 50
   Divaricella quadrisulcata ×1
   Oblique sculpture; ivory-white.

6. **CROSSHATCHED LUCINE,** sculpture ×5

7. **BUTTERCUP,** *Anodontia alba* ×1   p. 48
   Inflated; dull white; interior yellow to orange.

8. **WOVEN LUCINE,** *Lucina nassula* ×1   p. 47
   Sharp concentric and radiating lines; white.

9. **DECORATED LUCINE,** *Lucina amiantus* ×2   p. 46
   About 12 broad ribs, cut by concentric ridges; white.

10. **MANY-LINED LUCINE,** *Lucina multilineata* ×2   p. 47
    Circular, compressed; crowded radiating lines; white.

11. **SULCATE LUCINE,** *Lucina leucocyma* ×4   p. 47
    Somewhat triangular; broad folds; white.

12. **JAMAICA LUCINE,** *Lucina pectinatus* ×1   p. 47
    Solid; fold on posterior; yellowish white.

13. **FANCY LUCINE,** *Lucina muricata* ×1   p. 47
    Sharp concentric and radiating lines; white.

14. **FLORIDA LUCINE,** *Lucina floridana* ×1   p. 46
    Solid; fine concentric lines; white.

15. **PENNSYLVANIA LUCINE,** *Lucina pensylvanica* ×1   p. 48
    Strong posterior fold, concentric ridges; white.

16. **NORTHERN LUCINE,** *Lucina filosus* ×1   p. 46
    Circular; sharp concentric ridges; white.

CONTINUED FROM LEGEND PAGE FOR PLATE 20

22. **WAGNER'S DIPLODON,** *Diplodonta nucleiformis* ×1   p. 44
    Small, moderately solid; weak concentric lines; white.

23. **VERRILL'S DIPLODON,** *Diplodonta verrilli* ×2   p. 45
    Thin-shelled, inflated, nearly round; white.

Plate 22

LUCINES AND JEWEL BOXES

1. **LITTLE WHITE LUCINE,** *Codakia orbiculata* ×1   p. 49
   Radiating and concentric lines; lunule elongate; white.

2. **COSTATE LUCINE,** *Codakia costata* ×2   p. 49
   Radiating lines often in pairs; lunule small; white.

3. **GREAT WHITE LUCINE,** *Codakia orbicularis* ×1   p. 49
   Large; lunule heart-shaped; white, pinkish border inside.

4. **CORRUGATED JEWEL BOX**   p. 50
   Chama congregata ×1 (on coral)
   Wavy corrugations; upper valve often spiny, streaked with purple.

5. **LITTLE JEWEL BOX,** *Chama sarda* ×1   p. 51
   Wavy, tubular scales; white, sometimes with reddish rays.

6. **SPINY JEWEL BOX,** *Echinochama cornuta* ×1   p. 52
   Strong; spiny radiating ribs; white; red splashes inside.

7. **CARIBBEAN SPINY JEWEL BOX**   p. 51
   Echinochama arcinella ×1
   See text.

8. **LEFT-HANDED JEWEL BOX,** *Pseudochama radians* ×1   p. 51
   Right valve attached; whitish or grayish; interior often brownish.

9. **JEWEL BOX,** *Chama macerophylla* ×½   p. 50
   Scalelike foliations; color varies from pink and rose to yellow. Muscle scar and pallial line inked in.

10. **SMOOTH-EDGED JEWEL BOX,** *Chama sinuosa* ×½   p. 51
    See text. Muscle scar and pallial line inked in.

Plate 23

COCKLES

Plate 24

COCKLES AND HARD-SHELLED CLAMS

1. **POINTED VENUS,** *Anomalocardia cuneimeris* ×1   p. 62
   Wedge-shaped; concentric ribs; color variable.

2. **WEST INDIAN POINTED VENUS**   p. 61
   Anomalocardia brasiliana ×1
   See text.

3. **GREENLAND COCKLE,** *Serripes groenlandicus* ×1   p. 57
   Shell thin, moderately inflated, smooth; drab gray.

4. **SMALL RING CLAM,** *Cyclinella tenuis* ×1   p. 62
   Nearly orbicular, moderately inflated; white.

5. **SERENE GOULD CLAM,** *Gouldia cerina* ×3   p. 65
   Shiny; white, sometimes with brown markings.

6. **ABACO TIVELA,** *Tivela abaconis* ×2   p. 62
   Moderately inflated, polished; yellowish; rosy lavender at
   beaks.

7. **TRIGONAL TIVELA,** *Tivela mactroides* ×1   p. 62
   Solid, trigonal; clouded and rayed with brown.

8. **BROWN GEM CLAM,** *Parastarte triquetra* ×5   p. 67
   Triangular, solid; white, tinged with purple.

9. **WAVY CLAM,** *Liocyma fluctuosa* ×1   p. 67
   Concentric ridges; white; thin yellowish periostracum.

10. **RIGID VENUS,** *Ventricolaria rigida* ×1   p. 58
    Nearly round; yellowish gray, with brown mottlings.

11. **GEM SHELL,** *Gemma gemma* ×4   p. 67
    See text.

12. **QUEEN VENUS,** *Ventricolaria rugatina* ×1   p. 58
    Thick-shelled; concentric ribs; yellowish white, mottled.

13. **LISTER'S VENUS,** *Periglypta listeri* ×½   p. 58
    Solid; bladelike ribs at posterior; grayish white.

14. **QUAHOG,** *Mercenaria mercenaria* ×½   p. 59
    Sturdy; grayish; interior white, often with dark violet
    border.

Plate 25

HARD-SHELLED CLAMS

1. **MOTTLED CHIONE,** *Chione intepurpurea* ×1   p. 60
   Crowded ribs; white to cream; violet splotch on posterior portion of interior.

2. **CROSS-BARRED CHIONE,** *Chione cancellata* ×1   p. 59
   Elevated concentric and radiating ribs; yellowish white, sometimes banded.

3. **OLD VENUS CHIONE,** *Chione pubera* ×½   p. 61
   Large; gray, sometimes mottled with purplish brown.

4. **GRAY PYGMY CHIONE,** *Chione grus* ×2   p. 60
   Gray, sometimes with pinkish tinge; purple stain at both ends of hinge.

5. **WHITE PYGMY CHIONE,** *Chione pygmaea* ×2   p. 61
   Elongate; whitish; purple stain on posterior of hinge.

6. **KING CHIONE,** *Chione paphia* ×1   p. 61
   Broad ribs, pinched out at ends; mottled lilac and brown.

7. **BROAD-RIBBED CHIONE,** *Chione latilirata* ×1   p. 60
   Broad ribs, not pinched out at ends; mottled lilac and brown.

8. **CLENCH'S CHIONE,** *Chione clenchi* ×1   p. 60
   Posterior pointed; ribs distinct; gray, spotted with purplish brown.

9. **BEADED CHIONE,** *Chione granulata* ×1   p. 60
   Scaly radiating ribs; gray, with darker mottling.

10. **ATLANTIC RUPELLARIA,** *Rupellaria typica* ×1   p. 68
    Rough, plump; radiating lines; grayish white.

11. **CORAL PETRICOLA,** *Petricola lapicida* ×1   p. 68
    Inflated; crisscross sculpture; white.

12. **FALSE ANGEL WING,** *Petricola pholadiformis* ×1   p. 68
    Elongate; sharp sculpture on anterior end; white.

13. **ELEGANT DISK SHELL,** *Dosinia elegans* ×1   p. 66
    Circular, thin, with well-spaced concentric lines; white.

14. **WEST INDIAN DISK SHELL,** *Dosinia concentrica* ×1   p. 66
    See text.

15. **DISK SHELL,** *Dosinia discus* ×1   p. 66
    Circular, thin, with crowded concentric lines; white.

Plate 26

HARD-SHELLED CLAMS AND OTHERS

1. **CUBAN TRANSENNELLA,** *Transennella cubaniana* ×2   p. 63
   Somewhat triangular; white, flecked with brown.

2. **CONRAD'S TRANSENNELLA**   p. 63
   Transennella conradina ×2
   Shiny; yellowish gray, often zigzag scrawls of brown.

3. Inner margin of *Transennella* ×15

4. **STIMPSON'S TRANSENNELLA**   p. 63
   Transennella stimpsoni ×2
   Polished; creamy white, brown marks; interior purplish.

5. **LIGHTNING VENUS,** *Pitar fulminata* ×1   p. 64
   Plump; white, with brown or orange spots in a radial pattern.

6. **FALSE QUAHOG,** *Pitar morrhuana* ×1   p. 65
   Plump, smooth; rusty gray; interior white.

7. **PURPLE VENUS,** *Pitar circinata* ×1   p. 64
   White to purplish.

8. **CORDATE VENUS,** *Pitar cordata* ×1   p. 64
   Plump; posterior pointed; grayish white.

9. **WHITE VENUS,** *Pitar albida* ×1   p. 63
   Plump; beaks not prominent; white.

10. **WEST INDIAN VENUS,** *Pitar aresta* ×1   p. 63
    Plump; grayish white.

11. **ELEGANT VENUS,** *Pitar dione* ×1   p. 64
    Posterior ridge with long spines; lavender; interior white.

12. **TEXAS VENUS,** *Callocardia texasiana* ×1   p. 65
    Inflated, thin in substance; yellowish white.

13. **CORAL-BORING CLAM,**   p. 69
    Coralliophaga coralliophaga ×1
    Cylindrical; yellowish tan; interior white.

14. **WINGED SURF CLAM,** *Mactra alata* ×½   p. 69
    Inflated, with elevated ridge on posterior; white.

Plate 27

SURF CLAMS AND OTHERS

1. SURF CLAM, hinge ×1
2. STIMPSON'S SURF CLAM, hinge ×1
3. SURF CLAM, *Spisula solidissima* ×½ p. 70
   Sturdy; yellowish white; see text.
4. STIMPSON'S SURF CLAM, *Spisula polynyma* ×½ p. 70
   Sturdy; yellowish white; see text.
5. SOUTHERN SURF CLAM
   Spisula solidissima similis ×1
   See Remarks under Surf Clam (p. 70).
6. FRAGILE SURF CLAM, hinge ×2
7. FRAGILE SURF CLAM, *Mactra fragilis* ×1 p. 70
   Thin, with posterior ridge; white.
8. LINED DUCK, *Labiosa lineata* ×1 p. 71
   Posterior cordlike ridge; thin-shelled; white.
9. LITTLE SURF CLAM, *Mulinia lateralis* ×1 p. 70
   Triangular, smooth; yellowish white; thin periostracum.
10. PUERTO RICO SURF CLAM, *Mulinia portoricensis* ×1 p. 71
    Beaks prominent; whitish; brown periostracum.
11. CHANNELED DUCK, *Labiosa plicatella* ×1 p. 27
    Thin, fragile, gaping; concentric grooves; white.
12. ARCTIC WEDGE CLAM, *Mesodesma arctatum* ×1 p. 72
    Wedge-shaped, solid; yellowish periostracum.

Plate 28

TELLINS

1. **SAY'S TELLIN,** *Tellina sayi* ×1 p. 76
   Moderately inflated; glossy white.

2. **TAMPA TELLIN,** *Tellina tampaensis* ×1 p. 76
   Slightly inflated; white; interior often pinkish.

3. **CARIBBEAN TELLIN,** *Tellina caribaea* ×1 p. 73
   Elongate, thin; glossy white or pinkish.

4. **MARTINIQUE TELLIN,** *Tellina martinicensis* ×2 p. 75
   Moderately inflated; white.

5. **CRYSTAL TELLIN,** *Tellina cristallina* ×1½ p. 74
   Thin, translucent, with concentric ridges; white.

6. **TEXAS TELLIN,** *Tellina texana* ×1 p. 76
   Posterior short; white, sometimes pinkish, often iridescent.

7. **LINED TELLIN,** *Tellina alternata* ×1 p. 73
   Sharp concentric lines; white, pink, or yellow.

8. **DWARF TELLIN,** *Tellina agilis* ×2 p. 73
   Posterior short; white, yellowish, or pink.

9. **ANGULAR TELLIN,** *Tellina angulosa* ×1 p. 73
   Fine concentric lines; delicate pink.

10. **MERA TELLIN,** *Tellina mera* ×1 p. 75
    Anterior rounded, posterior bluntly pointed; white.

11. **DEKAY'S DWARF TELLIN,** *Tellina versicolor* ×1 p. 77
    Posterior slope; color variable, iridescent.

12. **IRIS TELLIN,** *Tellina iris* ×2 p. 74
    Compressed; glossy white; valves translucent.

13. **WEDGE TELLIN,** *Tellina candeana* ×2 p. 73
    Posterior and anterior nearly equal; white.

14. **CANDY STICK TELLIN,** *Tellina similis* ×1 p. 76
    Glossy white, often with pinkish rays.

15. **SMOOTH TELLIN,** *Tellina laevigata* ×1 p. 74
    Roundish, smooth; white, with pale orange rays.

16. **GREAT TELLIN,** *Tellina magna* ×1 p. 75
    Large; left valve white, right valve yellowish or orange.

17. **DALL'S DWARF TELLIN,** *Tellina sybaritica* ×2 p. 76
    Elongate, posterior slopes; pinkish or red.

18. **FAUST TELLIN,** *Arcophagia fausta* ×1 p. 77
    Roundish, solid; whitish; interior often tinged with yellow.

19. **GROOVED MACOMA,** *Apolymetis intastriata* ×1 p. 80
    Anterior twisted; well inflated; white.

Plate 29

MACOMAS, COQUINAS, AND OTHERS

1. **CONSTRICTED MACOMA,** *Macoma constricta* ×1   p. 78
   Feeble fold; white; yellowish periostracum.
2. **HENDERSON'S MACOMA,** *Macoma hendersoni* ×2   p. 78
   Thin, fragile; strong concentric undulations; white.
3. **BALTIC MACOMA,** *Macoma balthica* ×1   p. 77
   Beaks central; dull pinkish white; thin periostracum.
4. **SLENDER MACOMA,** *Macoma extenuata* ×1   p. 78
   Long anterior; polished white.
5. **SOUTHERN CUMINGIA,** *Cumingia antillarum* ×2   p. 85
   Posterior pointed; concentric lines; white.
6. **SHORT MACOMA,** *Macoma brevifrons* ×1   p. 77
   Dull white; interior buffy yellow; brownish periostracum.
7. **NARROWED MACOMA,** *Macoma tenta* ×2   p. 78
   Thin, delicate; pinkish white, iridescent.
8. **HANLEY'S MACOMA,** *Macoma aurora* ×1   p. 77
   Oval; short posterior; white, sometimes pinkish.
9. **CHALKY MACOMA,** *Macoma calcarea* ×1   p. 78
   Chalky white; dull gray periostracum.
10. **COMMON CUMINGIA,** *Cumingia tellinoides* ×1   p. 85
    Posterior pointed; concentric lines; white.
11. **LINEN TELLIN,** *Quadrans lintea* ×1   p. 80
    Posterior sloping; fine concentric lines; white.
12. **CRESTED TELLIN,** *Tellidora cristata* ×1   p. 80
    Compressed; saw-toothed lateral margins; white.
13. **CRENULATE TELLIN,** *Phylloda squamifera* ×1   p. 79
    Dorsal margins crenulate; white; sometimes yellowish or orange tinge.
14. **GREAT FALSE COQUINA,** *Iphigenia brasiliensis* ×1   p. 82
    Sturdy, broadly triangular; tan periostracum.
15. **FALSE COQUINA,** *Heterodonax bimaculata* ×1   p. 83
    Triangular-oval; bluish white; interior spotted with purple.
16. **PEA STRIGILLA,** *Strigilla pisiformis* ×2   p. 79
    Inflated; oblique wavy lines; white or pinkish.
17. **WHITE STRIGILLA,** *Strigilla mirabilis* ×2   p. 79
    Oblique wavy lines; white.
18. **ROSY STRIGILLA,** *Strigilla carnaria* ×1   p. 79
    Oblique wavy lines; pale rose; interior rosy pink.
19. **ROSY STRIGILLA,** sculpture ×10
20. **CARIBBEAN COQUINA,** *Donax denticulata* ×1   p. 81
    Wedge-shaped; varicolored.
21. **ROEMER'S COQUINA,** *Donax roemeri* ×1 (see text)   p. 81
22. **FOSSOR COQUINA,** *Donax fossor* ×1   p. 81
    Wedge-shaped, with fine radiating lines; bluish white.

(contd. on legend page for Plate 30)

Plate 30

SEMELES, RAZOR CLAMS, AND OTHERS

1. **ATLANTIC SANGUIN,** *Sanguinolaria cruenta* ×1   p. 82
   Oval; ends rounded; rosy pink to whitish.
2. **CANCELLATE SEMELE,** *Semele bellastriata* ×1   p. 84
   Round-oval; concentric and radiating lines; varicolored.
3. **TINY SEMELE,** *Semele nuculoides* ×3   p. 84
   Beaks near anterior end; white or yellowish.
4. **PURPLE SEMELE,** *Semele purpurascens* ×1   p. 84
   Oval, smooth; pale yellow, somewhat blotched with purple or brown or orange.
5. **GAUDY ASAPHIS,** *Asaphis deflorata* ×1   p. 83
   Radiating lines; color variable (see text).
6. **WHITE SEMELE,** *Semele proficua* ×1   p. 84
   Thin, compressed; creamy white; interior sometimes with purplish or pink spots.
7. **SHINING ERVILIA,** *Ervilia nitens* ×2   p. 72
   Plump, smooth; white, tinted with pinkish.
8. **WEDGE RANGIA,** *Rangia cuneata* ×1   p. 71
   Thick, solid; posterior slopes; gray-brown periostracum.
9. **RIBBED POD,** *Siliqua costata* ×1   p. 87
   Ovate-elliptical; low beaks, internal rib; yellowish-green periostracum.
10. **PROPELLER CLAM,** *Cyrtodaria siliqua* ×1   p. 88
    Thick; ends rounded, valves twisted; black periostracum.
11. **OVATE PARAMYA,** *Paramya subovata* ×1   p. 92
    Rather square; dull white or yellowish white.
12. **COMMON ATLANTIC ABRA,** *Abra aequalis* ×2   p. 85
    Rather plump, smooth; brownish.
13. **ELONGATED ABRA,** *Abra longicallis americana* ×1   p. 85
    Elongate, smooth, polished; pale brown.
14. **DALL'S LITTLE ABRA,** *Abra liocia* ×1   p. 85
    Small, shiny; buff.
15. **LITTLE GREEN RAZOR CLAM,** *Solen viridis* ×1   p. 86
    Thin, fragile, barely curved; shiny green periostracum.
16. **WEST INDIAN RAZOR CLAM,** *Solen obliquus* ×⅔   p. 86
    Sturdy, straight; white; brownish periostracum.
17. **DWARF RAZOR CLAM,** *Ensis minor* ×1   p. 87
    Thin, fragile; see text.
18. **COMMON RAZOR CLAM,** *Ensis directus* ×½   p. 86
    Thin, elongate, curved; glossy greenish periostracum.

CONTINUED FROM LEGEND PAGE FOR PLATE 29

23. **STRIATE COQUINA,** *Donax striatus* ×1   p. 81
    Posterior slope flat; bluish white.
24. **COQUINA,** *Donax variabilis* ×1   p. 81
    Wedge-shaped; varicolored.

Plate 31

SOFT-SHELLED AND BASKET CLAMS, OTHERS

1. **PURPLISH TAGELUS,** *Tagelus divisus* ×1   p. 83
   Thin, fragile; ends rounded; faint purple rays.

2. **STOUT TAGELUS,** *Tagelus plebeius* ×1   p. 83
   Elongate, gaping; yellowish-brown periostracum.

3. **ARCTIC ROCK BORER,** *Hiatella arctica* ×1   p. 88
   Oblong, irregular; dingy white.

4. **ST. MARTHA'S RAZOR CLAM**   p. 87
   Solecurtus sanctaemarthae ×2
   See text.

5. **CORRUGATED RAZOR CLAM**   p. 87
   Solecurtus cumingianus ×1
   Elongate; ends rounded; peculiar sculpture (see text).

6. Overlapping valves in *Corbula* ×8

7. **CARIBBEAN BASKET CLAM,** *Corbula caribaea* ×2   p. 91
   Posterior rostrate (beaked); white.

8. **FAT BASKET CLAM,** *Notocorbula operculata* ×3   p. 91
   Greatly inflated; glossy white; thin yellowish periostracum.

9. **COMMON BASKET CLAM,** *Corbula contracta* ×2   p. 91
   Solid, inflated; white; thin brownish periostracum.

10. **SNUB-NOSED BASKET CLAM,** *Corbula nasuta* ×2   p. 91
    Posterior drawn to blunt point; white or yellowish white.

11. **EQUAL-VALVED BASKET CLAM**   p. 91
    Corbula aequivalvis ×2
    Valves nearly equal; posterior ridge; white.

12. **SWIFT'S BASKET CLAM,** *Corbula swiftiana* ×3   p. 92
    Posterior ridge; white; thin yellowish periostracum.

13. **TRUNCATE SOFT-SHELLED CLAM,** *Mya truncata* ×1   p. 90
    Posterior truncate, flaring; yellowish-brown periostracum.

14. **ROUGH SOFT-SHELLED CLAM** *Panomya arctica* ×1   p. 89
    Rough, sturdy; posterior truncate; dull chalky white.

15. **SOFT-SHELLED CLAM,** *Mya arenaria* ×⅔   p. 90
    Both ends gape; dull gray or chalky white; thin grayish periostracum.

16. Chondrophore of *Mya* somewhat enlarged

17. **ATLANTIC GEODUCK,** *Panope bitruncata* ×½   p. 89
    Large, well inflated; dull grayish white.

Plate 32

BORING CLAMS

Plate 33

LYONSIAS, PANDORAS, SPOON SHELLS, ETC.

1. **GLASSY LYONSIA,** *Lyonsia hyalina* ×1   p. 98
   Fragile, translucent; pearly.

2. **SAND LYONSIA,** *Lyonsia arenosa* ×1   p. 97
   Blocky, fragile; silvery white; pearly within.

3. **FLORIDA LYONSIA,** *Lyonsia floridana* ×1   p. 98
   Unequal valves, fragile; thin grayish periostracum.

4. Sand entangled at margin of **GLASSY LYONSIA** ×10

5. **PEARLY LYONSIA,** *Lyonsia beana* ×1   p. 98
   Unequal valves, irregular; white to brownish.

6. **GOULD'S PANDORA,** *Pandora gouldiana* ×1   p. 99
   Wedge-shaped, very thin; white or grayish white; interior pearly.

7. **SAND PANDORA,** *Pandora arenosa* ×1   p. 99
   Unequal valves, weak posterior ridge; white; interior pearly.

8. **GRANULATE LYONSIELLA,** *Lyonsiella granulifera* ×1   p. 98
   Well inflated, smooth; white; interior pearly.

9. **THREE-LINED PANDORA,** *Pandora trilineata* ×1   p. 99
   Elongate; upturned posterior tip; white; interior pearly.

10. **WAVY SPOON SHELL,** *Periploma undulata* ×2   p. 102
    Fragile, posterior somewhat pointed; white.

11. **FRAGILE SPOON SHELL,** *Periploma fragile* ×1   p. 101
    Oval; prominent chondrophore; white.

12. **LEA'S SPOON SHELL,** chondrophore ×5

13. **LEA'S SPOON SHELL,** *Periploma leanum* ×1   p. 101
    Oval; valves unequal; pearly white; thin yellowish periostracum.

14. **PAPER SPOON SHELL,** *Periploma papyratium* ×1   p. 101
    Thin-shelled; short anterior; white; thin yellowish periostracum.

15. **UNEQUAL SPOON SHELL,** *Periploma inequale* ×2   p. 101
    Moderately inflated; chondrophore large; white.

16. **SHINING THRACIA,** *Thracia nitida* ×1   p. 100
    Broadly oval, gaping; whitish; greenish periostracum.

17. **CONRAD'S THRACIA,** *Thracia conradi* ×1   p. 100
    Oval, inflated; right beak perforated; thin brownish periostracum.

18. **NORTHERN THRACIA,** *Thracia septentrionalis* ×1   p. 100
    Oval, ends rounded; brownish; thin periostracum.

Plate 34

DIPPER SHELLS AND OTHERS

Plate 35

KEYHOLE LIMPETS

Plate 36

LIMPETS, KEYHOLE LIMPETS, ABALONES

1. **GREEN KEYHOLE LIMPET,** *Diodora viridula* ×1   p. 111
   Radiating lines; white and greenish; black around orifice.

2. **DYSON'S KEYHOLE LIMPET,** *Diodora dysoni* ×1   p. 110
   Orifice small; white, broad bands of black.

3. **SOWERBY'S KEYHOLE LIMPET,** *Lucapina sowerbii* ×1   p. 111
   Elongate; large orifice; white to buff, blotched with brown.

4. **CANCELLATED KEYHOLE LIMPET,** *Lucapina suffusa* ×1   p. 111
   Orifice round; radiating ribs; buffy or white.

5. **FILE KEYHOLE LIMPET,** *Lucapinella limatula* ×1¼   p. 111
   Oval; radiating ribs with scales; brownish, with whitish rays.

6. **RUFFLED EMARGINULA,** *Emarginula phrixodes* ×1   p. 112
   Moderately elevated; slit at margin; white.

7. **EMARGINATE LIMPET,** *Hemitoma emarginata* ×1   p. 112
   Highly elevated; slit at anterior margin (in young); whitish.

8. **POURTALÈS ABALONE,** *Haliotis pourtalesi* ×1   p. 107
   See text.

9. **BRIDLE RIMULA,** *Rimula frenulata* ×5   p. 113
   Elongate, canoe-shaped; apex hooked; white; interior glossy.

10. **PYGMY EMARGINULA,** *Emarginula pumila* ×2   p. 112
    Radiating and concentric lines; yellowish to whitish.

11. **DWARF LIMPET,** *Acmaea rubella* ×2   p. 114
    Small, oval; whitish or rusty brown.

12. **EIGHT-RIBBED LIMPET,** *Hemitoma octoradiata* ×1   p. 112
    Flattened to moderately elevated; nodular ribs; grayish white.

13. **NOAH'S PUNCTURED SHELL,** *Puncturella noachina* ×2   p. 113
    Oval base, steep summit, narrow slit in front of tip; dull white.

14. **TORTOISESHELL LIMPET,** *Acmaea testudinalis* ×1   p. 114
    Moderately arched; checkered brown marks.

15. **EELGRASS LIMPET,** *Acmaea testudinalis alveus* ×1
    Parallel sides; see Remarks under Tortoiseshell Limpet (p. 114).

16. **SOUTHERN LIMPET,** *Acmaea antillarum* ×1   p. 113
    Flattened; radiating black lines.

Plate 37

LIMPETS AND TOP SHELLS

Plate 38

TOP SHELLS, TURBANS, STARS, OTHERS

Plate 39

RISSOS, NERITES, AND OTHERS

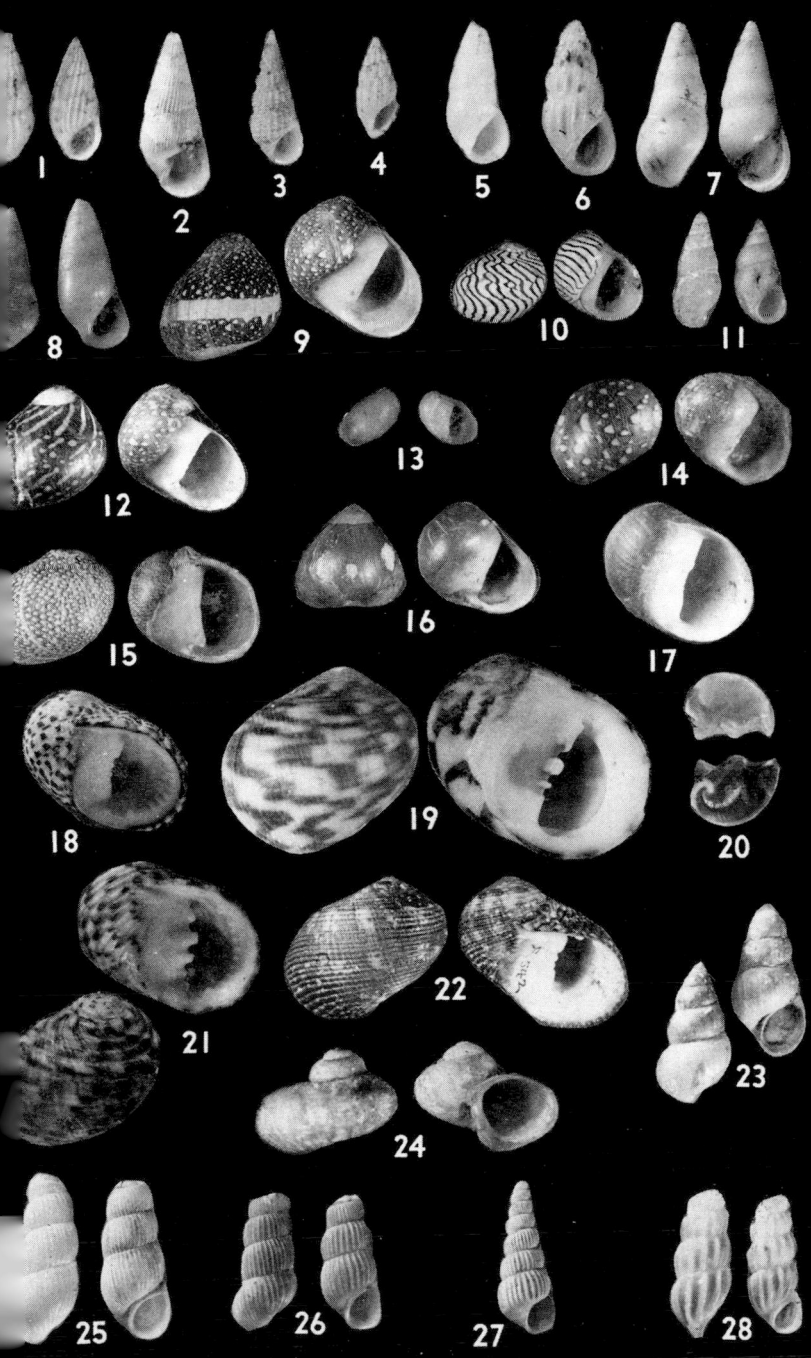

Plate 40

PERIWINKLES AND OTHERS

Plate 41

TURRET AND WORM SHELLS

1. **COMMON TURRET SHELL,** *Turritella exoleta* ×1   p. 140
   Whorls concave; creamy white, splashed with chocolate-brown.

2. **BORING TURRET SHELL,** *Turritella acropora* ×1   p. 140
   Whorls flat-sided; yellowish or brownish red.

3. **ERODED TURRET SHELL,** *Tachyrhynchus erosum* ×2   p. 141
   High-spired; blunt revolving furrows; brown.

4. **RETICULATE TURRET SHELL**   p. 141
   Tachyrhynchus reticulatum ×2
   High-spired; reticulate sculpture; brown.

5. **VARIEGATED TURRET SHELL**   p. 140
   Turritella variegata ×1
   Flattish whorls; reddish vertical streaks.

6. **COOPER'S ATLANTIC CAECUM,** *Caecum cooperi* ×10   p. 138
   Slightly curved; faint longitudinal lines; white or yellowish white.

7. **BEAUTIFUL CAECUM,** *Caecum pulchellum* ×10   p. 139
   Tubular, curved; rounded rings; yellowish white.

8. **DECUSSATE WORM SHELL,** *Serpulorbis decussata* ×1   p. 145
   Usually coiled; longitudinal ridges; yellowish brown.

9. **SLIT WORM SHELL,** *Siliquaria anguillae* ×1   p. 145
   Loosely coiled; slit along upper surface; brownish.

10. **ERECT WORM SHELL,** *Petaloconchus erectus* ×1   p. 144
    Ends of tubes free; smooth; whitish.

11. **FLORIDA WORM SHELL,** *Petaloconchus floridanus* ×1   p. 144
    Surface reticulate; yellowish brown.

12. **RIISE'S WORM SHELL,** *Serpulorbis riisei* ×1   p. 145
    Surface rather pebbly; pale brown.

13. **BLACK WORM SHELL,** *Petaloconchus nigricans* ×1   p. 144
    Wrinkled horizontally; reddish brown to black.

14. **IRREGULAR WORM SHELL**   p. 144
    Petaloconchus irregularis ×1
    Coarse revolving ridges; reddish brown.

15. **KNORR'S WORM SHELL,** *Vermicularia knorri* ×1   p. 141
    First few volutions white, most of shell buffy brown.

16. **FARGO'S WORM SHELL,** *Vermicularia fargoi* ×1   p. 141
    Longitudinal keels; reddish brown.

17. **COMMON WORM SHELL,** *Vermicularia spirata* ×1   p. 142
    Tubular, with longitudinal ridges; yellowish brown.

Plate 42

SUNDIALS AND CARRIER SHELLS

1. **SUNDIAL,** *Architectonica nobilis* ×1   p. 142
   Circular, somewhat flattened; marbled brown and purple.

2. **NOBLE FALSE DIAL,** *Pseudomalaxis nobilis* ×2   p. 143
   Flatly coiled; squarish ridges at shoulders; dull white.

3. **CYLINDER SUNDIAL,** *Heliacus cylindricus* ×1   p. 143
   Solid; blunt spire, weak revolving lines; brownish to black, dotted with white.

4. **CARIBBEAN CARRIER SHELL,** *Xenophora caribaea* ×1   p. 166
   Thin-shelled; wrinkle-like folds; only a few foreign objects on shell; white.

5. **KREBS' SUNDIAL,** *Philippia krebsi* ×1   p. 143
   Flatly coiled, smooth; yellowish white, with brownish specks at periphery.

6. **BISULCATE SUNDIAL,** *Heliacus bisulcatus* ×2   p. 143
   Flatly coiled; prominent beaded cords; yellowish brown.

7. **COMMON CARRIER SHELL**   p. 166
   Xenophora trochiformis ×1
   Well covered with foreign objects; yellowish brown.

CONTINUED FROM LEGEND PAGE FOR PLATE 43

21. **MIDDLE-SPINED HORN SHELL,** *Cerithium algicola* ×1½   p. 148
   Rugged, angled whorls, beaded; white or yellowish, blotched with brown.

22. **IVORY HORN SHELL,** *Cerithium eburneum* ×1½   p. 148
   Short canal turns left; white to deep brown, with reddish-brown blotches.

23. **VARIABLE HORN SHELL,** *Cerithium variabile* ×2   p. 149
   Stoutly conical; revolving beads, often 1 varix; color varies.

24. **FLORIDA HORN SHELL,** *Cerithium floridanum* ×1   p. 149
   Unequal spiraling ridges; white or gray, marked with brown.

25. **DOTTED HORN SHELL,** *Cerithium muscarum* ×1   p. 149
   Grayish white, with small chestnut dots.

26. **LETTERED HORN SHELL,** *Cerithium literatum* ×1   p. 149
   Stocky; white or gray, checked with brownish black.

Plate 43

HORN SHELLS

(contd. on legend page for Plate 42)

Plate 44

WENTLETRAPS AND VIOLET SNAILS

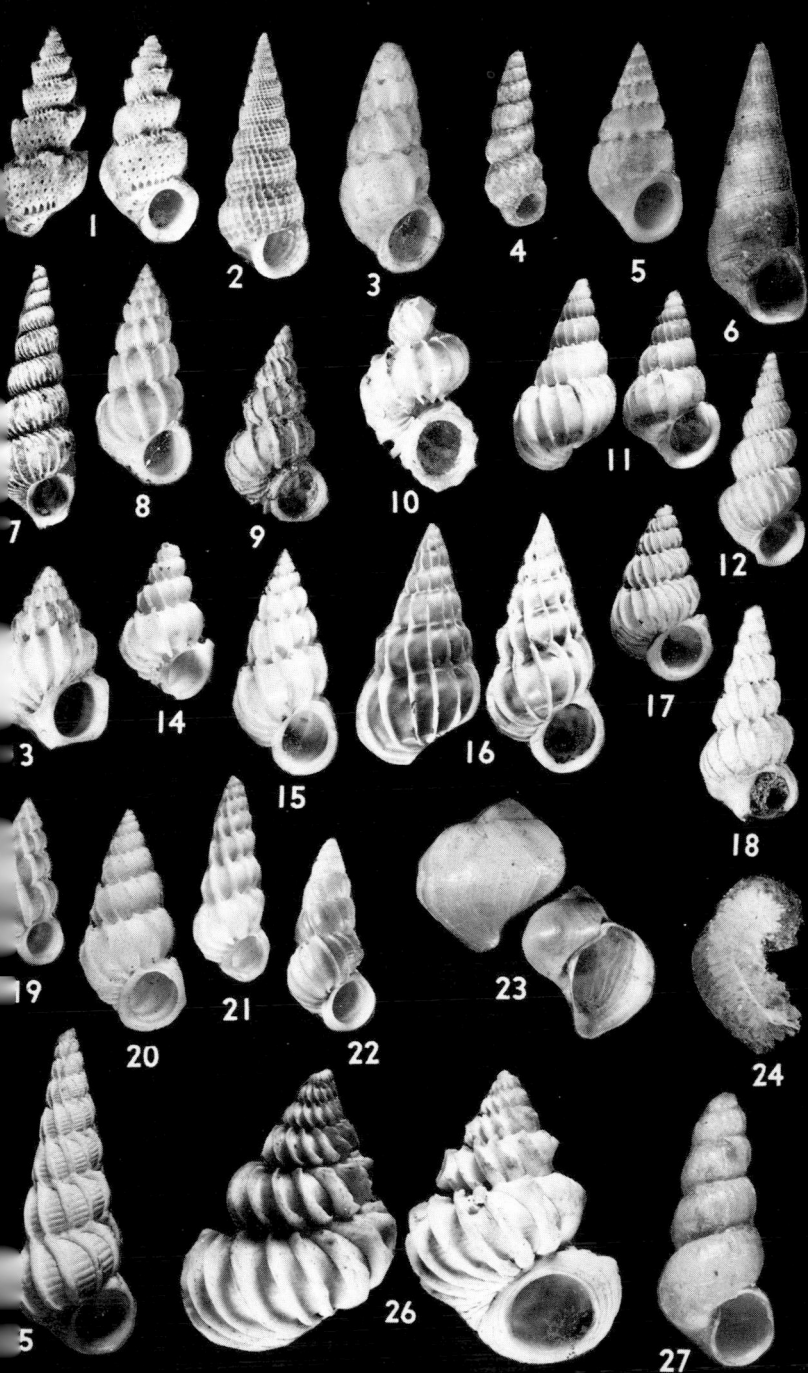

Plate 45

HOOF SHELLS, CUP-AND-SAUCER LIMPETS, SLIPPER SHELLS

1. **HOOF SHELL,** *Hipponix antiquatus* ×1   p. 161
   Caplike, variable; wrinkled surface; white or yellowish white.

2. **FALSE CUP-AND-SAUCER LIMPET,** *Cheilea equestris* × 1   p. 161
   Base irregular, circular; horseshoe-shaped plate within; grayish white.

3. **CIRCULAR CUP-AND-SAUCER LIMPET**   p. 163
   Calyptraea centralis ×2
   Cap-shaped; shelflike plate within; white.

4. **ROSY CUP-AND-SAUCER LIMPET**   p. 163
   Crucibulum auricula ×1
   Wavy margins; inner cup nearly free; grayish; interior pinkish.

5. **CUP-AND-SAUCER LIMPET,** *Crucibulum striatum* ×1   p. 164
   Fine radiating lines; inner cup fastened at one side; pinkish white, usually with brown streaks.

6. **INCURVED CAP SHELL,** *Capulus intortus* ×1   p. 162
   Large aperture; apex incurved; white.

7. **CAP SHELL,** *Capulus ungaricus* ×1   p. 163
   Cap-shaped; apex curved forward; whitish; grayish periostracum.

8. **CONVEX SLIPPER SHELL,** *Crepidula convexa* ×1   p. 164
   Oval, deeply convex; shelly plate deep; ashy brown, dotted or streaked.

9. **THORNY SLIPPER SHELL,** *Crepidula aculeata* ×1   p. 164
   Irregular thorny ridges; brownish or grayish, often rayed.

10. **COMMON SLIPPER SHELL,** *Crepidula fornicata* ×1   p. 164
    Platform slightly covex; pale gray, flecked with purplish chestnut.

11. **SPOTTED SLIPPER SHELL,** *Crepidula maculosa* ×1   p. 165
    Platform straight; pale gray surface, checked with chocolate marks.

12. **FLAT SLIPPER SHELL,** *Crepidula plana* ×1   p. 165
    Flat or concave; pure white.

Plate 46

STROMBS

1. **DUCKFOOT,** *Aporrhais occidentalis* ×1   p. 166
   Lip greatly expanded; rounded vertical folds; white or yellowish white.

2. **FIGHTING STROMB,** *Strombus pugilis* ×¾   p. 169
   Robust; spines on shoulders; lip notched below; yellowish brown. No purple as in Florida Stromb.

3. **FLORIDA STROMB,** *Strombus alatus* ×¾   p. 167
   Shoulders with or without spines; yellowish brown, clouded and sometimes striped with orange and purple.

4. **FLORIDA STROMB,** juvenile ×1

5. **HAWK WING,** *Strombus raninus* ×⅔   p. 169
   Knobby shoulders; streaked with brown and black.

6. **QUEEN STROMB,** juvenile ×1
   Zigzag axial stripes of brown.

7. **FLORIDA STROMB,** operculum ×1

8. **RIBBED STROMB,** *Strombus costatus* ×½   p. 167
   Solid; lip thick; mottled yellowish white.

9. **QUEEN STROMB,** *Strombus gigas* ×¼   p. 168
   Ponderous; wide lip; yellowish buff; interior rosy pink.

CONTINUED FROM LEGEND PAGE FOR PLATE 47

20. **SPOTTED EAR SHELL,** *Sinum maculatum* ×1   p. 174
    See text.

21. **GREENLAND MOON SHELL,** *Lunatia groenlandica* ×1   p. 172
    Small umbilicus; gray or white; greenish-yellow periostracum.

22. **NORTHERN MOON SHELL,** egg collar

23. **NORTHERN MOON SHELL,** *Lunatia heros* ×⅔   p. 172
    Large, globular; umbilicus deep; ashy brown.

Plate 47

MOON SHELLS AND OTHERS

1. NORTHERN HAIRY-KEELED SNAIL   p. 162
   Trichotropis borealis ×1
   Aperture large; stiff hairs on periostracum; brown.
2. SMOOTH VELUTINA, *Velutina laevigata* ×2   p. 170
   Substance very thin; thick brownish periostracum.
3. TRANSPARENT LAMELLARIA   p. 169
   Lamellaria pellucida ×2, with animal
   Tough periostracum conceals most of shell; pale yellow.
4. PERON'S ATLANTA, *Atlanta peroni* ×3   p. 170
   Flatly coiled, fragile; bladelike fins at periphery; white.
5. RANG'S LAMELLARIA, *Lamellaria rangi* ×2   p. 170
   Thin and delicate; aperture flaring; white.
6. MILKY MOON SHELL, *Polinices lacteus* ×1   p. 171
   Ovate, polished; callus partly over umbilicus; white.
7. TAN MOON SHELL, *Polinices hepaticus* ×1   p. 171
   Thick white callus over umbilicus; yellowish tan.
8. LOBED MOON SHELL, *Polinices duplicatus* ×1   p. 171
   Brown lobe over umbilicus; gray-brown, bluish tinge.
9. IMMACULATE MOON SHELL   p. 171
   Polinices immaculatus ×2
   Smooth; callus does not block umbilicus; white.
10. DWARF MILKY MOON SHELL *Polinices uberinus* ×2   p. 172
    See text.
11. SULCATE MOON SHELL, *Stigmaulax sulcata* ×1   p. 174
    Sturdy; operculum calcareous; yellowish, spotted brown.
12. ICELAND MOON SHELL, *Amauropsis islandica* ×1   p. 173
    Slitlike umbilicus; yellowish white; orange-brown periostracum.
13. COLORFUL MOON SHELL, *Natica canrena* ×1   p. 173
    Operculum ribbed, calcareous; spiral chestnut bars and zigzag marks.
14. MOROCCO MOON SHELL, *Natica marochiensis* ×1   p. 174
    Umbilicus deep; grayish brown, with obscure reddish-brown spots.
15. LIVID MOON SHELL, *Natica livida* ×1   p. 173
    Shiny; pearl-gray, with darker bands; callus brown.
16. ARCTIC MOON SHELL, *Natica clausa* ×1   p. 173
    Operculum calcareous; callus over umbilicus; pale brown.
17. MINIATURE MOON SHELL, *Natica pusilla* ×2   p. 174
    Small but solid; white or gray.
18. SPOTTED MOON SHELL, *Lunatia triseriata* ×2   p. 172
    Yellowish gray, usually with squarish brown spots; see text.
19. EAR SHELL, *Sinum perspectivum* ×1   p. 175
    Ovate, depressed; undulating lines; milky white.

Contd. on legend page for Plate 46

Plate 48

COWRIES AND THEIR RELATIVES

1. **COFFEE BEAN TRIVIA,** *Trivia pediculus* ×1   p. 177
   Solid, nearly spherical; ridged top; brown spots.

2. **FOUR-SPOTTED TRIVIA,** *Trivia quadripunctata* ×2   p. 177
   Nearly globular; pale pink, with 4 brownish spots on upper surface.

3. **ANTILLEAN TRIVIA,** *Trivia antillarum* ×3   p. 177
   Oval, well arched; narrow lines; dark purplish.

4. **WHITE GLOBE TRIVIA,** *Trivia nix* ×1   p. 177
   Nearly globular; strong ridges; white.

5. **PINK TRIVIA,** *Trivia suffusa* ×2   p. 178
   Small, transverse lines weakly beaded; rosy pink.

6. **SINGLE-TOOTHED SIMNIA,** *Neosimnia uniplicata* ×1   p. 178
   Thin, elongate, smooth; usually pink or purple.

7. **COMMON WEST INDIAN SIMNIA**   p. 178
   Neosimnia acicularis ×1
   Thin, elongate; ridged columella; yellowish or purplish.

8. **MAUGER'S ERATO,** *Erato maugeriae* ×2   p. 176
   Pear-shaped, smooth; pale tan often with rosy tint.

9. **DWARF RED OVULA,** *Primovula carnea* ×2   p. 178
   Aperture larger than shell; pinkish white or yellowish.

10. **FLAMINGO TONGUE,** *Cyphoma gibbosum* ×1   p. 179
    Elongate-oval; transverse dorsal ridge; white or yellowish.

11. **McGINTY'S FLAMINGO TONGUE**   p. 179
    Cyphoma macgintyi ×1
    See text.

12. **YELLOW COWRY,** *Cypraea spurca acicularis* ×1   p. 176
    Yellowish tan, spotted with yellow; violet spots along sides.

13. **DEER COWRY,** *Cypraea cervus* ×½   p. 175
    Large; brown, with many small whitish spots that are not in rings.

14. **GRAY COWRY,** *Cypraea cinerea* ×1   p. 175
    Plump; teeth rather small; grayish brown, paler bands.

15. **MEASLED COWRY,** *Cypraea zebra* ×½   p. 176
    Purplish brown, with small whitish spots, often in rings.

16. **MEASLED COWRY,** juvenile ×1

17. **MEASLED COWRY** ×½, cut to show early whorls

Plate 49

BONNETS, HELMETS, AND OTHERS

1. **WOOD LOUSE,** *Morum oniscus* ×1 p. 180
   Warty surface; soiled white, mottled brown and gray.

2. **PAPER FIG SHELL,** *Ficus communis* ×⅔ p. 188
   Thin-shelled, flat-topped; pinkish gray, with weak brown dots.

3. **ROYAL BONNET,** *Sconsia striata* ×1 p. 180
   Usually 1 or 2 varices on last volution; buffy gray, with chestnut spots.

4. **SCOTCH BONNET,** *Phalium granulatum* ×1 p. 180
   Inner lip pustulate; surface ribbed; squarish brown spots.

5. **SCOTCH BONNET,** operculum ×1

6. **CROSSE'S TUN,** *Eudolium crosseanum* ×1 p. 188
   Short spire, large aperture; buffy white.

7. **FLAME HELMET,** *Cassis flammea* ×½ p. 181
   Body whorl large, surface smooth; yellowish, with brown streaks and splashes.

8. **EMPEROR HELMET,** *Cassis madagascariensis*
   Apertural view greatly reduced. See Remarks under Clench's Helmet (p. 181).

9. **CLENCH'S HELMET** p. 181
   Cassis madagascariensis spinella ×¼
   Ventral surface oval; grayish yellow, lip not spotted.

10. **KING HELMET,** *Cassis tuberosa* ×⅓ p. 182
    Ventral surface triangular; mottled buff and brown; lip spotted.

11. **KING HELMET,** apertural view greatly reduced

Plate 50

TRITONS, TUNS, AND OTHERS

Plate 51

FROG SHELLS, MUREX AND TRUMPET SHELLS

1. **CUBAN FROG SHELL,** *Bursa cubaniana* ×1   p. 186
   2 varices on each whorl; beaded surface; reddish brown, white blotches.

2. **CORRUGATED FROG SHELL,** *Bursa corrugata* ×1   p. 186
   2 varices on each whorl; knobby surface; yellowish brown, clouded darker.

3. **CHESTNUT FROG SHELL,** *Bursa spadicea* ×1   p. 186
   Aperture notched above; surface beaded; yellowish brown, banded with orange-brown.

4. **SHORT-FROND MUREX,** *Murex brevifrons* ×½   p. 190
   3 varices; grayish, mottled with white and brown.

5. **LACE MUREX,** *Murex florifer* ×1   p. 191
   Frondlike spines; deep brownish black, aperture pink.

6. **ST. THOMAS FROG SHELL,** *Bursa thomae* ×1   p. 187
   Squat; 2 varices; yellowish white, spotted with reddish.

7. **RED MUREX,** *Murex recurvirostris rubidus* ×1   p. 193
   3 rounded varices; mottled gray, brown, pink, or reddish.

8. **PITTED MUREX,** *Murex cellulosus* ×1   p. 191
   Solid, squat; grayish; aperture purplish.

9. **APPLE MUREX,** *Murex pomum* operculum ×1   p. 192

10. **TRUMPET SHELL,** operculum ×½

11. **CABRIT'S MUREX,** *Murex cabriti* ×1½   p. 190
    Long canal, 3 varices; pinkish buff.

12. **TAWNY MUREX,** *Murex fulvescens* ×½   p. 191
    3 varices; spiny; buffy white.

13. **TRUMPET SHELL,** *Charonia variegata* ×¼   p. 185
    Large; variegated buff, brown, purple, and red.

Plate 52

ROCK OR DYE SHELLS, OYSTER DRILLS, ETC.

1. CIBONEY MUREX, *Murex ciboney* ×1 — p. 191
   Moderate canal, short spines; white, with thin brown lines.
2. TRYON'S MUREX, *Murex tryoni* ×1½ — p. 193
   Long canal, small aperture, long spines; yellowish white.
3. HANDSOME MUREX, *Murex pulcher* ×1 — p. 192
   Short sharp apex; yellowish white, banding of reddish brown.
4. BEQUAERT'S MUREX, *Murex bequaerti* ×1 — p. 190
   3 weblike varices, canal short; white.
5. PAZ'S MUREX, *Murex pazi* ×1 — p. 192
   Varices with long spines at shoulders; canal short; white.
6. WOODRING'S MUREX, *Murex woodringi* ×1 — p. 193
   Canal long, nearly closed; short spines; yellowish white.
7. McGINTY'S MUREX, *Murex macgintyi* ×1 — p. 192
   Solid, squat; canal short; revolving ribs; grayish white.
8. CAILLET'S MUREX, *Murex cailleti* ×1 — p. 191
   3 rugged varices, canal closed; yellowish white, with brownish bands.
9. AGUAYO'S MUREX, *Murex aguayoi* ×1 — p. 189
   Canal long, closed, curved; nodulous ridges; yellowish white.
10. BEAU'S MUREX, *Murex beaui* ×1 — p. 190
    Webs at base of spines; yellowish or brownish.
11. TAMPA OYSTER DRILL, *Urosalpinx tampaensis* ×1 — p. 195
    Strong sculpture; mottled grayish brown.
12. GULF OYSTER DRILL, *Urosalpinx perrugata* ×1 — p. 195
    Solid, with vertical folds; yellowish gray.
13. OYSTER DRILL, *Urosalpinx cinerea* ×1 — p. 194
    Solid; dingy gray; aperture dark purple.
14. FALSE DRILL, *Pseudoneptunea multangula* ×1 — p. 194
    Color variable, usually creamy white flecked with brown.
15. THICK-LIPPED OYSTER DRILL, *Eupleura caudata* ×1 — p. 195
    Lip thickened, ridges knobby; reddish brown to bluish white.
16. WRETCHED ASPELLA, *Aspella paupercula* ×1½ — p. 194
    Thin varices; grayish, clouded with brown or black.
17. FRILLY DWARF ROCK SHELL, *Ocenebra intermedia* ×1 — p. 200
    Revolving ribs and frilly varices; white, sometimes banded.
18. CARIBBEAN CORAL SNAIL, *Coralliophila caribaea* ×1 — p. 197
    Slanting vertical ribs; whitish; aperture purplish.
19. GLOBULAR CORAL SNAIL, *Coralliophila aberrans* ×1 — p. 197
    Knobby shoulders, aperture pear-shaped; white; purple inside.

(contd. on legend page for Plate 53)

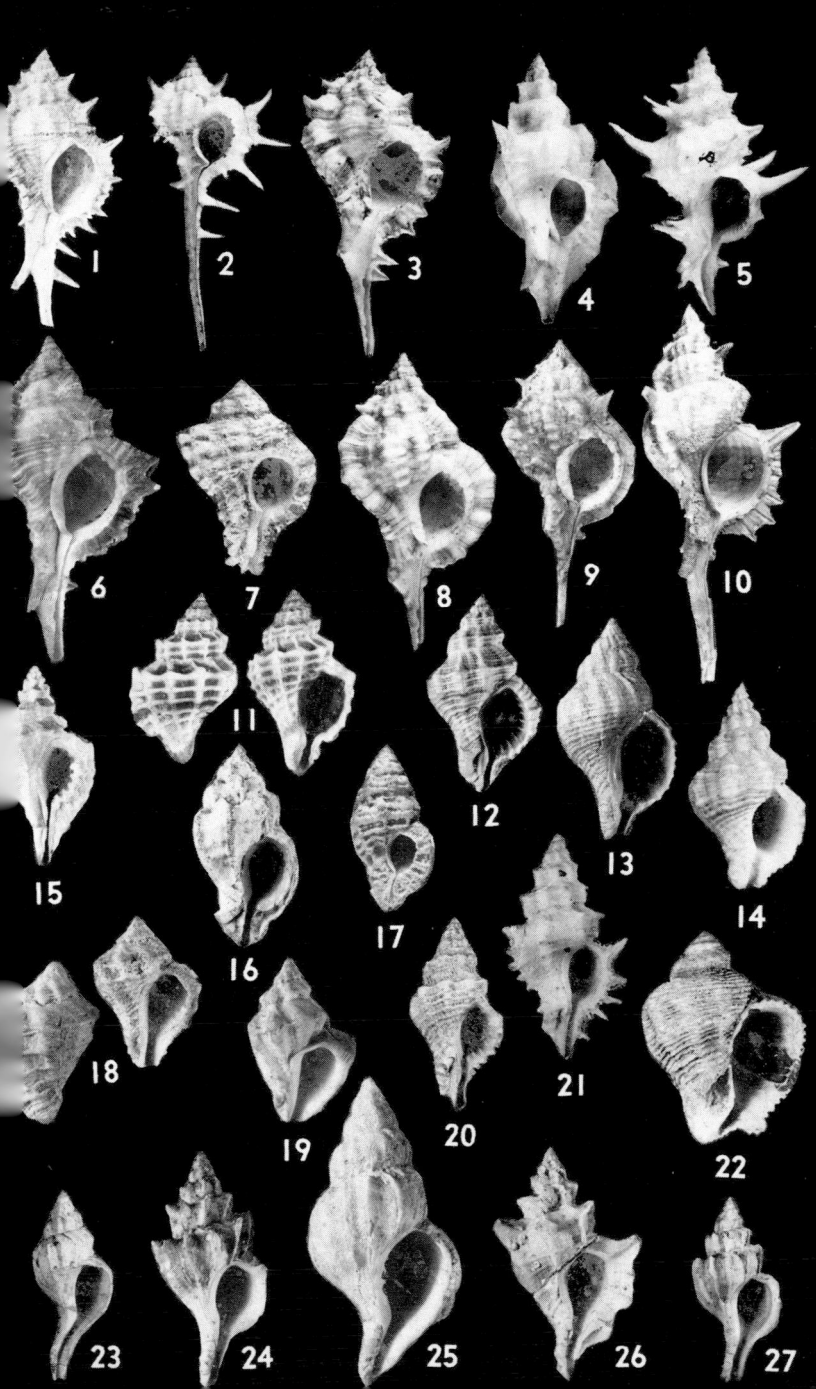

Plate 53

ROCK OR DYE SHELLS

1. **BLACKBERRY SNAIL,** *Drupa nodulosa* ×1   p. 198
   Prominent nodules; grayish green, nodules shiny black.
2. **DELTOID ROCK SHELL,** *Thais deltoidea* ×1   p. 198
   Stubby; grayish or pinkish white, blotched with brown and purple.
3. **RUSTIC ROCK SHELL,** *Thais rustica* ×1   p. 200
   Shoulders nodular; gray or brown, sometimes with white bands.
4. **FLORIDA ROCK SHELL**   p. 199
   Thais haemastoma floridana ×1
   See text.
5. **HAYS' ROCK SHELL,** *Thais haemastoma haysae* ×1   p. 199
   See text.
6. **Operculum of** *Thais* ×1
7. **DOGWINKLE,** *Thais lapillus* ×1   p. 199
   Thick-shelled; color variable, ranges from white to lemon-yellow and purplish brown; some banded with white or yellow.
8. **WIDE-MOUTHED ROCK SHELL,** *Purpura patula* ×1   p. 198
   Flaring aperture; grayish green or brown; interior salmon-pink.

Continued from legend page for Plate 52

20. **MAUVE-MOUTHED MUREX,** *Muricopsis ostrearum* ×1   p. 193
    Spire elevated; strong sculpture; aperture mauve.
21. **HEXAGONAL MUREX,** *Muricopsis oxytatus* ×1   p. 193
    Surface spiny; grayish white, tinged with reddish.
22. **CORAL SNAIL,** *Coralliophila abbreviata* ×1   p. 197
    Aperture flares; grayish; aperture pinkish violet.
23. **CURVED TROPHON**
    Boreotrophon clathratum rostratum ×1
    See Remarks under Clathrate Trophon (p. 196).
24. **CLATHRATE TROPHON,** *Boreotrophon clathratum* ×1   p. 196
    Moderately large; sharp vertical ridges; chalky white.
25. **STAIRCASE TROPHON**   p. 196
    Boreoptrophon scalariformis ×1
    Thin varices; grayish brown.
26. **LATTICED TROPHON,** *Boreotrophon craticulatum* ×1   p. 196
    Widely spaced thin varices; grayish white.
27. **GUNNER'S TROPHON,** *Boreotrophon gunneri* ×1   p. 196
    Shouldered whorls, bladelike varices; chalky white.

Plate 54

DOVE SHELLS AND DOG WHELKS

1. MOTTLED DOVE SHELL, *Columbella mercatoria* ×1   p. 200
   Color variable, usually gray clouded with purplish brown, rusty red, and white.
2. OVATE DOVE SHELL, *Pyrene ovulata* ×1½   p. 201
   Surface smooth, spire blunt; reddish brown, with whitish scrawls.
3. SPOTTED DOVE SHELL, *Columbella rusticoides* ×1   p. 201
   Well-elevated spire; yellowish, heavily marked with brown.
4. GREEDY DOVE SHELL, *Anachis avara* ×2   p. 201
   Well-spaced vertical ribs; brownish yellow.
5. WELL-RIBBED DOVE SHELL, *Anachis translirata* ×2   p. 202
   See text.
6. CHAIN DOVE SHELL, *Anachis catenata* ×2   p. 201
   Apex rather blunt; yellowish white, scattered brown dots.
7. FAT DOVE SHELL, *Anachis obesa* ×2   p. 202
   Stocky, spindle-shaped; yellowish gray.
8. HALIAECT'S DOVE SHELL, *Anachis haliaecti* ×2   p. 202
   Distinct sutures, vertical ribs; yellowish brown.
9. SPRINKLED DOVE SHELL, *Anachis sparsa* ×2   p. 202
   Glossy; yellowish white, with splashes of orange-brown.
10. BEAUTIFUL DOVE SHELL, *Anachis pulchella* ×2   p. 202
    Sutures indistinct; yellowish white, with reddish-brown scrawls.
11. GLOSSY DOVE SHELL, *Nitidella nitida* ×2   p. 203
    Polished; pale yellow, mottled with chestnut-brown.
12. SMOOTH DOVE SHELL, *Nitidella laevigata* ×2   p. 203
    Smooth and glossy; usually yellowish orange, with white streaks.
13. CRESCENT MITRELLA, *Mitrella lunata* ×4   p. 204
    Smooth; yellowish tan, with reddish-brown markings.
14. BANDED MITRELLA, *Mitrella zonalis* ×4   p. 204
    Well polished; reddish brown.
15. LITTLE WHITE MITRELLA, *Mitrella albella* ×4   p. 203
    Weak vertical folds; whitish, with darker markings.
16. TWO-COLORED DOVE SHELL, *Nitidella dichroa* ×2   p. 203
    Smooth, shiny; whitish, with vertical streaks of brown.
17. WHITE-SPOTTED DOVE SHELL, *Nitidella ocellata* ×2   p. 203
    Elongate; yellowish tan to nearly black, with round white spots.
18. MANY-SPOTTED DOVE SHELL   p. 204
    Parastola monilifera ×4
    Elongate; revolving ridges; whitish, with broken brown bands.
19. VARIABLE DOG WHELK, *Nassarius albus* ×1   p. 205
    Well shouldered; color variable, usually white.

(*contd. on legend page for Plate 55*)

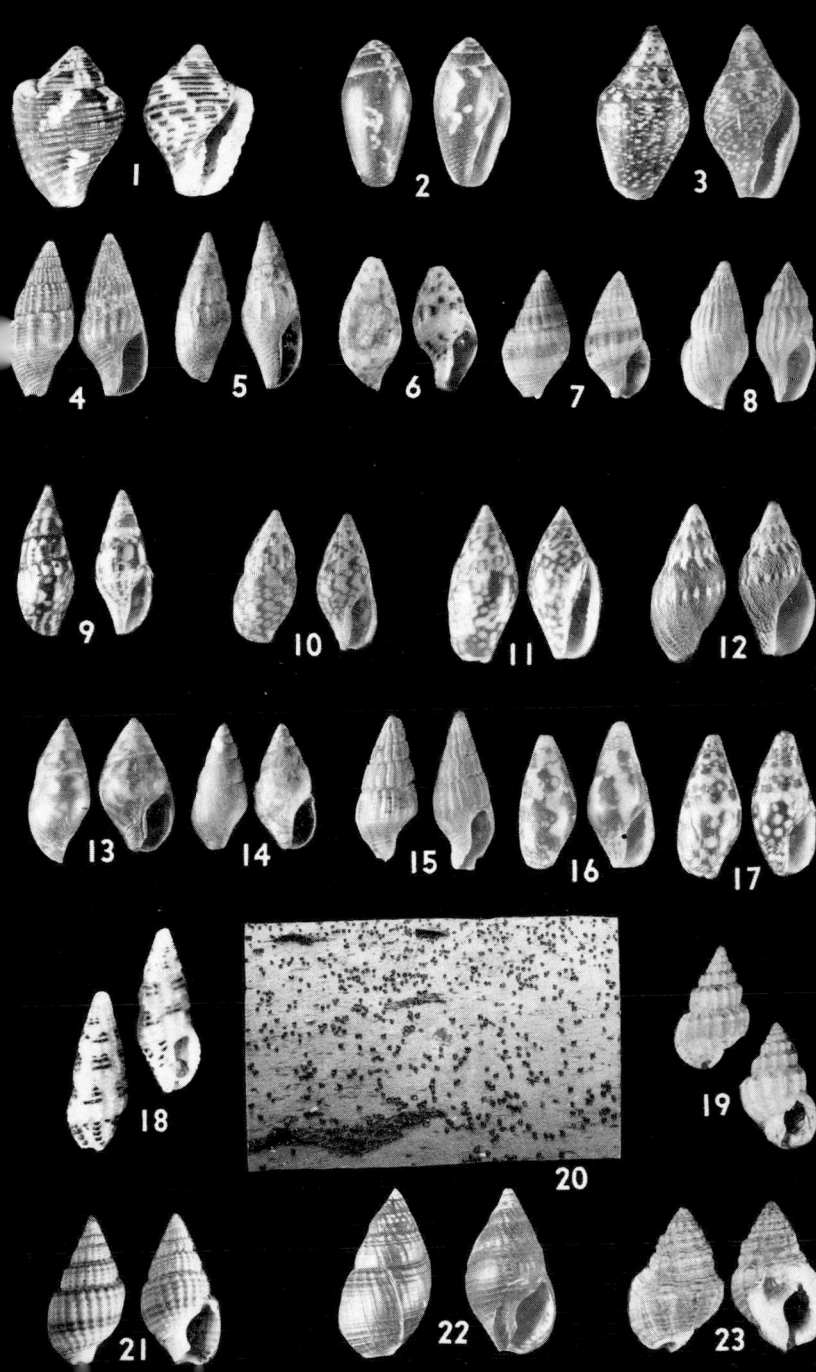

Plate 55

WHELKS

1. **INTRICATE BAILY SHELL,** *Bailya intricata* ×1   p. 206
   Ridge on outer lip; surface cancellate; grayish white.

2. **DEEP-SEA WHELK,** *Buccinum abyssorum* ×1   p. 206
   Keeled whorls; dull yellowish white.

3. **LITTLE BAILY SHELL,** *Bailya parva* ×1   p. 206
   Strong vertical and weaker revolving lines; yellowish
   white, with orange-brown bands.

4. **WAVED WHELK,** operculum ×1

5. **TOTTEN'S WHELK,** *Buccinum totteni* ×1   p. 207
   Surface smooth; thin yellowish periostracum.

6. **DONOVAN'S WHELK,** *Buccinum donovani* ×1   p. 207
   Sharp sutures, weak revolving lines; yellowish brown.

7. **WAVED WHELK,** *Buccinum undatum* ×1   p. 207
   Wavy vertical folds; pale reddish or yellowish brown.

8. **WAVED WHELK,** eggs

9. **BLUISH WHELK,** *Buccinum cyaneum* ×1   p. 207
   Sharp sutures; dull bluish white.

10. **SILKY WHELK,** *Buccinum tenue* ×1   p. 207
    Vertical folds; yellowish brown.

CONTINUED FROM LEGEND PAGE FOR PLATE 54

20. **MUD DOG WHELK** on mudflats

21. **NEW ENGLAND DOG WHELK**   p. 205
    Nassarius trivittatus ×1
    Shoulders flattened; white or yellowish white, sometimes
    banded.

22. **MUD DOG WHELK,** *Nassarius obsoletus* ×1   p. 205
    Apex commonly corroded; dark reddish purple to almost
    black.

23. **MOTTLED DOG WHELK,** *Nassarius vibex* ×1   p. 205
    Inner lip plastered onto body whorl; white, sometimes
    banded.

Plate 56

WHELKS

Plate 57

WHELKS, SPINDLES, AND OTHERS

1. **CANDÉ'S PHOS,** *Antillophos candei* ×2   p. 211
   Beaded surface; brownish yellow, lightly banded.

2. **GUADELOUPE PHOS,** *Engoniophos guadelupensis* ×1   p. 211
   Stocky; strong vertical ribs; whitish.

3. **MOTTLED SPINDLE,** *Cantharus tinctus* ×1½   p. 213
   Inner lip toothed above; reddish brown, mottled.

4. **CROSS-BARRED SPINDLE,** *Cantharus cancellaria* ×1½   p. 213
   Cancellate sculpture; reddish brown, mottled with white.

5. **PISA SHELL,** *Pisania pusio* ×1½   p. 212
   Polished; purplish brown, with irregular dark and light marks.

6. **SWIFT'S DWARF TRITON,** *Colubraria swifti* ×1   p. 213
   Slender, no varices on spire; yellowish gray, brown-blotched.

7. **GAUDY SPINDLE,** *Cantharus auritula* ×1   p. 213
   Stocky; knobby shoulders; mottled gray, brown, and black.

8. **WHITE-SPOTTED ENGINA,** *Engina turbinella* ×1½   p. 212
   Spindle-shaped; white knobs encircle volutions.

9. **LEANING DWARF TRITON,** *Colubraria obscura* ×1   p. 212
   Usually 2 varices each whorl; pale brown, with scattered orange-brown spots.

10. **KROYER'S WHELK,** *Plicifusus kroyeri* ×1   p. 210
    High-spired, smooth; brownish periostracum.

11. **ARROW TRITON,** *Colubraria lanceolata* ×2   p. 212
    2 riblike varices; pale brown or yellowish buff, with scattered orange-brown spots.

12. **LARGILLIERT'S WHELK,** *Volutopsius largillierti* ×1   p. 210
    Large aperture; yellowish white; greenish-brown periostracum.

13. **OSSIAN'S WHELK,** *Beringius ossiani* ×1   p. 209
    High-spired, with revolving riblets; yellowish brown; brownish periostracum.

14. **UNFINISHED WHELK,** *Neptunea despecta tornata* ×1   p. 211
    See text.

Plate 58

WHELKS AND CONCHS

1. **KNOBBED WHELK,** *Busycon carica* ×½   p. 215
   Knobbed shoulders; yellowish gray; aperture orange-red.

2. **CHANNELED WHELK,** *Busycon canaliculatum* ×½   p. 214
   Deep channel at suture; buffy gray; yellowish-brown periostracum.

3. **FIG WHELK,** *Busycon spiratum* ×⅔   p. 216
   Large aperture; flesh-colored, with reddish-brown streaks.

4. **CHANNELED WHELK,** egg ribbon

5. **CHANNELED WHELK,** young just hatched

6. **CROWN CONCH,** *Melongena corona*   p. 214
   Operculum ×½

7. **PERVERSE WHELK,** *Busycon perversum* ×⅕   p. 216
   Sinistral (may be dextral); swollen body whorl; grayish, with violet-brown streaks.

8. **HIGH-SPIRED CROWN CONCH**
   Melongena corona altispira ×1
   See Remarks under Crown Conch (p. 214).

9. **BROWN CROWN CONCH,** *Melongena melongena* ×½   p. 214
   Spines on shoulders; dark brown, with white or yellowish bands.

Plate 59

HORSE CONCH

1. **HORSE CONCH,** juveniles ×1

2. **HORSE CONCH,** *Pleuroploca gigantea* ×¼   p. 219
   Ponderous; revolving ridges; brown; aperture orange-red.

3. **HORSE CONCH,** operculum, showing both sides

Plate 60

LATIRUS, NUTMEGS, VASE SHELLS, OTHERS

1. McGINTY'S LATIRUS, *Latirus macgintyi* ×1   p. 218
   Rugged; vertical knobs; yellowish, dark brown between knobs.

2. SHORT-TAILED LATIRUS, *Latirus brevicaudatus* ×1   p. 217
   Stocky; light brown, with encircling dark brown lines.

3. KEY WEST LATIRUS, *Latirus cayohuesonicus* ×1   p. 217
   Stocky; indistinct sutures; dull gray; aperture brownish.

4. VIRGIN ISLAND LATIRUS, *Latirus virginensis* ×1   p. 218
   Solid; canal short; yellowish brown, knobs whitish.

5. WHITE-SPOTTED LATIRUS, *Leucozonia ocellata* ×1   p. 217
   Spindle-shaped; shoulders knobby; brown, knobs whitish.

6. CHESTNUT LATIRUS, *Leucozonia nassa* ×1   p. 217
   Wavy tubercles at shoulders; brown to blackish, pale band at base.

7. DISTINCT LATIRUS, *Latirus distinctus* ×⅔   p. 218
   Rugged; canal short; brownish gray.

8. AGASSIZ'S NUTMEG, *Trigonostoma agassizi* ×2   p. 230
   Columella grooved, shoulders knobby; yellowish white.

9. COUTHOUY'S NUTMEG, *Admete couthouyi* ×1   p. 231
   Revolving lines, nodular shoulders; yellowish brown.

10. RIDGED TULIP SHELL, *Ptychatractus ligatus* ×1   p. 220
    Spindle-shaped, with revolving lines; grayish brown.

11. NUTMEG, *Cancellaria reticulata* ×1   p. 230
    Solid; pleated columella; bright or pale orange, often banded.

12. PHILIPPI'S NUTMEG, *Trigonostoma tenerum* ×2   p. 231
    Flattened shoulders; pale orange, sometimes with chocolate spots.

13. SPINY VASE, *Vasum capitellus* ×1   p. 227
    Spiny, solid; deep umbilicus; yellowish brown.

14. VASE SHELL, *Vasum muricatum* ×1   p. 227
    Rugged; folds on columella; tough brownish periostracum.

15. TULIP SHELL, *Fasciolaria tulipa* ×½   p. 219
    Spindle-shaped, smooth; pinkish gray to reddish orange, streaked with reddish brown.

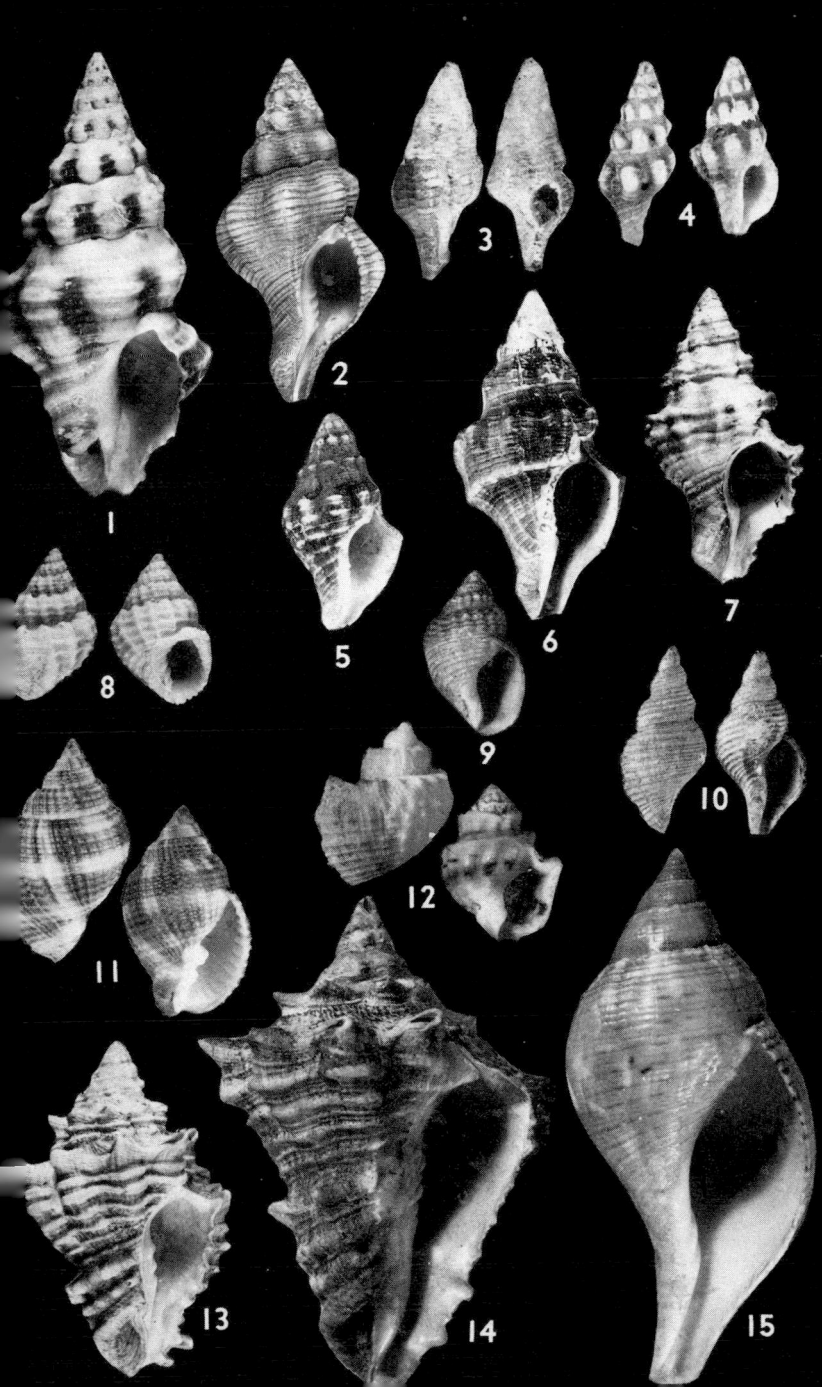

Plate 61

OLIVE AND SPINDLE SHELLS

1. **NETTED OLIVE,** *Oliva reticularis* ×1   p. 221
   Polished; white or grayish, with purplish-brown reticulations.

2. **GOLDEN OLIVE,** *Oliva sayana citrina* ×1
   Pale yellow to nearly golden; see Remarks under Lettered Olive (p. 222).

3. **LETTERED OLIVE,** *Oliva sayana* ×1   p. 222
   Cylindrical, polished; bluish gray, marked with chestnut and pink.

4. **CARIBBEAN OLIVE,** *Oliva caribaeensis* ×1   p. 221
   Glossy; mottled grayish purple, with white specks; aperture purplish.

5. **RICE DWARF OLIVE,** *Olivella floralia* ×1½   p. 222
   Bluish white, darker markings at sutures; apex often orange.

6. **DWARF OLIVE,** *Olivella mutica* ×1½   p. 222
   Color variable; often with revolving bands of purplish brown.

7. **CARIBBEAN DWARF OLIVE,** *Olivella petiolita* ×1½   p. 223
   Whitish, with irregular, wavy, vertical purplish-brown lines.

8. **TINY DWARF OLIVE,** *Olivella perplexa* ×2   p. 223
   Polished; base rather square; white.

9. **JASPER DWARF OLIVE,** *Jaspidella jaspidea* ×1½   p. 223
   Yellowish gray, with narrow chocolate band at sutures.

10. **WEST INDIAN DWARF OLIVE,** *Olivella nivea* ×1½   p. 223
    Slender; white, clouded with orange-brown.

11. **MINUTE DWARF OLIVE,** *Olivella minuta* ×2   p. 222
    Solid, polished; bluish gray, whitish at base; usually a white or brown band at sutures.

12. **ORNAMENTED SPINDLE,** *Fusinus eucosmius* ×1   p. 220
    Long canal, convex whorls; orange-white to pure white.

13. **COUE'S SPINDLE,** *Fusinus couei* ×½   p. 220
    Long canal, spiral ridges; white.

14. **SLENDER SPINDLE,** *Fusinus amphiurgus* ×1   p. 220
    Long canal, vertical folds; grayish white.

15. **TURNIP SPINDLE,** *Fusinus timessus* ×⅔   p. 221
    Long canal, stocky; white, tinged with orange-yellow.

Plate 62

MITERS AND LAMP SHELLS

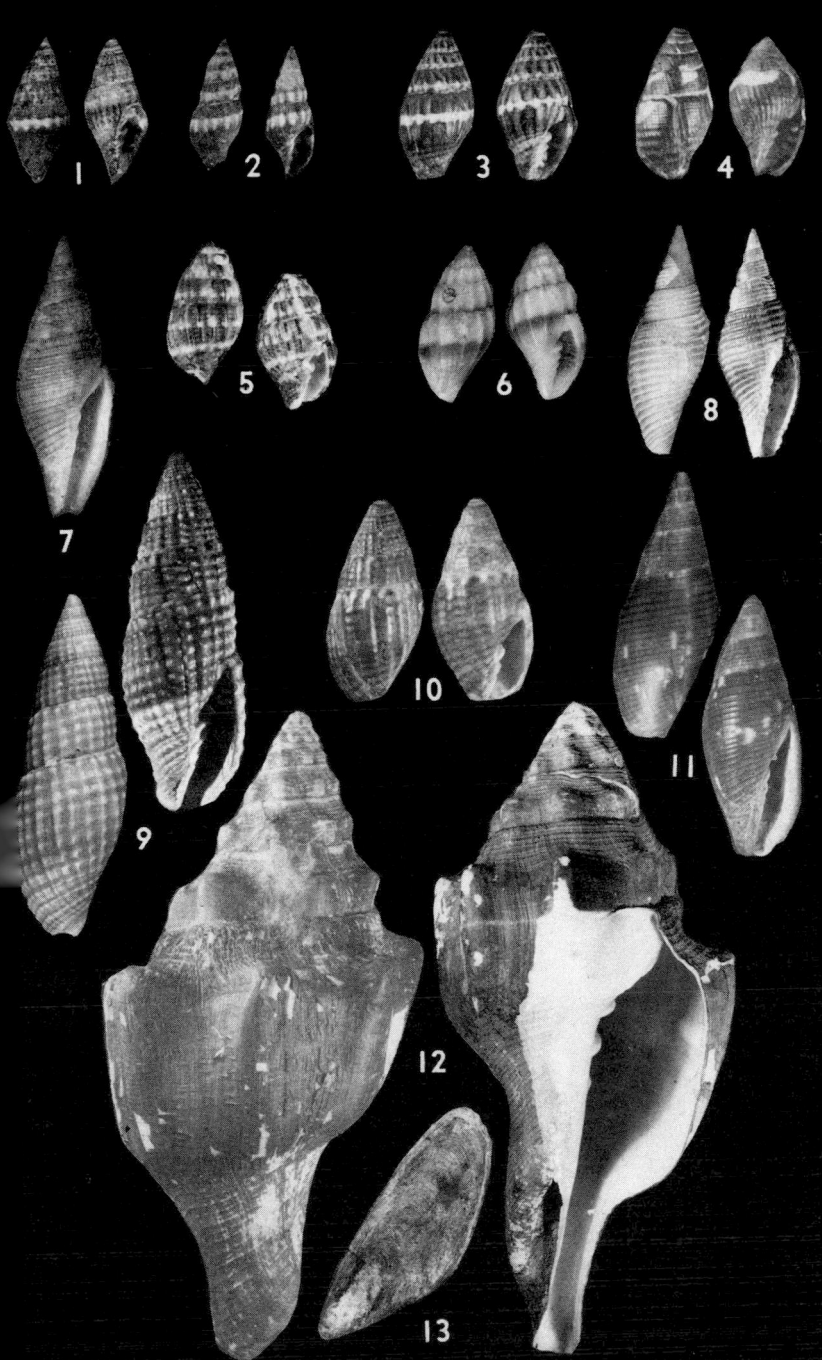

Plate 63

VOLUTES

Plate 64

MARGINELLAS

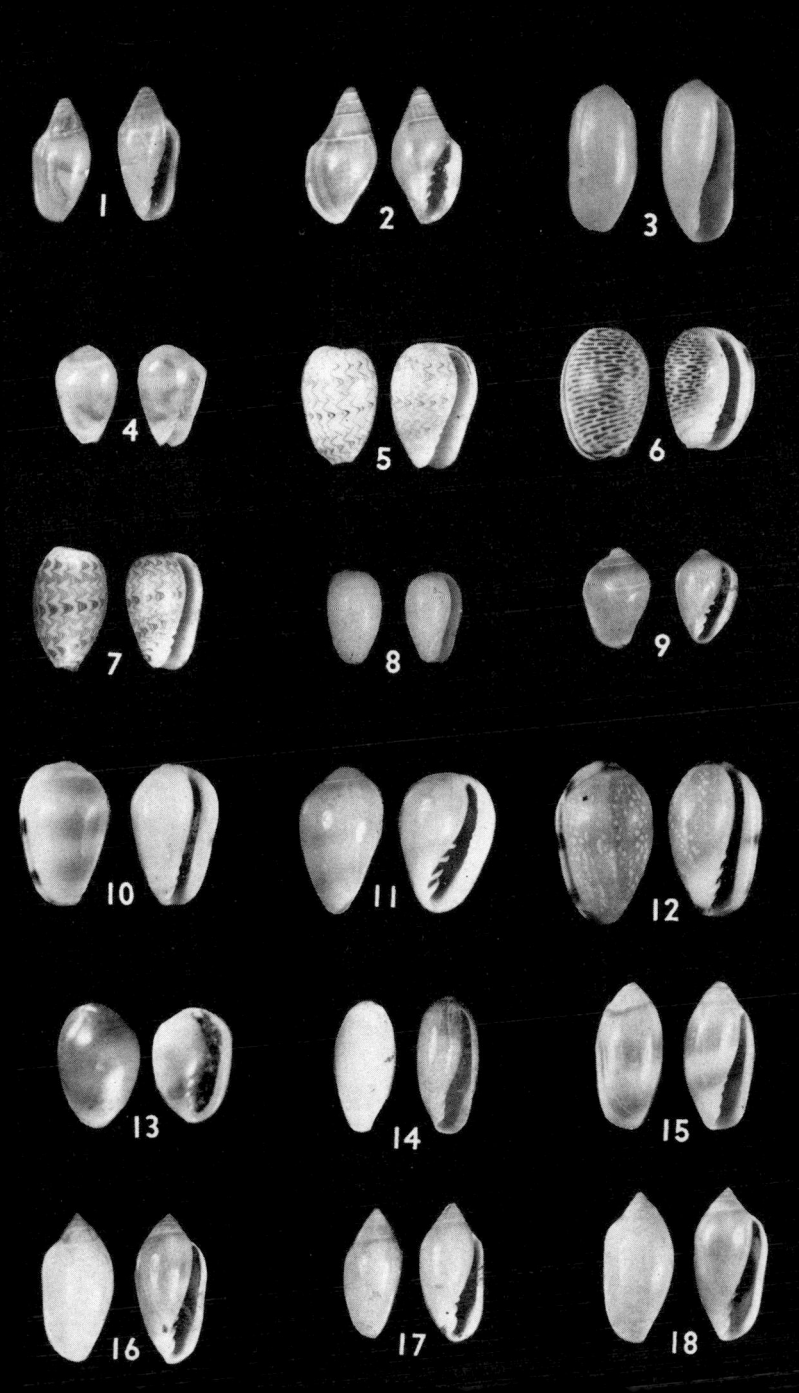

Plate 65

CONES

1. **GOLDEN-BANDED CONE,** *Conus aureofasciatus* ×1 p. 236
   Reddish brown, with pale yellow bands.

2. **CARROT CONE,** *Conus daucus* ×1 p. 236
   Orange to yellow; weak central band of yellowish.

3. **ANTILLEAN CONE** detail ×4

4. **ANTILLEAN CONE,** *Conus dominicanus* ×⅔ p. 236
   Yellowish tan, with pinkish-white blotches; see text.

5. **SOZON'S CONE,** *Conus sozoni* ×⅔ p. 238
   Sharp spire; rich orange, with white spotted bands.

6. **AGATE CONE,** *Conus ermineus* ×⅔ p. 236
   Grayish or bluish white, with deep brown blotches.

7. **FLORIDA CONE,** *Conus floridanus* ×1 p. 237
   Buffy yellow, with brown and white bands.

8. **ALPHABET CONE,** *Conus spurius spurius* ×⅔
   Markings are in broad bands; see Remarks under Atlantic Alphabet Cone (p. 239).

9. **ATLANTIC ALPHABET CONE** p. 239
   Conus spurius atlanticus ×⅔
   Creamy white, with orange-brown blotches; see text.

10. **CROWN CONE,** *Conus regius* ×⅔ p. 238
    Mottled chocolate-brown and purplish.

Plate 66

CONES

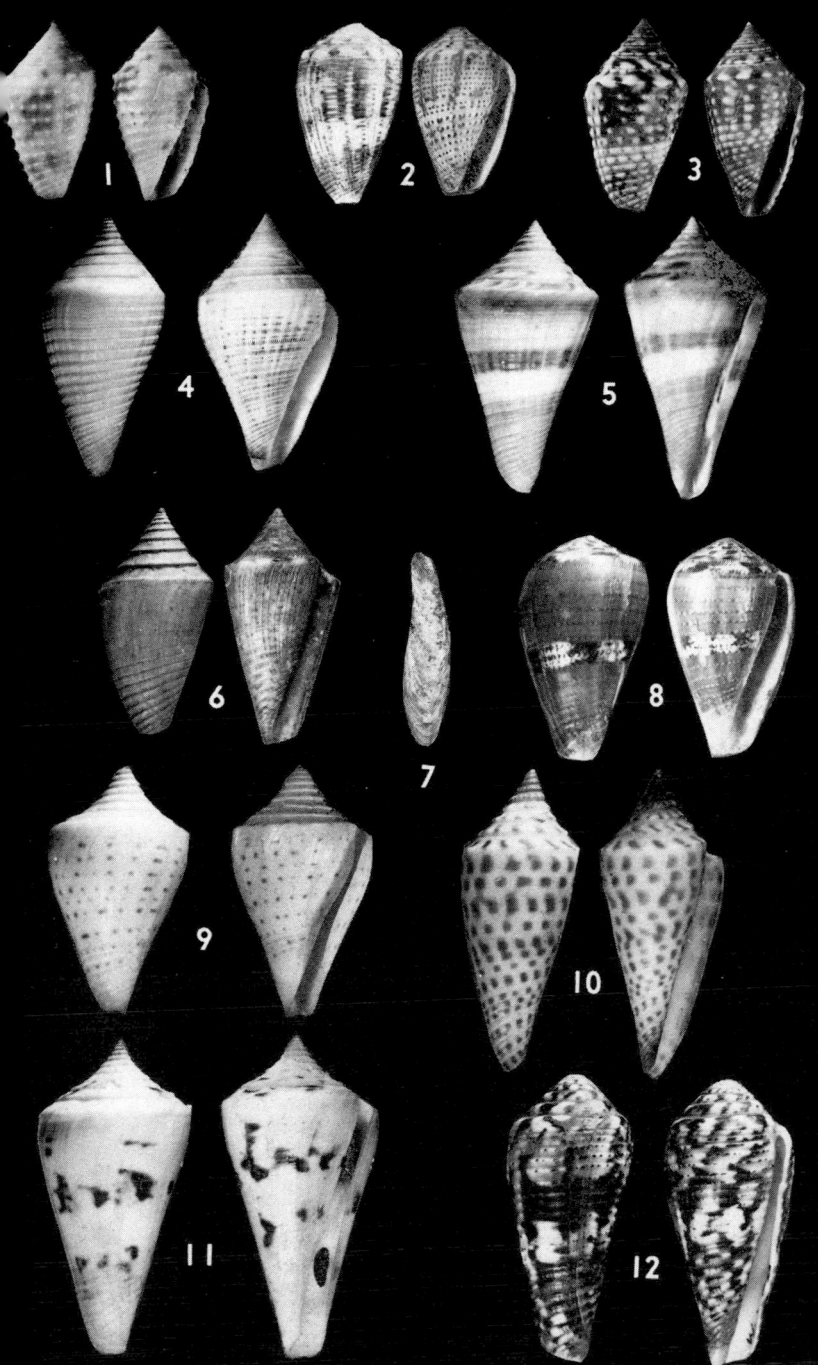

Plate 67

AUGERS AND TURRETS

1. **SHINY AUGER/** *Terebra hastata* ×1   p. 241
   Rather stout, polished; creamy white, with broad pale orange bands.

2. **CONCAVE AUGER,** *Terebra concava* ×1   p. 240
   Slender, whorls concave; gray, sometimes tinged with yellow.

3. **BLACK AUGER,** *Terebra protexta* ×1   p. 241
   Slender, with sharp vertical folds; deep chocolate-brown.

4. **COMMON AUGER,** *Terebra dislocata* ×1   p. 240
   Spikelike; impressed line below suture; ashy gray to pale brown, sometimes almost white.

5. **FLORIDA AUGER,** *Terebra floridana* ×1   p. 241
   Tall, with thin line at center of each whorl; yellowish white.

6. **SALLÉ'S AUGER,** *Terebra salleana* ×1   p. 241
   Dark gray or brown, with paler band; apex purple.

7. **FLAME AUGER,** *Terebra taurinum* ×⅔   p. 242
   Tall, slender; yellowish white, streaked with reddish brown.

8. **GRAY AUGER,** *Terebra cinerea* ×1   p. 240
   Sutures indistinct; gray or brown, with whitish band.

9. **GEM TURRET,** *Gemmula periscelida* ×1   p. 242
   Channeled above sutures; gray, with yellowish periostracum.

10. **PEBBLED STAR TURRET,** *Ancistrosyrinx elegans* ×1   p. 243
    Spire terraced, spines short, surface pebbled; yellowish white.

11. **STAR TURRET,** *Ancistrosyrinx radiata* ×4   p. 244
    Spines on angled suture; long open canal; yellowish gray.

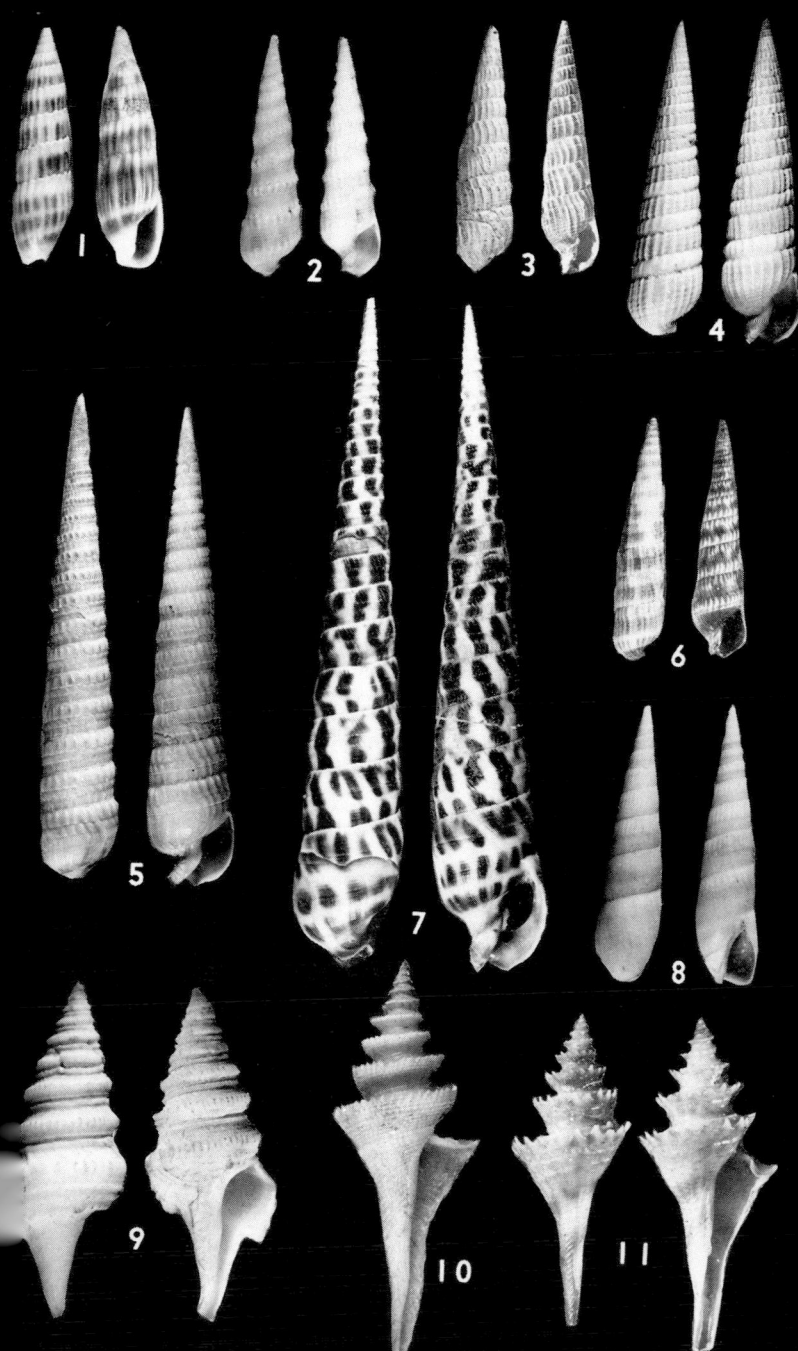

Plate 68

TURRET OR SLIT SHELLS

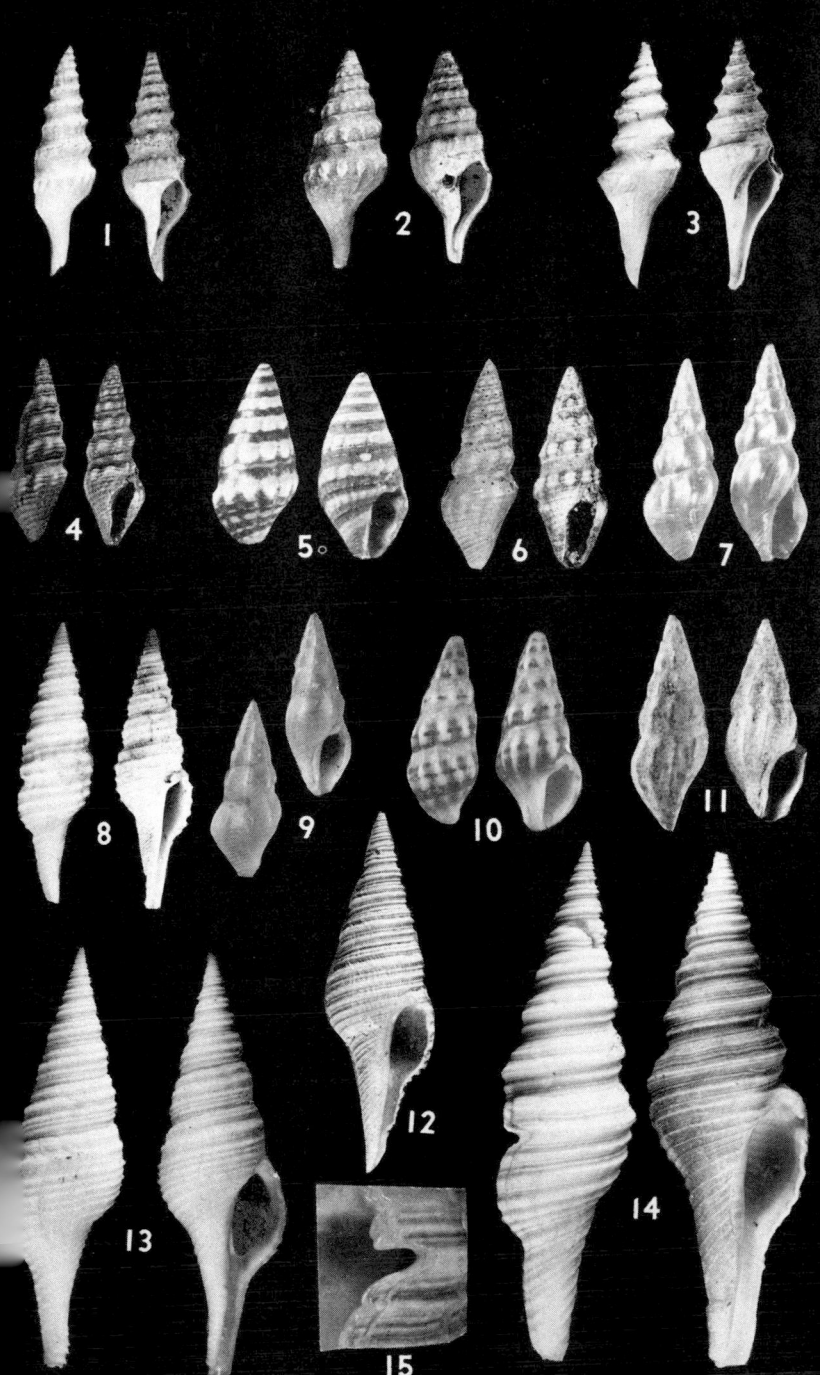

Plate 69

TURRET OR SLIT SHELLS

1. **SPINDLE TURRET,** *Crassispira alesidota* ×1   p. 244
   Elongate; crease above shoulders; ashy brown.
2. **BLACK TURRET,** *Crassispira fuscescens* ×1   p. 245
   Solid, with ridged encircling band; dark brown or black.
3. **TAMPA TURRET,** *Crassispira tampaensis* ×1   p. 245
   Ridge at sutures; reddish brown.
4. **SANIBEL TURRET,** *Crassispira sanibelensis* ×1   p. 245
   Distinct vertical ribs; pale brown or reddish brown.
5. **OYSTER TURRET,** *Crassispira ostrearum* ×1   p. 245
   Elevated ridges at sutures; reddish brown.
6. **STRIPED TURRET,** *Crassispira mesoleuca* ×1   p. 245
   Dark brown or black; white knobs commonly at shoulders.
7. **TALL-SPIRED TURRET,** *Inodrillia aepynota* ×2   p. 247
   Slender; strong vertical ribs; pinkish white.
8. **MIAMI TURRET,** *Inodrillia miama* ×2   p. 247
   Rather stout, with vertical ribs; pinkish white.
9. **FROSTED TURRET**   p. 252
   Glyphoturris quadrata rugirima ×2
   Angled whorls, vertical ribs; white.
10. **SHORT TURRET,** *Gymnobela curta* ×2   p. 252
    Stout, sutures deep; grayish or greenish white.
11. **GRAY-PILLARED TURRET,** *Kurtziella atrostyla* ×2   p. 248
    Small, slender, with sharp vertical ribs; grayish white.
12. **VOLUTE TURRET,** *Daphnella lymneiformis* ×2   p. 252
    Thin-shelled; yellowish white, blotched orange-brown.
13. **SLUGLIKE TURRET,** *Daphnella limacina* ×4   p. 252
    Smooth, shiny; ivory-white.
14. **CYDIA TURRET,** *Drillia cydia* ×2   p. 246
    Sturdy, with vertical ribs; white.
15. **BRIEF TURRET,** *Gymnobela brevis* ×2   p. 251
    Stout; obscure nodes at shoulders; yellowsh white.
16. **RATHBUN'S PLEUROTOMELLA**   p. 251
    Pleurotomella rathbuni ×1
    Surface weakly reticulate; yellowish white.
17. **PACKARD'S PLEUROTOMELLA**   p. 251
    Pleurotomella packardi ×1
    Stoutly spindle-shaped, with oblique ribs; pale flesh-color.
18. **JEFFREY'S PLEUROTOMELLA,** *Pleurotomella jeffreysi* ×1   p. 251
    Shoulders sharply angled and nodular; lustrous white.
19. **AGASSIZ'S PLEUROTOMELLA**   p. 250
    Pleurotomella agassizi ×1
    Strong shoulders, slanting ribs; yellowish brown.
20. **BAIRD'S PLEUROTOMELLA,** *Pleurotomella bairdi* ×1   p. 250
    Slanting shoulders, weak nodules; pale yellowish white.

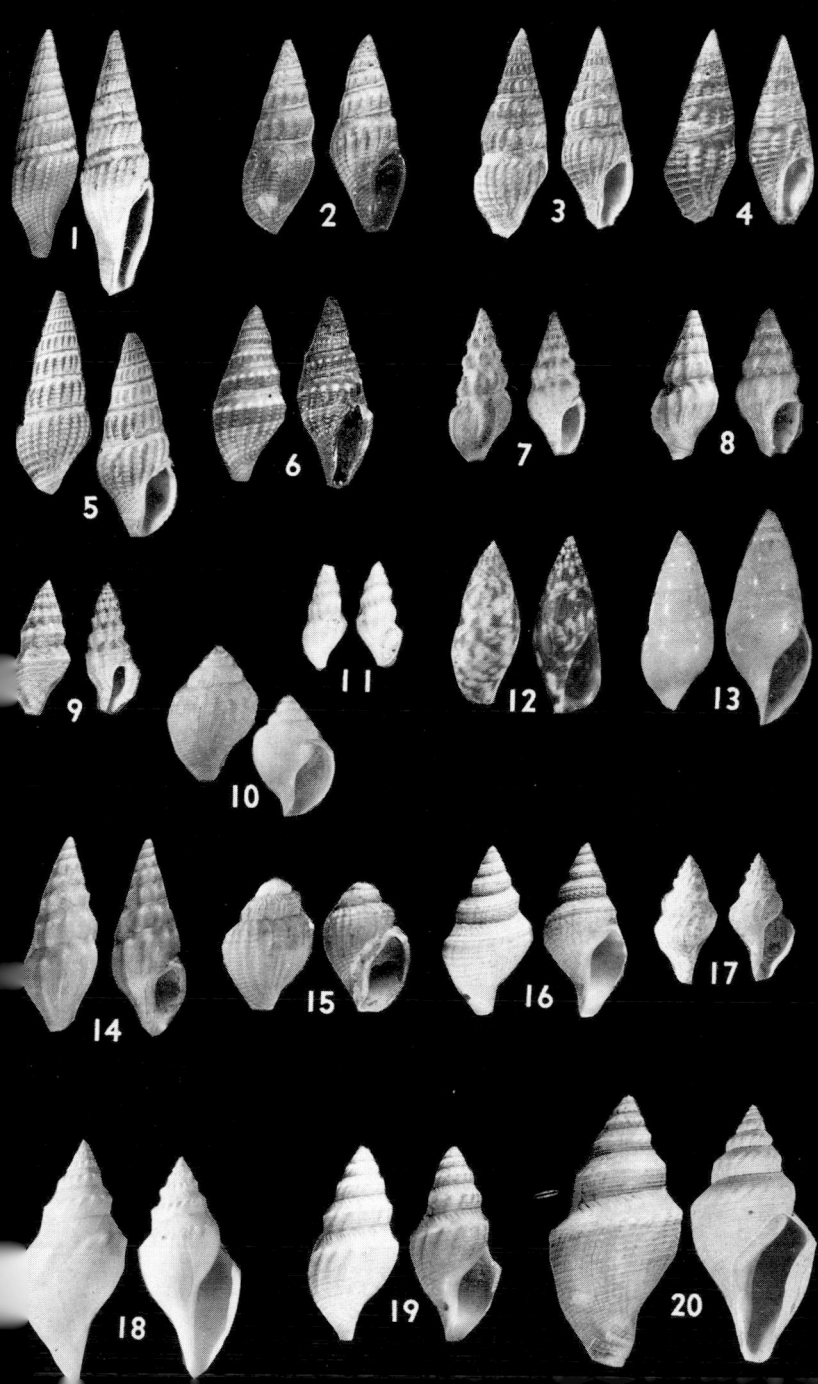

Plate 70

LORAS, OBELISKS, AND OTHERS

1. **NOBLE LORA,** *Lora nobilis* ×1   p. 249
   Shoulders flattened, spire turreted; yellowish white.
2. **HARP LORA,** *Lora harpularia* ×2   p. 248
   Oblique rounded ribs; buffy flesh color.
3. **CANCELLATE LORA,** *Lora cancellata* ×2   p. 248
   Cancellate sculpture; pinkish white.
4. **STAIRCASE LORA,** *Lora scalaris* ×1   p. 249
   Angled shoulders, vertical ridges; yellowish brown.
5. **LITTLE WAX-COLORED MANGELIA**   p. 249
   Mangelia cerinella ×3
   Slender, with broad vertical folds; yellowish white.
6. **WAX-COLORED MANGELIA,** *Mangelia cerina* ×3   p. 249
   Rounded vertical ribs; pale yellowish white.
7. **BLACK MANGELIA,** *Mangelia melanitica* ×4   p. 250
   Blunt vertical ribs, encircling lines; white.
8. **DARK MANGELIA,** *Mangelia fusca* ×2   p. 250
   Stubby, with vertical ribs; reddish brown.
9. **RIBBED MANGELIA,** *Mangelia plicosa* ×3   p. 250
   Network sculpture; dark brown.
10. **GOLDEN-BANDED EULIMA,** *Eulima auricincta* ×4   p. 253
    Slender, flat-sided, glossy; grayish white, with brownish bands.
11. **WALLER'S ACLIS,** *Aclis walleri* ×3   p. 255
    High-spired, glossy; white.
12. **STRIATED ACLIS,** *Aclis striata* ×5   p. 255
    Stocky; impressed sutures; yellowish white.
13. **INTERRUPTED NISO,** *Niso interrupta* ×2   p. 255
    Flattened whorls; pale brown; encircling reddish-brown line at sutures.
14. **TALL EULIMA,** *Eulima hypsela* ×2   p. 253
    Slender, spikelike, glossy; milky white.
15. **NARROWMOUTH EULIMA,** *Eulima stenotrema* ×4   p. 253
    Tall, slender, glassy; amber-colored.
16. **CONELIKE BALCIS,** *Balcis conoidea* ×3   p. 254
    Rather flat whorls, weak sutures; glossy white.
17. **TWO-LINED BALCIS,** *Balcis bilineata* ×4   p. 254
    Stoutly conical, shiny; milky white.
18. **TWO-BANDED EULIMA,** *Eulima bifasciata* ×4   p. 253
    Slender, flat-sided; yellowish white, 2 brown bands on each volution.
19. **INTERMEDIATE BALCIS,** *Balcis intermedia* ×2   p. 254
    Slender, tapering; glossy white, sometimes flecked with yellowish brown.
20. **HEMPHILL'S BALCIS,** *Balcis hemphilli* ×5   p. 254
    Flattened whorls; glossy white.
21. **JAMAICAN BALCIS,** *Balcis jamaicensis* ×4   p. 254
    Sutures plain; glossy white.

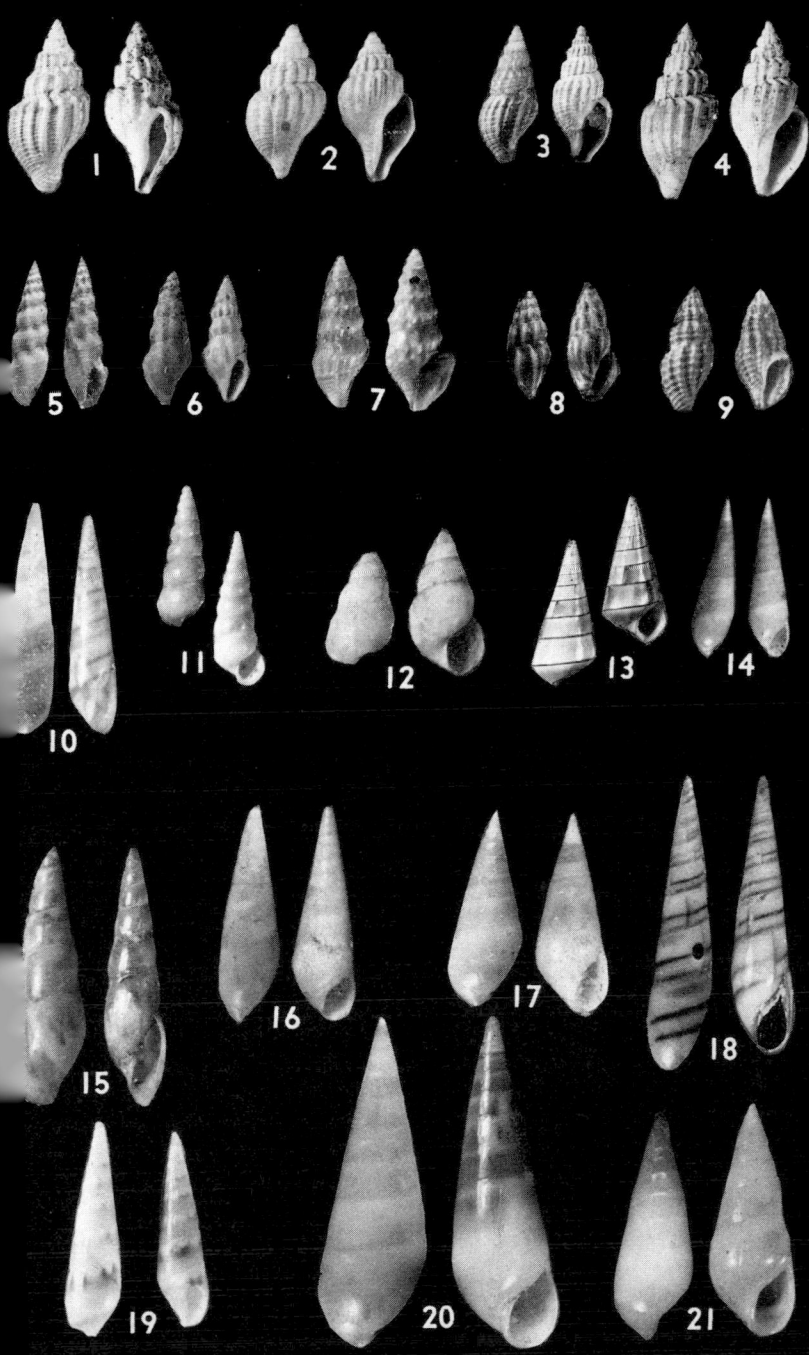

Plate 71

PYRAMID SHELLS, TURBONILLES, STYLUSES

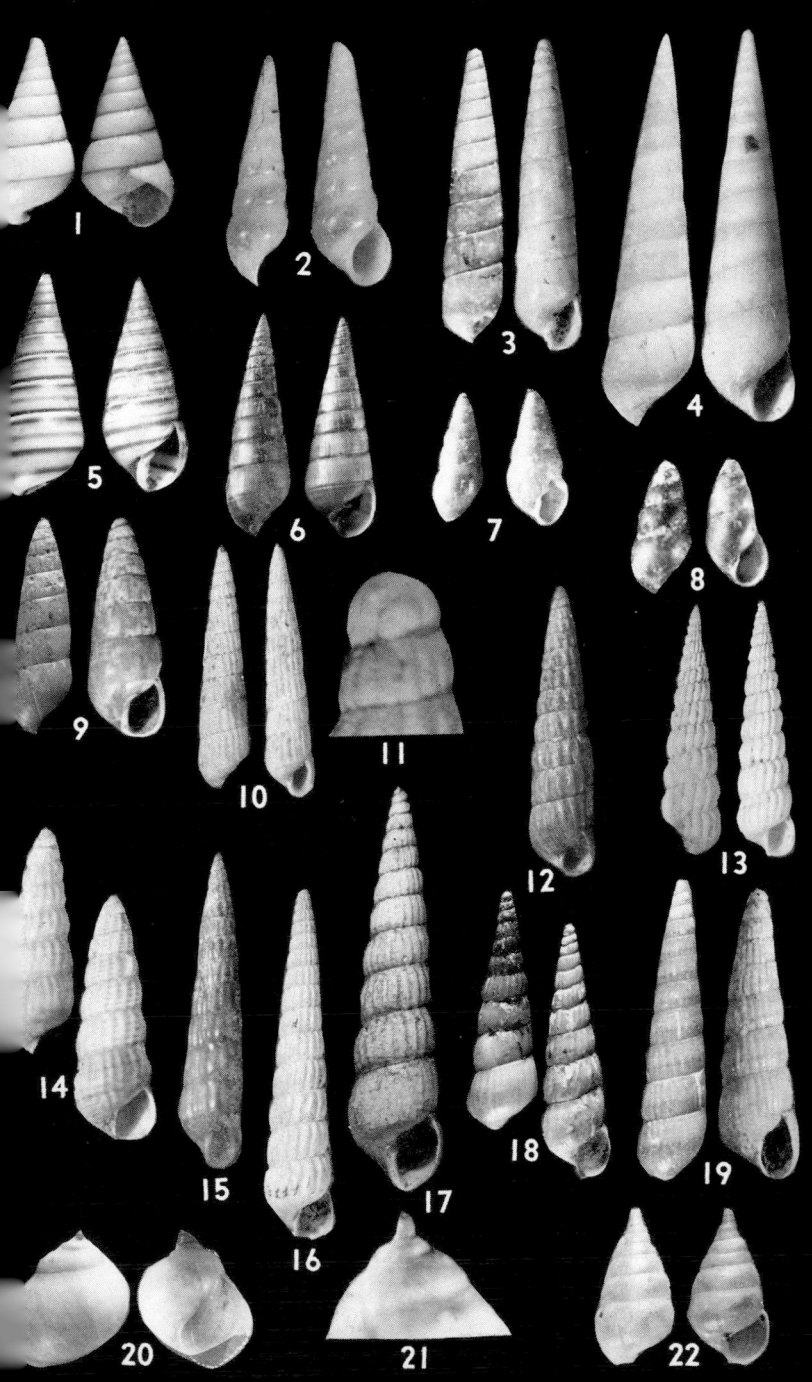

Plate 72

ODOSTOMES AND BUBBLE SHELLS

1. **HALF-SMOOTH ODOSTOME,** *Odostomia seminuda* ×5   p. 260
   Channeled sutures; surface appears beaded; white.
2. **TRIPARTITE ODOSTOME,** *Odostomia trifida* ×5   p. 261
   Impressed lines below suture; pale greenish white.
3. **INCISED ODOSTOME,** *Odostomia impressa* ×5   p. 260
   Channeled sutures, spiral grooves; milky white.
4. **DOUBLE-SUTURED ODOSTOME**   p. 259
   Odostomia bisuturalis ×5
   Double line at suture; dull white.
5. **GEMLIKE ODOSTOME,** *Odostomia gemmulosa* ×5   p. 260
   Nodular ridges, rather blunt apex; white.
6. **CHANNELED ODOSTOME,** *Odostomia canaliculata* ×5   p. 260
   Sutures slightly channeled; white.
7. **ANGULATED ODOSTOME,** *Odostomia engonia* ×5   p. 260
   Flattish whorls, sutures deeply impressed; white.
8. **STRIATE HELMET BUBBLE,** *Ringicula semistriata* ×5   p. 262
   Stout; blunt spire, twisted columella; polished white.
9. **ADAMS' BABY BUBBLE,** *Acteon punctostriatus* ×4   p. 261
   Stout; moderate spire; punctate at base; white.
10. **VERRILL'S HELMET BUBBLE,** *Ringicula nitida* ×5   p. 261
   Stout; pointed spire, twisted columella; white.
11. **LINED BUBBLE,** *Hydatina vesicaria* ×1   p. 262
   Thin-shelled; wide aperture; thin wavy brown lines.
12. **ARCTIC BUBBLE,** *Diaphana minuta* ×4   p. 263
   Globose; aperture wide below; translucent tan.
13. **MINIATURE MELO,** *Micromelo undata* ×2   p. 262
   Semiglobular, fragile; reddish-brown lines, vertical streaks.
14. **SHARP'S GLASSY BUBBLE,** *Atys sharpi* ×2   p. 265
   Elongate; long aperture; milky white.
15. **CARIBBEAN GLASSY BUBBLE,** *Atys caribaea* ×2   p. 264
   Stoutly elongate; pure white.
16. **SOLITARY GLASSY BUBBLE,** *Haminoea solitaria* ×2   p. 265
   Small depression at top of shell; bluish white.
17. **ANTILLEAN GLASSY BUBBLE,** *Haminoea antillarum* ×1   p. 265
   Thin and fragile; translucent greenish yellow.
18. **AMBER GLASSY BUBBLE,** *Haminoea succinea* ×2   p. 265
   Moderately slender; translucent amber.
19. **ELEGANT GLASSY BUBBLE,** *Haminoea elegans* ×1   p. 265
   Fragile, semitransparent; greenish yellow.
20. **COMMON BUBBLE,** *Bulla occidentalis* ×1½   p. 263
   Reddish gray, mottled with purplish brown.
21. **SOLID BUBBLE,** *Bulla solida* ×1   p. 264
   Solid, well inflated; mottled gray and brown.
22. **STRIATE BUBBLE,** *Bulla striata* ×1   p. 264
   See text.
23. **BEAU'S GLASSY BUBBLE,** *Cylindrobulla beaui* ×4   p. 266
   Broadly cylindrical; narrow aperture; soiled white.

Plate 73

BUBBLE SHELLS

1. **ARCTIC BARREL BUBBLE,** young ×4
   Flat-topped, shiny; yellowish gray.
2. **CANDÉ'S BARREL BUBBLE,** *Retusa candei* ×5   p. 267
   Cylindrical; short spire; milky white.
3. **IVORY BARREL BUBBLE,** *Retusa eburnea* ×6   p. 267
   Oval; rounded apex; ivory-white.
4. **CHANNELED BARREL BUBBLE,** *Retusa canaliculata* ×5   p. 266
   Short spire; dull chalky white.
5. **GOULD'S BARREL BUBBLE,** *Retusa gouldi* ×4   p. 267
   Oval; rounded apex; pale brown.
6. **ARCTIC BARREL BUBBLE,** *Retusa obtusa* ×6   p. 267
   Cylindrical, smooth; dingy white.
7. **BROWN'S BUBBLE,** *Cylichna alba* ×5   p. 269
   Cylindrical, spire sunken; rusty-brown periostracum.
8. **RICE BUBBLE,** *Cylichna oryza* ×5   p. 270
   Stoutly oval, shiny; yellowish gray.
9. **ORBIGNY'S BUBBLE,** *Cylichna bidentata* ×5   p. 269
   Elongate-oval; yellowish white.
10. **DEPRESSED BUBBLE,** *Cylichna vortex* ×3   p. 270
    Bluntly oval; white or yellowish white.
11. **VERRILL'S BUBBLE,** *Cylichna verrilli* ×5   p. 270
    Cylindrical; thin yellowish periostracum.
12. **CONCEALED BUBBLE,** *Cylichna occulta* ×4   p. 269
    Oval, flat-topped; yellowish-brown periostracum.
13. **AUBER'S BUBBLE,** *Cylichna auberi* ×4   p. 269
    Oval, flat-sided; white.
14. **OVATE LITTLE PEAR SHELL,** *Pyrunculus ovatus* ×5   p. 267
    Pear-shaped, with pit at apex; brownish gray.
15. **SPINDLE BUBBLE,** *Rhizorus oxytatus* ×5   p. 268
    Cylindrical, with pointed apex; pale yellow periostracum.
16. **FINMARKIAN PAPER BUBBLE,** *Philine finmarchia* ×5   p. 271
    Top flattened, shiny; ivory-white.
17. **FUNNEL-LIKE PAPER BUBBLE,** *Philine infundibulum* ×1   p. 271
    Large and fragile; flaring aperture; translucent white.
18. **QUADRATE PAPER BUBBLE,** *Philine quadrata* ×5   p. 271
    Flaring aperture; translucent grayish or whitish.
19. **FILE PAPER BUBBLE,** *Philine lima* ×5   p. 271
    Elongate-oval, with wide aperture; translucent yellowish white.
20. **GIRDLED PAPER BUBBLE,** *Philine cingulata* ×5   p. 270
    Fragile, with wide aperture; yellowish white.

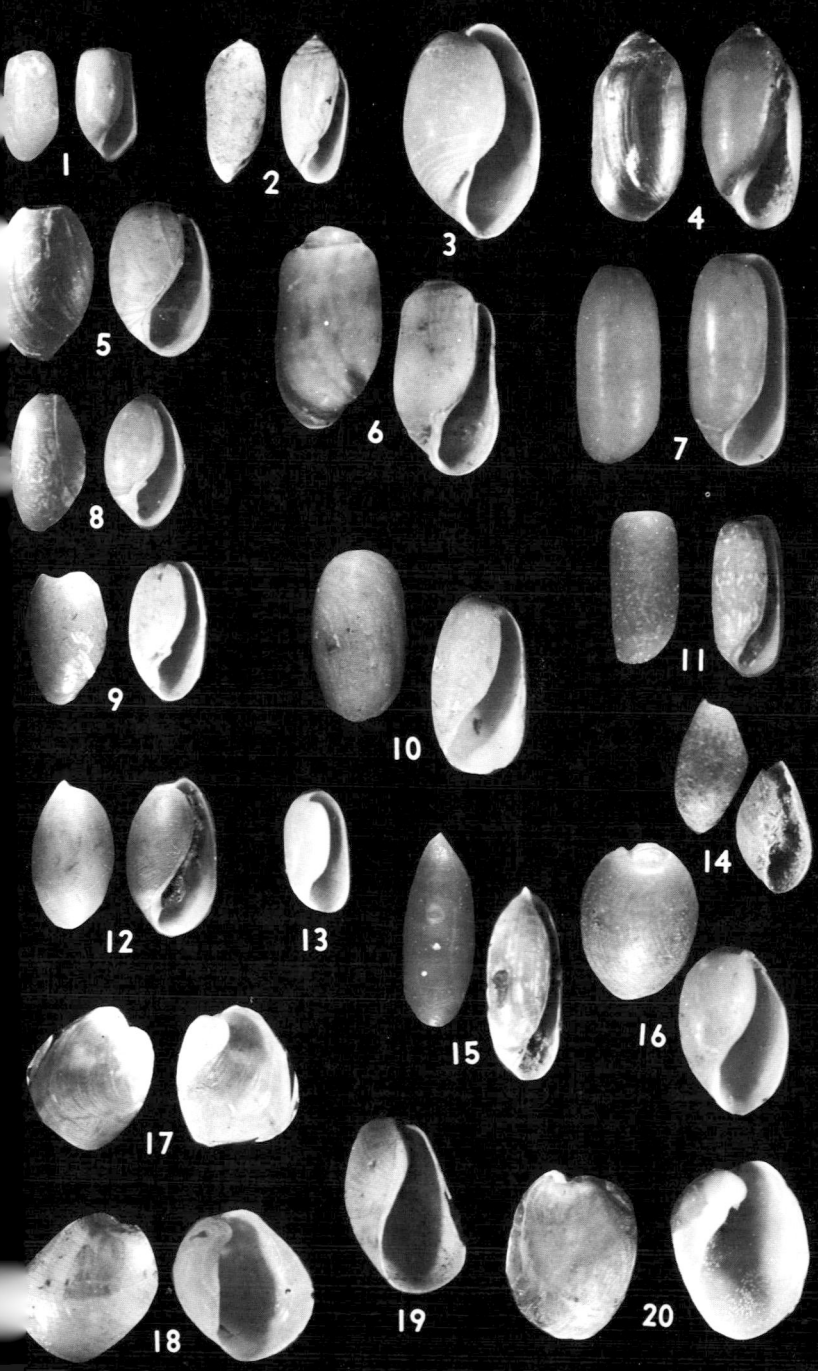

Plate 74

CANOE AND TUSK SHELLS, OTHERS

(Contd. on legend page for Plate 75)

Plate 75

MISCELLANEOUS MOLLUSKS

1. **TRIDENTATE CAVOLINE,** *Cavolina telemus* ×1   p. 275
   Translucent amber; see text.

2. Animal of *Cavolina* ×1

3. **THREE-SPINED CAVOLINE,** *Cavolina trispinosa* ×2   p. 275
   Translucent smoky white; see text.

4. **LONG-SNOUT CAVOLINE,** *Cavolina longirostris* ×3   p. 275
   Milky white; see text.

5. **INFLEXED CAVOLINE,** *Cavolina inflexa* ×3   p. 275
   Milky white; see text.

6. **CIGAR PTEROPOD,** *Herse columnella* ×2   p. 277
   White, with smooth surface; see text.

7. **HUMPBACK CAVOLINE,** *Cavolina gibbosa* ×2   p. 275
   Milky white; see text.

8. **UNCINATE CAVOLINE,** *Cavolina uncinata* ×2   p. 276
   Pale amber; see text.

9. **CUSPIDATE CLIO,** *Clio cuspidata* ×1   p. 276
   Opaque, white; see text.

10. **STRAIGHT NEEDLE PTEROPOD,** *Creseis acicula* ×2   p. 276
    White or yellowish white; see text.

11. **KEELED CLIO,** *Styliola subula* ×2   p. 276
    Whitish; see text.

12. **PYRAMID CLIO,** *Clio pyramidata* ×2   p. 276
    Translucent white; see text.

13. Female *Argonauta* with egg case

14. Animal of *Spirula*

15. **RAM'S HORN,** *Spirula spirula* ×1   p. 285
    Chambered; pure white.

16. **PAPER ARGONAUT,** *Argonauta argo* ×⅓   p. 286
    Not chambered; milky white.

17. **WILLCOX'S SEA HARE,** *Aplysia willcoxi* ×1   p. 277
    Glossy white on attached side, outer surface covered with
    a thin yellowish periostracum.

CONTINUED FROM LEGEND PAGE FOR PLATE 74

20. **CAROLINA CADULUS,** *Cadulus carolinensis* ×2   p. 283
    Slightly swollen at middle, polished; milky white.

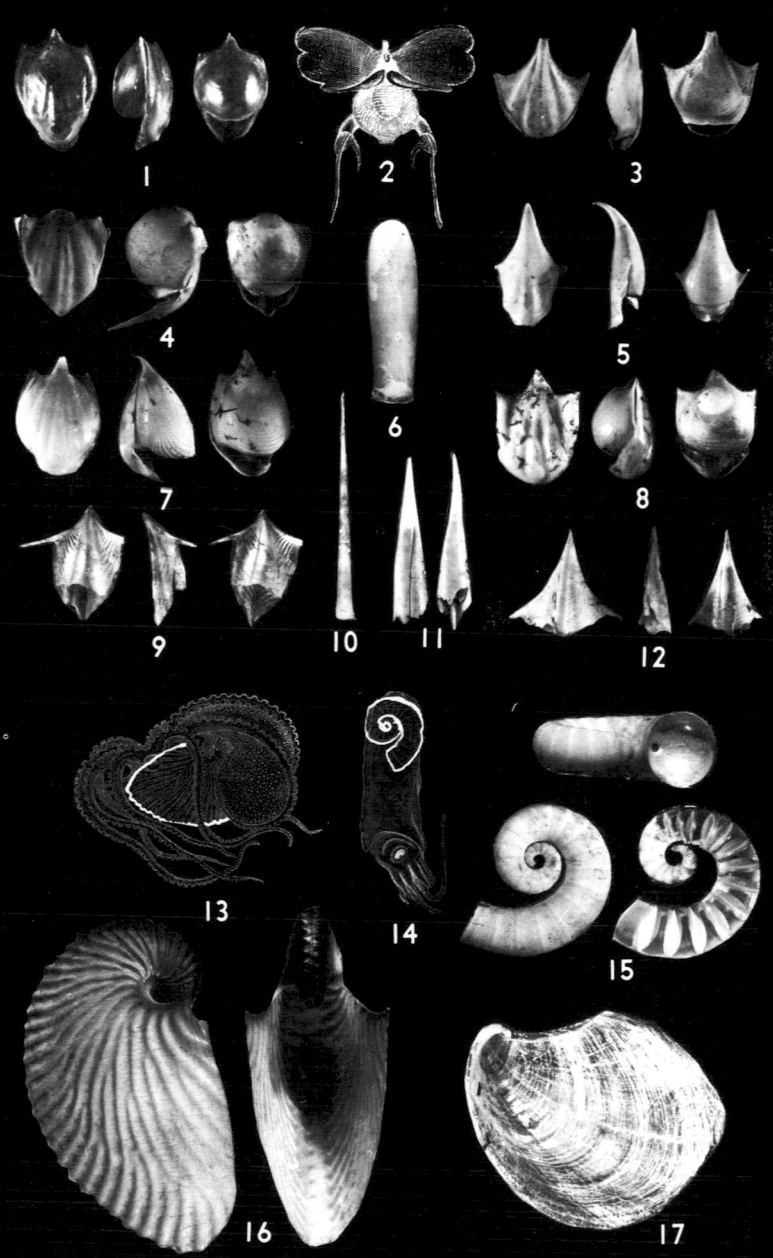

Plate 76

CHITONS

1. **SQUAMOSE CHITON,** *Chiton squamosus* ×1   p. 281
   Girdle scaly; greenish gray, with brownish streaks.

2. **WHITE CHITON,** *Ischnochiton albus* ×1   p. 280
   Elongate, with sandpapery girdle; dead specimens pale cream to white; surface of live specimens bluish black, easily rubbed off.

3. **MARBLED CHITON,** *Chiton marmoratus* ×1   p. 281
   Smooth plates, scaly girdle; color variable.

4. **RIO JANEIRO CHITON,** *Calloplax janeirensis* ×2   p. 280
   Strongly sculptured; girdle with silky hairs; greenish brown.

5. **MOTTLED CHITON,** *Tonicella marmorea* ×2   p. 279
   Plates keeled, girdle leathery; tan, sometimes speckled with red.

6. **RED CHITON,** *Ischnochiton ruber* ×1   p. 281
   Plates strongly keeled; girdle scaly, reddish brown.

7. **PYGMY CHITON,** *Acanthochitona pygmaea* ×2   p. 279
   Plates minutely dotted; leathery, hairy girdle; color variable.

8. **SLENDER CHITON,** *Ischnochiton floridanus* ×1   p. 280
   Elongate, with smooth girdle; whitish or pale green, mottled with gray.

9. **TUBERCULATE CHITON,** *Chiton tuberculatus* ×1   p. 281
   Plates strongly sculptured, girdle scaly; clouded olive-green.

10. *Chiton* ×1, showing animal

11. **BEE CHITON,** *Chaetopleura apiculata* ×2   p. 280
    Oval; plates grooved, girdle smooth; grayish or pale chestnut.

12. **TUBERCULATE CHITON** ×1, showing underside; animal removed

Genus *Planaxis* Lamarck 1822

LINED PLANAXIS *Planaxis lineatus* (Da Costa) **Pl. 40**
Range: S. Florida to West Indies.
Habitat: Shallow water; under stones.
Description: About ¼ in. high. Quite solid and well inflated.
3 or 4 whorls, decorated with evenly spaced spiral grooves.
Aperture oval, notched at lower margin. Outer lip thick.
Yellowish brown, with spiral brown bands.

BLACK PLANAXIS *Planaxis nucleus* (Brug.) **Pl. 40**
Range: S. Florida to West Indies.
Habitat: Shallow water; under stones.
Description: Height about ¼ in. A stubbier shell than Lined
Planaxis. 3 or 4 whorls, the last one making up most of the
shell, sutures distinct. Aperture large and oval, notched below, with shiny area on columella. Outer lip strongly crenulate within. Color rich brown, with rather thick grayish periostracum.

Modulus: Family Modulidae

FLATTISH, top-shaped shells, the whorls grooved and tubercu-
lated. There is a narrow umbilicus. The columella ends below
in a sharp tooth. There is but a single genus in this family.
Found in warm seas.

Genus *Modulus* Gray 1842

ANGLED MODULUS **Pl. 40**
Modulus carchedonius (Lam.)
Range: S. Florida through West Indies to northern coast of
S. America.
Habitat: Shallow water.
Description: Height about ½ in. 4 or 5 whorls, rather
pointed apex. Periphery sharply angled. Small umbilicus,
distinct tooth on lower columella. Sculpture of revolving
beaded lines. Color yellowish tan, beaded lines reddish brown.

ATLANTIC MODULUS *Modulus modulus* (Linn.) **Pl. 40**
Range: Florida to Texas; West Indies.
Habitat: Shallow water; among weeds.
Description: A knobby little shell about ½ in. high and
about the same in diam. Spire rather low. 3 or 4 whorls,
body whorl large, with sloping shoulders. Periphery keeled.

Sculpture of low revolving ridges and stout vertical ribs separated by deep grooves. Aperture nearly round, outer lip thin and crenulate. Operculum horny, small umbilicus. Color yellowish white, spotted and marked with brown.

Remarks: A somewhat more knobby form used to be called a subspecies, *M. m. floridanus* Conrad, but now all of the specimens found in Florida are considered as one single, somewhat variable species.

Horn Shells: Family Potamididae

THESE are mud dwellers, with elongated, many-whorled shells, usually with vertical ribs or grooves. The aperture is round, the lip commonly flaring. Distributed in warm seas.

Genus *Cerithidea* Swainson 1840

COSTATE HORN SHELL Pl. 43
Cerithidea costata (Da Costa)
   Range: Florida to West Indies.
   Habitat: Shallow water.
   Description: Height ½ in. About 12 rounded whorls, sutures well defined. Each volution sculptured with rather thin and sharp vertical ribs, which tend to fade somewhat on body whorl. Aperture nearly round, lip thin. Color yellowish brown.

PLICATE HORN SHELL Pl. 43
Cerithidea pliculosa (Menke)
   Range: Louisiana to Texas; Cuba.
   Habitat: Shallow water.
   Description: Height 1 in. From 10 to 12 well-rounded whorls, sutures well defined. Sculpture is of curved vertical ribs crossed by fine spiral lines; usually 1 or 2 pronounced varices. Color brown, sometimes with paler band midway of each volution.

LADDER HORN SHELL Pl. 43
Cerithidea scalariformis (Say)
   Range: S. Carolina to West Indies.
   Habitat: Shallow water.
   Description: About 1 in. high. 10 to 13 well-rounded whorls, sutures very distinct. Sculptured with closely spaced vertical ribs, no varices. Base of body whorl bears spiral riblets. Aperture large and circular, outer lip partially reflected. Operculum horny. Color pale to buffy gray.

Genus *Batillaria* Benson 1842

BLACK HORN SHELL *Batillaria minima* (Gmelin) **Pl. 43**
Range: S. Florida to West Indies.
Habitat: Shallow water.
Description: About ½ in. high. Elongate, with 6 to 8 whorls, sutures distinct. Apex sharp. Sculpture of low vertical ribs, broken by unequally knobby ridges. Aperture oval, canal short and turned to left. Black or deep brown, commonly with a paler band encircling each volution just below the suture.
Remarks: An extremely variable shell in its markings. Many individuals are sharply black and white. A jet-black form is sometimes listed as a subspecies, *B. m. nigrescens* (Menke).

Horn Shells: Family Cerithiidae

A LARGE family of generally elongate, many-whorled snails living in moderately shallow water, mostly on grasses and seaweeds, in tropical and semitropical seas. The aperture is small and oblique, and there is a short anterior canal, commonly somewhat twisted.

Genus *Cerithium* Bruguière 1789

MIDDLE-SPINED HORN SHELL **Pl. 43**
Cerithium algicola C. B. Adams
Range: S. Florida to West Indies.
Habitat: Shallow water.
Description: About 1½ in. high, a rugged shell of 8 or 9 angled whorls. Surface bears revolving rows of beads, those at periphery pointed. Aperture with notch at upper angle, short recurved canal below. Operculum horny. Color white or yellowish, blotched with brown.

IVORY HORN SHELL *Cerithium eburneum* Brug. **Pl. 43**
Range: Florida to West Indies.
Habitat: Shallow water.
Description: About 1 in. high. Elongate, spire tapering to sharp apex. 6 or 7 whorls, sutures distinct. Sculpture of revolving bands of small tubercles gives surface a beaded appearance. Aperture oval, canal short and turned to left. White to deep brown, usually with reddish-brown blotches.

FLORIDA HORN SHELL Pl. 43
Cerithium floridanum Mörch
    Range: N. Carolina to Florida.
    Habitat: Shallow water.
    Description: Height to 1½ in. Sturdy, apex sharp, about 10 whorls with distinct sutures. Sculpture of elevated, nodular ribs, sharply angled at edge of volutions, and many unequal ridges and fine lines spiraling over entire shell. Aperture small, oval, oblique. White or gray, with spiral pattern of brown.

LETTERED HORN SHELL Pl. 43
Cerithium literatum Born
    Range: Bermuda; s. Florida to West Indies.
    Habitat: Shallow water.
    Description: About 1 in. high. Shell stout, apex sharply pointed. 7 whorls, sutures somewhat indistinct. 2 rows of prominent knobs encircle each volution (those on shoulders the largest), and between them are spiral lines more or less beaded. Aperture moderately large and oval, lip thickened. Canal short and only partially recurved. Color varies from white to gray, checked with brownish black.

DOTTED HORN SHELL *Cerithium muscarum* Say Pl. 43
    Range: S. Florida to West Indies.
    Habitat: Shallow water.
    Description: Height 1 in. Elongate-conical, apex sharp. 9 or 10 whorls, sutures rather distinct. Sculpture of prominent vertical ribs, about 11 on body whorl, crossed by impressed revolving lines. Aperture oval and oblique, canal short and turned to left. Color grayish white, with small chestnut dots.

VARIABLE HORN SHELL Pl. 43
Cerithium variabile C. B. Adams
    Range: Florida to West Indies; Texas.
    Habitat: Shallow water.
    Description: Generally less than ½ in. high. Stoutly conical, apex sharp. 7 or 8 whorls, sutures indistinct. Revolving rows of beads decorate volutions, and commonly there is a varix present on the last whorl. Color varies from dark brown to whitish with darker mottlings.

Genus *Bittium* Gray 1847

ALTERNATE BITTIUM *Bittium alternatum* (Say) Pl. 43
    Range: Massachusetts to Virginia.
    Habitat: Shallow water.

Description: Height about ⅕ in. 6 to 8 rounded whorls, well-impressed sutures. Sculpture a granulated network of elevated spiral lines crossed by rounded vertical folds. Aperture obliquely rounded, lip sharp. Canal a mere notch or fissure. Operculum horny. Color bluish black or slate-gray, the lower whorls sometimes paler in tone. See next species for differentiation.

Remarks: Young specimens are reddish brown, and sometimes occur in such numbers that the sand appears to be alive with them.

VARIABLE BITTIUM   *Bittium varium* (Pfr.)   **Pl. 43**
Range: Maryland to Texas.
Habitat: Shallow water; on eelgrass.
Description: Much like *B. alternatum,* but a trifle smaller and slimmer, usually with a thickened rib (varix) on the body whorl. Sculpture of weak vertical folds and spiral lines. Grayish or slate-colored.

Genus *Litiopa* Rang 1829

SARGASSUM SNAIL   *Litiopa melanostoma* Rang   **Pl. 40**
Range: Massachusetts to Florida.
Habitat: Pelagic.
Description: About ¼ in. high. A thin and delicate shell, its shape rather stoutly conic. 6 or 7 roundish whorls, sutures well impressed. Sharply pointed apex, elongate-oval aperture, lip thin and sharp. Volutions bear weak revolving lines, but surface appears smooth and glossy. Color pale yellowish gray, often with rows of small brownish dots.

Remarks: This is a pelagic gastropod living in floating masses of sargassum weed, and specimens are often found in clumps blown ashore.

Genus *Alaba* H. & A. Adams 1853

DISHEVELED ALABA SHELL   **Pl. 43**
Alaba incerta (Orb.)
Range: Florida Keys to West Indies.
Habitat: Shallow water; weeds.
Description: About ¼ in. high. A slender elongate shell of about 7 whorls, sutures fairly well indented. Aperture oval, lip thin. Surface glossy, color white or gray.
Remarks: Formerly listed as *A. tervaricosa* Adams.

Genus *Alabina* Dall 1902

MINIATURE HORN SHELL   Pl. 43
Alabina cerithidioides Dall
  Range: Bermuda, Florida, West Indies.
  Habitat: Shallow water.
  Description: About 3/16 in. high. 8 or 9 rounded whorls, sutures distinct, apex pointed. Surface with weak curving ribs, usually lacking on first few volutions. Color yellowish white.

Genus *Cerithiopsis* Forbes & Hanley 1849

CRYSTALLINE CERITH   Pl. 43
Cerithiopsis crystallinum Dall
  Range: Gulf of Mexico.
  Habitat: Moderately deep water.
  Description: A slender spikelike shell ½ in. high. 9 or 10 whorls, sutures moderately distinct. Decorated with revolving rows of tubercles. Lower portion of body whorl bears spiral grooves. Aperture small. Color white.

EMERSON'S CERITH   Pl. 43
Cerithiopsis emersoni (C. B. Adams)
  Range: Massachusetts to West Indies.
  Habitat: Moderately shallow water.
  Description: About ½ in. high. A slender shell of 10 to 14 flattish whorls, sutures indistinct. Sculpture of spiraling rows of tiny beads, base of last whorl smooth, with cordlike ridges. Aperture quite small, outer lip somewhat thickened, inner lip slightly twisted. Color chocolate-brown.
  Remarks: *C. emersoni* may be a synonym of *C. subulata* (Montagu).

GREEN'S CERITH   Pl. 43
Cerithiopsis greeni (C. B. Adams)
  Range: Massachusetts to w. Florida.
  Habitat: Shallow water.
  Description: Nearly ¼ in. high. A shiny shell of about 8 whorls. Volutions rather flat, sutures indistinct. Ornamented by revolving rows of tiny beads. Aperture quite small, with smooth area on the slightly arched inner lip. Color brown.

BROAD CERITH   *Cerithiopsis latum* (C. B. Adams)   Pl. 43
  Range: West Indies.
  Habitat: Shallow water.

Description: About ¼ in. high. 9 to 10 rather flattish whorls, sutures moderately distinct. Aperture small, canal short and twisted. Sculpture of fine revolving lines. Color white, with pale brown band below suture.

PIKE CERITH *Cerithiopsis matara* Dall   **Pl. 43**
Range: S. Florida.
Habitat: Deep water.
Description: A high-spired shell somewhat more than ¼ in. tall. About 16 flattish whorls, sutures deeply impressed. Surface sculptured with revolving rows of beads connected by weak threads. Aperture small, canal short and slightly twisted. Color brown.

Genus *Cerithiella* Verrill 1882

WHITEAVES' LITTLE HORN SHELL   **Pl. 43**
Cerithiella whiteavesi (Verrill)
Range: Gulf of St. Lawrence to Massachusetts.
Habitat: Deep water.
Description: Height ½ in. A rather stout shell, composed of 8 angled whorls. A weak row of beads produces the angle, with the main sculpture consisting of vertical striations. Apex quite blunt, aperture small, short canal considerably twisted. Color brown.

Genus *Seila* A. Adams 1861

WOOD SCREW SHELL *Seila adamsi* (Lea)   **Pl. 43**
Range: Massachusetts to West Indies; Texas.
Habitat: Shallow water.
Description: Height about ½ in. Slender, elongate, with about 12 whorls, sutures indistinct. Sculpture of 3 or 4 elevated revolving ridges on each volution, with fine vertical lines between them. Aperture small, canal short and twisted. Color brown.
Remarks: Formerly listed as *Cerithiopsis terebralis* (C. B. Adams).

Left-handed Snails: Family Triphoridae

AN interesting family of very small, elongate, left-handed (sinistral) gastropods. There are many whorls. Widely distributed in warm seas.

Genus *Triphora* Blainville 1828

WHITE TRIPHORA   **Pl. 43**
Triphora melanura (C. B. Adams)
  Range: West Indies.
  Habitat: Shallow water.
  Description: About ¼ in. high. A slender shell of some 12 rather flat whorls, decorated with revolving rows of small beads. Apex sharp, aperture small. Sinistral (left-handed). Color white, sometimes with a few scattered reddish-brown marks.

BLACK TRIPHORA   **Pl. 43**
Triphora nigrocincta (C. B. Adams)
  Range: Massachusetts to West Indies; Texas.
  Habitat: Shallow water; weeds.
  Description: About ¼ in. high. Shell elongate, 10 to 12 whorls, sutures slightly excavated. Surface bears revolving rows of beadlike tubercles, 4 rows on body whorl. Spiral turns to left (sinistral) instead of to right as with most gastropods. Color pale brown, occasionally with lighter band near suture.
  Remarks: Sometimes considered a subspecies of the European *T. perversa* (Linn.).

ORNATE TRIPHORA   *Triphora ornata* (Desh.)   **Pl. 43**
  Range: Florida to West Indies.
  Habitat: Shallow water.
  Description: Height ¼ in. A rather sturdy little shell of about 12 whorls, sutures indistinct. Sculpture of relatively large revolving beads. Aperture small, sinistral. Color dark brown, with encircling line of white beads; a few scattered white beads in the brown area.

BEAUTIFUL TRIPHORA   **Pl. 43**
Triphora pulchella (C. B. Adams)
  Range: S. Florida to West Indies.
  Habitat: Moderately shallow water.
  Description: Nearly ¼ in. high. As many as 15 moderately rounded whorls, sutures indistinct. Spiral turns to left (sinistral). Sculpture consists of 3 rows of tiny beads on each whorl, the beads connected by threadlike lines. Upper portions of each volution rich brown, lower part white.

THOMAS' TRIPHORA   **Pl. 43**
Triphora turristhomae (Holten)
  Range: West Indies.
  Habitat: Shallow water.

Description: Height ¼ in. Slender and high-spired, 12 to 16 whorls. Sinistral. Sculpture of 2 revolving rows of beads, upper row white, the lower with a brownish band.

Wentletraps: Family Epitoniidae

THESE are predatory, carnivorous snails, occurring in all seas, and popularly known as wentletraps or staircase shells. They are high-spired, usually white and polished, and consist of many rounded whorls. The outer lip is thickened considerably during rest periods in shell growth, and this thickened lip becomes a new varix as the mollusk increases in size. Among the most delicately graceful of all marine shells and great favorites with collectors.

Genus *Cirsotrema* Mörch 1852

PITTED WENTLETRAP *Cirsotrema dalli* Rehder **Pl. 44**
  Range: N. Carolina to West Indies and south to Brazil.
  Habitat: Deep water.
  Description: About 1 in. high. A slender, high-spired shell of about 8 whorls, sutures well indented. Aperture round, bordered by a thickened lip, and as growth takes place successive varices are left on the earlier whorls at regular intervals in the form of thickened ribs. Surface between varices heavily pitted. Color dusky white.

Genus *Acirsa* Mörch 1857

NORTHERN WENTLETRAP **Pl. 44**
Acirsa costulata (Mighels & Adams)
  Range: Arctic Ocean to Massachusetts.
  Habitat: Moderately shallow water.
  Description: Height to 1¼ in. 8 or 9 whorls, surface smooth except for very tiny revolving lines. This wentletrap completely lacks the characteristic vertical ribs of this group. Aperture circular, outer lip thin. Color yellowish to grayish white.
  Remarks: Formerly listed as *Scalaria borealis* Beck.

Genus *Opalia* H. & A. Adams 1853

CRENULATED WENTLETRAP **Pl. 44**
Opalia crenata (Linn.)

Range: Florida Keys, West Indies, and south to Trinidad; also e. Atlantic.
Habitat: Moderately shallow water.
Description: About ¾ in. high. Approximately 7 whorls, each crenulate at the shoulder. Upper volutions with weak vertical ribs, remainder relatively smooth. Aperture round, lip thickened. Color white.

NOTCHED WENTLETRAP   Pl. 44
Opalia hotessieriana (Orb.)
Range: S. Florida to West Indies.
Habitat: Moderately shallow water.
Description: About ½ in. high. 7 whorls, the vertical ribs only faintly present, but along the top of each whorl is a series of large squarish indentations, or notches. Aperture round. Color grayish white.

PUMILIO WENTLETRAP   *Opalia pumilio* (Mörch)   Pl. 44
Range: Florida to West Indies.
Habitat: Moderately shallow water.
Description: About ½ in. high. 7 or 8 whorls, sutures distinct. Sculpture of curving vertical ribs, about 15 on body whorl. Aperture round, lip thickened. Color white or yellowish.

Genus *Amaea* H. & A. Adams 1853

RETICULATE WENTLETRAP   Pl. 44
Amaea retifera Dall
Range: N. Carolina to Florida.
Habitat: Moderately deep water.
Description: About 1 in. high. A tall and graceful shell of about 16 rounded whorls, sutures deeply impressed. Sculpture consists of strong revolving lines, about 5 to a volution, crossed at regular intervals by equally strong and sharp vertical lines; consequently surface bears distinct network pattern. Aperture round, outer lip wavy. Color pale brown, sometimes with darker bands.

Genus *Epitonium* Röding 1798

BLADED WENTLETRAP   Pl. 44
Epitonium albidum (Orb.)
Range: S. Florida to West Indies.
Habitat: Moderately shallow water.
Description: Height about ¾ in. A stocky shell of 7 or 8 rounded whorls, sutures distinct. Sculpture of sharp bladelike

ribs, about 12 on body whorl. Substance thin, the ribs visible through shell in aperture. Color white.

ANGLED WENTLETRAP   Pl. 44
Epitonium angulatum (Say)
   Range: Long Island to Texas.
   Habitat: Moderately shallow water.
   Description: About ¾ in. high. Stoutly elongate, some 6 whorls that do not touch each other in the coil. About 10 vertical bladelike varices on each volution, each with a more or less blunt angle (shoulder) above, near the suture. Color white.

CHAMPION'S WENTLETRAP   Pl. 44
Epitonium championi Clench & Turner
   Range: Massachusetts to N. Carolina.
   Habitat: Shallow water.
   Description: Height about ½ in. About 10 whorls that are not as convex as in most of the wentletraps. Volutions bear thickened varices, about 9 to a whorl, and between them the surface is marked with distinct revolving lines. No ridge at base of body whorl. Color dull white.
   Remarks: This species was not discovered and named until 1951. Specimens seen before that time were thought to be young examples of the Greenland Wentletrap (below).

DALL'S WENTLETRAP   Pl. 44
Epitonium dallianum (Verrill & Smith)
   Range: New Jersey to S. Carolina.
   Habitat: Deep water.
   Description: About ½ in. high. A slender shell of about 10 convex whorls. Volutions decorated with numerous closely spaced bladelike varices, as many as 20 on body whorl, each varix bearing a minute point, or hook, at the shoulder. Color grayish white.

DENTICULATE WENTLETRAP   Pl. 44
Epitonium denticulatum Sby.
   Range: Cape Hatteras to West Indies.
   Habitat: Moderately deep water.
   Description: About ½ in. high. About 7 whorls, expanding rapidly to a relatively large body whorl. About 12 thin and sharp vertical ribs on last volution, angled at suture. Color white.

WAVY-RIBBED WENTLETRAP   Pl. 44
Epitonium echinaticostum (Orb.)
   Range: Bermuda; s. Florida to West Indies.

Habitat: Moderately deep water.
Description: Less than ½ in. high. About 8 very convex whorls, sculptured with sharp varices, about 7 on body whorl. Varices thin, commonly waved or fluted, so that the shell has an elaborate appearance. Aperture round, bordered by thickened lip. Small but well-defined umbilicus. Color pure white.
Remarks: In occasional specimens the last few whorls become free and well separated (see Plate 44, Fig. 10).

GREENLAND WENTLETRAP Pl. 44
Epitonium greenlandicum (Perry)
Range: Circumpolar; south to Long Island; Alaska, Norway.
Habitat: Deep water.
Description: A large wentletrap, attaining a height of nearly 2 in., although most are closer to 1 in. Shell elongate, tapering regularly to a rounded point. About 12 moderately rounded whorls, barred with about 12 stout, rather flat varices. Spaces between varices sculptured with rounded revolving ridges. Generally a well-marked spiral ridge at base. Aperture circular, bordered by a stoutly thickened lip. Color dull yellowish white.
Remarks: This is one of the showy shells of the n. Atlantic, making up in bizarre sculpture what it lacks in size.

HUMPHREY'S WENTLETRAP Pl. 44
Epitonium humphreysi (Kiener)
Range: Massachusetts to Texas.
Habitat: Moderately shallow water.
Description: About ¾ in. high. 8 or 9 rounded whorls, decorated with well-spaced thickened varices; spaces between smooth and glossy, sometimes showing faint spiral lines. Color pure white.
Remarks: Formerly listed as *Scalaria sayana* Dall.

KREBS' WENTLETRAP *Epitonium krebsi* (Mörch) Pl. 44
Range: S. Florida, Bahamas, and south through West Indies to Barbados.
Habitat: Moderately shallow water.
Description: About ½ in. high. 5 or 6 convex whorls that are hardly joined at sutures. Body whorl relatively large, giving shell a somewhat squat appearance; there is a noticeable umbilicus. About 10 varices to a volution, slightly toothed at upper ends. Color white.

TRELLIS WENTLETRAP Pl. 44
Epitonium lamellosum (Lam.)
Range: S. Florida to West Indies and south to Tobago; also e. Atlantic.

Habitat: Moderately shallow water.
Description: Height to 1¼ in. high. 7 or 8 whorls, each with about 11 thin bladelike varices, spaces between them smooth and polished. Single revolving line at base. Shell elongate, rather stout at base. Creamy to pure white, sometimes with irregular markings of pale brown.
Remarks: Formerly listed as *E. clathrum* (Sby.).

CROWDED WENTLETRAP Pl. 44
Epitonium multistriatum (Say)
Range: Massachusetts to Texas.
Habitat: Moderately deep water.
Description: About ½ in. high. Well elevated and sharply pointed, with 7 or 8 convex whorls, sutures well indented. Varices closely crowded, particularly on earlier whorls; about 19 on last volution, but on a section 2 or 3 volutions from apex there may be as many as 40. Color dull white.

NEW ENGLAND WENTLETRAP Pl. 44
Epitonium novangliae (Couthouy)
Range: Virginia, West Indies, and south to Brazil.
Habitat: Moderately shallow water.
Description: About ½ in. high. A delicate shell of 8 or 9 whorls. Some 12 robust varices to a volution, the spaces between marked with weak spiral lines, and still weaker cross-lines, exhibiting a faint network pattern. Color white, sometimes lightly banded on later whorls.
Remarks: The first specimen (type) of this species was taken from the stomach of a cod in Massachusetts — hence the name *novangliae;* however, no specimens have ever been taken alive north of Virginia.

WEST ATLANTIC WENTLETRAP Pl. 44
Epitonium occidentale (Nyst)
Range: S. Carolina to West Indies.
Habitat: Deep water.
Description: Nearly 1 in. high. 7 or 8 rounded whorls, sutures deep. About 14 thin varices on body whorl, well angled at suture. Surface between varices smooth and glossy. Color white.

FISHHAWK WENTLETRAP Pl. 44
Epitonium pandion Clench & Turner
Range: New Jersey to Cape Hatteras.
Habitat: Deep water.
Description: A tiny shell, ¼ in. high, composed of 8 or 9 rounded whorls. The usual varices of this group are almost lacking in this species, especially on the later whorls, although

traces of them may be discernible at the shoulders. There is a weak sculpture of revolving lines. Color pale brown.

Remarks: Originally described as *Acirsa gracilis* by Verrill in 1880.

POURTALÈS' WENTLETRAP Pl. 44
Epitonium pourtalesi (Verrill & Smith)
Range: New Jersey to West Indies.
Habitat: Deep water.
Description: Nearly 1 in. high. A sturdy and rather stout shell of 9 or 10 convex whorls, sutures well indented. Expansion is rapid, so body whorl is wide in proportion to shell's height. About 14 thin varices on last volution, developing sharp nodes at shoulder. Color white.

LINED WENTLETRAP *Epitonium rupicola* (Kurtz) Pl. 44
Range: Massachusetts to Texas.
Habitat: Moderately shallow to deep water.
Description: Height ½ in. About 10 rounded whorls, each with some 10 to 16 delicate varices, spaces between smooth and polished. Sutures distinct. Color pinkish white, often with 1 or 2 brownish bands on later volutions.
Remarks: Formerly listed as *E. lineatum* (Say).

TOLLIN'S WENTLETRAP Pl. 44
Epitonium tollini Bartsch
Range: Gulf of Mexico.
Habitat: Moderately shallow water.
Description: About ½ in. high. 8 to 10 rounded whorls, each with about 12 varices that are rounded at the shoulders, spaces between them smooth and polished. Color pure white.

Genus *Sthenorhytis* Conrad 1862

NOBLE WENTLETRAP Pl. 44
Sthenorhytis pernobilis (Fischer & Bern.)
Range: N. Carolina to West Indies.
Habitat: Deep water.
Description: Height to 1¾ in. A sturdy shell of 6 or 7 rounded whorls, sutures well indented. Taper rapid, body whorl wide. Sculpture of strong curving bladelike varices, 12 on last volution; surface in between with weak revolving lines. Aperture round. Color white.
Remarks: Probably our largest wentletrap, but unfortunately quite rare.

Violet Snails:
Family Janthinidae

THESE are floating pelagic mollusks (living miles from land). They are very delicate lavender or purple shells, shaped much like land snails. There is no operculum. The animal is capable of ejecting a purplish fluid when disturbed. The mollusk is fastened to a raft of frothy bubbles of its own making, to which the eggs are attached. These snails float about in tremendous quantities; sometimes huge numbers are blown ashore, where the purple color may stain the beach for considerable distances.

Genus *Janthina* Röding 1798

GLOBE VIOLET SNAIL   *Janthina globosa* Swain.   **Pl. 44**
   Range: Both coasts of U.S.
   Habitat: Pelagic; warm waters.
   Description: Diam. about ¾ in. Shell globular, thin, very fragile, with about 3 whorls. Aperture large and moderately elongate. Color pale violet all over, usually darker toward base.

VIOLET SNAIL   *Janthina janthina* (Linn.)   **Pl. 4**
   Range: Both coasts of U.S.
   Habitat: Pelagic; warm waters.
   Description: Diam. about 1½ in. This graceful shell is thin and fragile, and consists of 3 or 4 sloping whorls. Surface bears very fine lines (striae). Aperture large, lip thin and sharp. No operculum or umbilicus. A two-toned shell: pale violet above, deep purple below.

Hoof Shells:
Family Hipponicidae

SHELL obliquely conical and generally thick and heavy, the apex hooked backward but not spirally. Surface usually rough, wrinkled, and yellowish or grayish white. The animal secretes a shelly plate between itself and the object on which it lives. This plate was once believed to be a 2nd valve, and several species were described as bivalve mollusks.

Genus *Hipponix* Defrance 1819

HOOF SHELL *Hipponix antiquatus* (Linn.) **Pl. 45**
 Range: Florida to West Indies; w. Mexico.
 Habitat: Shallow water; on stones.
 Description: Height ½ to ¾ in. Shell thick, conical, cap-shaped, concave at base. Apex bluntly pointed and curved backward. Surface variable, sometimes fairly smooth but often wrinkled by coarse laminations. Variable in shape too, since it lives attached to a rock or shell and grows to conform to its particular shape of seat; secretes a calcareous plate between itself and object to which it adheres. Color white or yellowish white.

Genus *Cheilea* Modeer 1793

FALSE CUP-AND-SAUCER LIMPET **Pl. 45**
Cheilea equestris (Linn.)
 Range: S. Florida to West Indies.
 Habitat: Shallow water.
 Description: Diam. about 1 in. Orbicular at base, although circumference may be irregular because of wavy rim; outer edge rises to a blunt point situated somewhat to the rear and directed backward. Exterior with rough corrugated appearance, interior with thin horseshoe-shaped plate. Color dull grayish white.
 Remarks: This snail lives attached to other shells and usually excavates a shallow cavity at the place of attachment.

Fossarus: Family Fossaridae

SMALL, short, and stubby shells, the surface strongly sculptured. Occur in cold seas, generally at considerable depths.

Genus *Fossarus* Philippi 1841

ELEGANT FOSSARUS **Pl. 38**
Fossarus elegans Verrill & Smith
 Range: Massachusetts to N. Carolina.
 Habitat: Deep water.
 Description: Height ⅛ in. Shell ovate, stout, and rugged, with acutely pointed spire. 5 sharply carinated whorls, elegantly latticed with sharp bladelike lines between the keels (carinae). Aperture nearly round, outer lip thickened, with projections where the carinae terminate. Color white.

Hairy-keeled Snails:
Family Trichotropidae

SMALL thin shells with widely flaring outer lips. The angles of the whorls are keeled, and there is a small umbilicus. Found in cool seas.

Genus *Trichotropis*
Broderip & Sowerby 1829

NORTHERN HAIRY-KEELED SNAIL   **Pl. 47**
Trichotropis borealis Brod. & Sby.
    Range: Labrador to Massachusetts; n. Pacific.
    Habitat: Moderately shallow water.
    Description: About ½ in. high. 4 whorls with deeply channeled sutures. Body whorl relatively large, encircled by 2 prominent keels, as well as 2 or 3 less conspicuous ones. Aperture broad and rounded above and somewhat pointed below. Outer lip flares considerably. Color brown, with yellowish periostracum that rises like a bristly fringe along the keels.
    Remarks: A common shell in the stomachs of fishes taken off the New England coast.

Cap Shells:
Family Capulidae

SHELL conical, cap-shaped, without internal plate or cup. The apex is spiral, the base open, and the whole shell is considerably curved.

Genus *Capulus* Montfort 1810

INCURVED CAP SHELL *Capulus intortus* Lam.   **Pl. 45**
    Range: Florida to West Indies.
    Habitat: Shallow water; on stones.
    Description: Height ½ in. Caplike, nearly all body whorl. Apex incurved. Aperture large and round. Sculpture of revolving and vertical lines, producing a cancellate pattern. Color glossy white.
    Remarks: The dark markings on the specimen illustrated on Plate 45 are algal growths.

CAP SHELL *Capulus ungaricus* (Linn.)   **Pl. 45**
Range: Greenland to West Indies.
Habitat: Deep water.
Description: Height 1 to 1½ in. and about 1 in. across at base. Conical, apex coiled and curved forward. Surface bears fine lines that radiate, and these are crossed by less frequent lines of growth. Open base shaped to fit the object to which the gastropod is attached, for the cap shell is limpetlike in habits. It forms a shallow excavation at the place of attachment, and sometimes deposits a shelly floor. Whitish, with grayish periostracum.
Remarks: The conical shell reminds one of a jester's hat. It is a modern representative of a very ancient group — fossil specimens of the same genus being found in rocks of Silurian age some 300,000,000 years old.

Cup-and-Saucer Limpets and Slipper Shells: Family Calyptraeidae

LIMPETLIKE gastropods, cap-shaped, with a shelly plate (platform) or cup-shaped process on inner side of shell. They live attached to other shells or to stones. Found in nearly all seas, from shallow water to moderate depths.

Genus *Calyptraea* Lamarck 1799

CIRCULAR CUP-AND-SAUCER LIMPET   **Pl. 45**
Calyptraea centralis Conrad
Range: Cape Hatteras to West Indies.
Habitat: Moderately shallow water.
Description: Diam. about ½ in. Cap-shaped, circular at base. Apex blunt, centrally situated, only moderately elevated. Small shelflike plate on inside. Color pure white.

Genus *Crucibulum* Schumacher 1817

ROSY CUP-AND-SAUCER LIMPET   **Pl. 45**
Crucibulum auricula (Gmelin)
Range: S. Florida to West Indies.
Habitat: Moderately shallow water.
Description: Cap-shaped, diam. at base about 1 in. Summit moderately elevated, rim of aperture wavy. Inner cup stands nearly free, whereas in the Cup-and-Saucer Limpet that

structure is fastened by almost half of its area. Color grayish; interior usually pinkish.

CUP-AND-SAUCER LIMPET Pl. 45
Crucibulum striatum (Say)
Range: Nova Scotia to Florida.
Habitat: Moderately shallow water.
Description: Moderately solid and cap-shaped, diam. at base 1 in. Surface bears numerous slightly elevated radiating lines. Summit usually smooth and bluntly pointed. Inner partition cup-shaped and attached by one side to shorter end of shell. Color pinkish white, generally streaked with brown.

Genus *Crepidula* Lamarck 1799

THORNY SLIPPER SHELL Pl. 45
Crepidula aculeata (Gmelin)
Range: N. Carolina to West Indies; Pacific.
Habitat: Shallow water.
Description: About 1 in. long. Apex turned considerably to one side. Outer surface bears irregular thorny or spiny ridges, radiating from apex. Inner cavity with shelflike shelly plate. Color brownish or grayish, quite variable in its markings; commonly there are broad rays of paler tint. Interior polished and often rayed with brown, shelly plate white.

CONVEX SLIPPER SHELL Pl. 45
Crepidula convexa Say
Range: Massachusetts to West Indies; Texas.
Habitat: Shallow water.
Description: Length about ½ in. Shell obliquely oval and deeply convex. Apex prominent and separate from the body whorl, turning down nearly to the plane of the aperture, occasionally beyond it. Shelly plate deeply situated. Outer surface minutely wrinkled. Color ashy brown, with streaks or dots of reddish brown.

COMMON SLIPPER SHELL Pl. 45
Crepidula fornicata Say
Range: Nova Scotia to Gulf of Mexico.
Habitat: Shallow water.
Description: Length 1½ in. Shell obliquely oval, apex prominent and turned to one side, not separate from body of shell. Moderately convex, according to the object on which it is seated. Cavity partially divided by a horizontal white plate (platform). Color pale gray, flecked with purplish chestnut. See next species for differentiation.
Remarks: Also called Boat Shell and Quarterdeck. These

are among the first objects collected by children at the seashore, since they make excellent miniature boats to sail in tide pools and also serve as tiny scoops for digging in the sand. The empty shells have a commercial value; under the name Quarterdecks many tons are annually scattered over the ocean floor for embryo oysters to settle upon.

SPOTTED SLIPPER SHELL Pl. 45
Crepidula maculosa Conrad
Range: W. Florida.
Habitat: Shallow water.
Description: This slipper shell is much like the Common Slipper Shell, but the edge of the shelly plate is straight instead of convex. The exterior is usually checked with small brown or chocolate marks.

FLAT SLIPPER SHELL *Crepidula plana* Say Pl. 45
Range: Nova Scotia to Florida.
Habitat: Shallow water; on shells.
Description: Length 1 in. General shape may be flat, slightly curved, or even concave, according to the outline of the object to which it is attached. Platform less than half the length of shell. Outer surface wrinkled by concentric lines of growth, inner surface highly polished, sometimes iridescent. Color pure white, semitransparent when young.
Remarks: Generally found flattened against the inside of the aperture of dead shells, particularly large examples of *Busycon* and *Polinices*. Often a large king (horseshoe) crab (*Limulus*), will be found to have several dozen on the lower side of its domelike shell.

Carrier Shells:
Family Xenophoridae

SHELL top-shaped, somewhat flattened. All but the deep-sea forms camouflage their shells by cementing pebbles and broken shell fragments to them. From above they look like small piles of debris. Early in life the snail fastens a bit of shell or a tiny pebble to its back, at the upper edge of its aperture. As the shell grows the foreign object is firmly anchored in place, and a larger piece is then added. This results in a spiral of foreign material following the suture line. The snail usually sticks to one kind of object, rarely mixing them. Fossil specimens of this genus are found as far back as the Cretaceous period (100,000,000 years), showing that this curious habit of self-ornamentation has persisted for countless thousands of generations.

Genus *Xenophora* Waldheim 1807

CARIBBEAN CARRIER SHELL   **Pl. 42**
Xenophora caribaea Petit
  Range: Florida Keys and West Indies.
  Habitat: Deep water.
  Description: Diam.. 3½ in. A large but thin-shelled, top-shaped species. About 6 whorls, the surface bearing fine radiating lines and oblique wrinkle-like folds. Color white, the shell having a curious translucent appearance.
  Remarks: This species does not cover its back with an assortment of objects, but is content with a single row of very small bits of shell or pebbles at the suture line.

COMMON CARRIER SHELL   **Pl. 42**
Xenophora trochiformis (Born)
  Range: N. Carolina, West Indies, and south to Brazil.
  Habitat: Moderately shallow water.
  Description: Diam. about 2 in. Shell top-shaped, apex low but sharp. 6 or 7 whorls, rather flattened. Surface bears prominent growth lines, although they may be visible only from below. Aperture large and oblique, lip thin. No umbilicus, operculum horny. Color yellowish brown.
  Remarks: This species does a good job in concealing the upperpart of its shell with marine trash. It appears to prefer bivalve shells, particularly those of the Cross-barred Chione, *Chione cancellata* (p. 59), and it is noteworthy that the bits of shell are so disposed as not to curve downward beyond edge of the shell (this would impede progress of the animal) but are almost invariably placed with their concave sides uppermost.

Duckfoot Shells:
Family Aporrhaidae

STRONG and solid shells, with a high spire and a long and narrow aperture. The outer lip is greatly thickened and flaring, forming a winglike expansion. The best-known example of this family is the Pelican's Foot, *Aporrhais pespelicani* (Linn.), of European waters.

Genus *Aporrhais* Da Costa 1778

DUCKFOOT   *Aporrhais occidentalis* Beck   **Pls. 1, 46**
  Range: Labrador to N. Carolina.

Habitat: Moderately deep water.
Description: About 2 in. high. Shell thick and strong, with 8 to 10 whorls, each decorated with numerous smooth, rounded, crescent-shaped folds, about 20 on body whorl. Surface ornamented by closely spaced revolving lines. Aperture semilunar; outer lip expanded into a wide 3-cornered wing. Color white or yellowish white.

Strombs: Family Strombidae

AN interesting family of active snails, widely distributed in warm seas. The shells are thick and solid, with greatly enlarged body whorls. The aperture is long and narrow, with a notch at each end, and the outer lip in adults is usually thickened and expanded. The operculum is clawlike, and does not close the aperture.

Genus *Strombus* Linnaeus 1758

FLORIDA STROMB   *Strombus alatus* Gmelin   **Pl. 46**
Range: N. Carolina to Texas.
Habitat: Shallow water.
Description: Height 3 to 4 in. Shell solid, about 7 whorls, spire pointed. Early volutions decorated with revolving ribs, body whorl smooth. Shoulders of later whorls may or may not bear blunt spines. Aperture long, lip thickened and flaring, notched below. Operculum clawlike. Color yellowish brown, clouded and sometimes striped with orange and purple. Interior commonly dark brown. Juvenile, or half-grown individuals, do not have the flaring lip, and look considerably like cone shells (*Conus*).
Remarks: There has been some confusion regarding the identity of this species and the Fighting Stromb, *S. pugilis* (below). Since the original figure of *S. alatus* showed a specimen without shoulder spines, it is commonly believed that the spineless form is *S. alatus* and the form with well-developed spines is *S. pugilis*. However, *S. alatus* also has spines in many cases. It is very common in Florida and is usually quite mottled; the shoulder spines, when present, are confined to the last whorl, whereas the typical *S. pugilis* is a West Indies form, uncommon in Florida, with prominent spines on the last 2 whorls. Furthermore, *S. pugilis* is a somewhat heavier shell, and is generally of a more solid, orange color.

RIBBED STROMB   *Strombus costatus* Gmelin   **Pl. 46**
Range: Florida to West Indies.

Habitat: Shallow water.
Description: A large and solid shell, 4 to 6 in. high. About 10 whorls, producing a fairly tall spire. Body whorl greatly expanded. A series of large blunt nodes on last whorl, with a series of smaller ones twisting up the spire. There are also a number of rather distinct horizontal corrugations on the body whorl. Aperture long and narrow, outer lip expanded and much thickened in old specimens. Color yellowish white, somewhat mottled; interior pure white, with a noticeable aluminumlike glaze on columella and outer lip.

COCK STROMB *Strombus gallus* Linn. **Pl. 8**
Range: S. Florida to West Indies.
Habitat: Moderately shallow water.
Description: Height 4 to 7 in. Strong and solid, with sharp spire of about 7 whorls. Sculpture of strong revolving ribs; body whorl bears blunt spines, or nodes, at shoulders. Outer lip expanded, its upper tip extending above top of spire. Color a mottled brown, white, and orange.
Remarks: Relatively uncommon throughout its range.

QUEEN STROMB *Strombus gigas* Linn. **Pl. 46**
Range: Bermuda; s. Florida to West Indies.
Habitat: Shallow water.
Description: Height 8 to 12 in. Heavy and solid, with short conical spire. Mostly all body whorl, the aperture moderately narrow and channeled at both ends. Outer lip thickened and greatly flaring when mollusk is fully grown. 8 to 10 whorls, with blunt nodes on shoulders. Operculum clawlike, horny. Color yellowish buff; interior bright rosy pink; young specimens have zigzag axial strips of brown.
Remarks: This is the shell that for generations has been used by families of seafaring men as a doorstop or for decorating the borders of flower beds. It is one of our largest and heaviest gastropods, individuals sometimes weighing more than 5 lbs. It is a commercial shell and large numbers are exported from the Bahamas for cutting into cameos; the scrap material is ground into powder for manufacturing porcelain. The flesh is eaten in the West Indies, where the aborigines formerly made scrapers, chisels, and various other tools from the shell. Semiprecious pearls are occasionally found within the mantle fold. When traveling through Florida one often sees huge piles of this shell heaped up beside a gas station or tourist lodge. They sell for around a dollar each and make popular Florida souvenirs, even though most of them are probably imported from one of the West Indian islands.

FIGHTING STROMB *Strombus pugilis* Linn. **Pl. 46**
Range: S. Florida (uncommon), West Indies, and south to Brazil.
Habitat: Shallow water.
Description: A robust shell of 7 whorls, maximum height about 5 in. and most specimens between 3 and 4 in. Early whorls sculptured with revolving ribs. Prominent spines on later whorls. Body whorl enlarged, aperture quite narrow, outer lip widely flaring and deeply notched below. Operculum clawlike. Color deep yellowish brown, commonly with a paler band midway of body whorl. Aperture deep orange.
Remarks: See Remarks under Florida Stromb (above).

HAWK WING *Strombus raninus* Gmelin **Pl. 46**
Range: S. Florida to West Indies.
Habitat: Shallow water.
Description: Height 4 to 5 in. About 8 whorls, well-developed spire. Shoulders of body whorl bear small knobs, and there is a single large knob on the back, with a smaller knob between it and the margin. Outer lip thick and flaring, deeply notched below. Color yellowish white, streaked and blotched with brown and black.
Remarks: Formerly listed as *S. bituberculatus* Lam.

Wide-mouthed Snails: Family Lamellariidae

SMALL, thin-shelled mollusks living chiefly in cold seas, and commonly at considerable depths. Most of them feature a very heavy periostracum.

Genus *Lamellaria* Montagu 1815

TRANSPARENT LAMELLARIA **Pl. 47**
Lamellaria pellucida Verrill
Range: Massachusetts to Delaware.
Habitat: Deep water.
Description: A globular shell with few whorls, the diam. about ½ in. Enclosed in a tough skinlike periostracum that conceals all but a small portion at the base. Color pale yellow, sometimes lightly flecked with brown, and the shell substance is thin, delicate, and almost transparent.

RANG'S LAMELLARIA   *Lamellaria rangi* Bergh   **Pl. 47**
Range: Florida, Gulf of Mexico, and West Indies.
Habitat: Moderately deep water.
Description: Diam. about ½ in. Thin and delicate, about 8 whorls, the body whorl making up most of the shell. Very fine growth lines the only sculpture. Aperture flaring. Color white.
Remarks: In life the shell is covered by the soft parts.

Genus *Velutina* Fleming 1821

SMOOTH VELUTINA   *Velutina laevigata* (Linn.)   **Pl. 47**
Range: Labrador to Massachusetts; also Alaska to California.
Habitat: Moderately deep water.
Description: A thin and fragile shell of 3 whorls, diam. about ½ in. Mostly all body whorl, no spire. Aperture large and flaring. Surface shows minute revolving threadlike lines, and there is a thick brownish periostracum that is usually ragged and frayed. Color brown or light tan.
Remarks: This snail gives the impression that it is made up of periostracum with a thin layer of shell on the inside.

Atlantas: Family Atlantidae

SMALL flatly coiled shells, thin and translucent. They are pelagic snails, living a floating life in tropical and semitropical seas.

Genus *Atlanta* Lesueur 1827

PERON'S ATLANTA   *Atlanta peroni* Lesueur   **Pl. 47**
Range: Massachusetts to West Indies; West Coast.
Habitat: Pelagic.
Description: Diam. ½ in. A fragile and glassy shell of 3 whorls. It is very flatly coiled, like a compressed ram's horn, with fairly large aperture. There is a thin bladelike keel, or fin, around the periphery of last whorl. Color white.
Remarks: Specimens are now and then washed up on the beaches after storms.

Moon Shells: Family Naticidae

THESE are carnivorous snails, found in all seas. The shell is usually globular, sometimes depressed, smooth, and often polished. The snails burrow into the sand, and have no eyes. Foot of the animal very large and often conceals entire shell when mollusk is extended. Operculum may be calcareous or horny.

Genus *Polinices* Montfort 1810

LOBED MOON SHELL *Polinices duplicatus* (Say) **Pl. 47**
Range: Massachusetts to Florida and Gulf of Mexico.
Habitat: Sand and mud flats.
Description: About 2 in. high, diam. nearly 3 in. Shell solid, 4 or 5 whorls, their upper portions compressed so as to give the shell a somewhat pyramidal outline. Umbilicus irregular, and covered wholly or in part by a thick, chestnut-brown lobe. Operculum horny. Color grayish brown, tinged with bluish.

TAN MOON SHELL *Polinices hepaticus* (Röding) **Pl. 47**
Range: S. Florida to West Indies.
Habitat: Shallow water.
Description: About 1½ in. high. An ovate shell of some 4 whorls, the body whorl constituting most of shell. Polished white callus over the umbilicus. Operculum horny. Color yellowish tan, aperture white.
Remarks: Formerly listed as *P. brunneus* (Link).

IMMACULATE MOON SHELL **Pl. 47**
Polinices immaculatus (Totten)
Range: Gulf of St. Lawrence to N. Carolina.
Habitat: Moderately shallow water.
Description: Not quite ½ in. high. 3 or 4 whorls, short pointed apex. Surface smooth and shiny. Thickened callus does not block the umbilicus. Operculum horny. Color white, with thin yellowish periostracum.

MILKY MOON SHELL *Polinices lacteus* (Guild.) **Pl. 47**
Range: Florida to West Indies; Texas.
Habitat: Shallow water.
Description: About 1 in. high. Shell obliquely oval, smooth and polished. 3 or 4 whorls, less flattened than with most of the genus. Umbilicus partly filled by a callus. Operculum horny, the color amber or claret. Shell milky white.
Remarks: See next species for similarities.

DWARF MILKY MOON SHELL Pl. 47
Polinices uberinus (Orb.)
Range: Florida to West Indies.
Habitat: Shallow water.
Description: Height ½ in. This shell resembles the Milky Moon Shell and may easily be only a form of that species. The umbilicus is proportionally larger, the callus is small, and there is a slight ridge extending from it into the umbilicus. Color usually white, shining.

Genus *Lunatia* Gray 1847

GREENLAND MOON SHELL Pl. 47
Lunatia groenlandica (Möller)
Range: Greenland to New Jersey.
Habitat: Moderately deep water.
Description: About 1 in. high. Approximately 5 well-rounded whorls, apex blunt. Small umbilicus, nearly closed by an enlargement at upperpart of inner lip. Operculum horny. Color gray or white, with thin greenish-yellow periostracum. Aperture white.

NORTHERN MOON SHELL *Lunatia heros* (Say) Pl. 47
Range: Gulf of St. Lawrence to N. Carolina.
Habitat: Sand and mud flats.
Description: Height about 4 in. 5 very convex whorls, somewhat flattened at the top. Aperture large and oval, operculum horny. Umbilicus large, rounded, coarsely wrinkled, and extending through to top of shell. Outline somewhat globular. Color ashy brown, with thin yellowish periostracum. See also the next species.
Remarks: This is a voracious feeder, drilling into and sucking the contents from many a luckless bivalve encountered in its subterranean wanderings. The eggs of this and most moon shells are laid in a mass of agglutinated sand grains that is molded over the shell, and, upon hardening, form the fragile "sand collars" often found lying on the beach during the summer months.

SPOTTED MOON SHELL *Lunatia triseriata* (Say) Pl. 47
Range: Gulf of St. Lawrence to N. Carolina.
Habitat: Shallow water.
Description: Less than 1 in. high. 4 or 5 whorls, general shape almost exactly like a miniature specimen of Northern Moon Shell. Color yellowish gray, sometimes uniformly so and sometimes with oblique chestnut-brown squarish spots arranged in bands.

Remarks: Some writers have contended that this is indeed merely the juvenile stage of the Northern Moon Shell; but certain characters, particularly a relatively thick callus on the inner lip, seem to indicate a mature shell.

Genus *Amauropsis* Mörch 1857

ICELAND MOON SHELL Pl. 47
Amauropsis islandica (Gmelin)
 Range: Arctic Ocean to Virginia.
 Habitat: Deep water.
 Description: About 1 in. high. A rather plump shell of 4 whorls. Shell thin in substance but well inflated. Aperture semicircular, outer lip thin and sharp. Inner lip overspread with white callus. A small slitlike umbilicus. Operculum horny. Yellowish white, with orange-brown periostracum.

Genus *Natica* Scopoli 1777

COLORFUL MOON SHELL Pls. 4, 47
Natica canrena (Linn.)
 Range: N. Carolina to West Indies.
 Habitat: Shallow water.
 Description: Height 1 to 2 in. Shape globular, somewhat flattened at the top. About 4 whorls, body whorl large and expanded, apex small. Operculum calcareous, thick, white, with several deeply cut grooves or channels following its outside curvature. Surface smooth and polished. Color pale bluish gray, with spiral chestnut bars or stripes and zigzag markings, along with various mottlings of dark brown and purple.

ARCTIC MOON SHELL Pl. 47
Natica clausa Brod. & Sby.
 Range: Circumpolar; to N. Carolina; also from Arctic Ocean to California.
 Habitat: Moderately deep water.
 Description: Height 1½ in. at times but usually smaller. Shell nearly spherical, and consists of 4 or 5 whorls, spire only slightly elevated. Aperture oval, lip sharp, thickened, and rounded as it ascends to umbilicus, which is completely closed by a shiny ivory-white callus. Operculum calcareous, bluish white. Shell pale brown.

LIVID MOON SHELL *Natica livida* Pfr. Pl. 47
 Range: Bermuda; s. Florida to West Indies.
 Habitat: Shallow water.

Description: About ½ in. high. A smooth and shiny shell of 3 or 4 whorls, apex rounded. Lip thin and sharp. Umbilicus deep, partly shielded by a deep brown callus. Operculum calcareous. Color pearl-gray, often with faint darker markings; brownish around aperture.

MOROCCO MOON SHELL Pl. 47
Natica marochiensis Gmelin
 Range: Bermuda, Florida, West Indies.
 Habitat: Shallow water.
 Description: About 1 in. high. 4 rounded whorls. Umbilicus deep, with sturdy white plug. Operculum shelly. Surface smooth, color grayish brown, with encircling rows of obscure reddish-brown spots.

MINIATURE MOON SHELL *Natica pusilla* Say Pl. 47
 Range: Massachusetts to Florida; West Indies.
 Habitat: Shallow water to moderately deep water.
 Description: About ¼ in. high. A small but sturdy shell of about 3 whorls, the body whorl making up most of the shell. Umbilicus a slit in back of rolled callus. Operculum calcareous. Color white or gray, sometimes faintly spotted or banded with brown.

Genus *Stigmaulax* Mörch 1852

SULCATE MOON SHELL Pl. 47
Stigmaulax sulcata (Born)
 Range: West Indies.
 Habitat: Moderately shallow water.
 Description: Height ¾ in. 3 or 4 rounded whorls, short spire. Aperture large and high, umbilicus broad and deep, with strong plug. Sculpture of revolving ridges cut by vertical lines. Color yellowish, with pale brown blotches; base white.

Genus *Sinum* Röding 1798

SPOTTED EAR SHELL *Sinum maculatum* (Say) Pl. 47
 Range: N. Carolina to Florida.
 Habitat: Shallow water.
 Description: This species is very much like the Ear Shell, but the top is slightly more elevated and the shell is not quite so flat; also, the sculpture is more delicate. Color white or buffy white, somewhat spotted with weak yellowish-brown marks.

EAR SHELL *Sinum perspectivum* (Say)   **Pl. 47**
Range: Virginia to West Indies.
Habitat: Shallow water.
Description: Diam. about 1½ in., height less than ½ in. Shell elongate-ovate, greatly depressed. 3 or 4 whorls, the enormously expanded body whorl makes up more than ¾ of entire shell; aperture wide and flaring. Surface sculptured with numerous impressed, transverse, slightly undulating lines. No umbilicus, operculum rudimentary. Color milky white. See Spotted Ear Shell for comparisons.
Remarks: This curious gastropod looks like a moon shell that has been squeezed flat. In life the shell is almost buried in the mantle of the mollusk.

Cowries: Family Cypraeidae

COWRIES are a large family of brightly polished and often brilliantly colored snails that have always been great favorites with collectors. The shell is more or less oval, well inflated, with the spire usually covered by the body whorl, and the aperture — lined with teeth on both sides — running the full length of the shell. There is no operculum. This is predominantly a tropical family, with scores of richly colored representatives distributed all around the world. Only a few species, however, are hardy enough to live as far north as Florida.

The genus *Cypraea* has been split into numerous subgenera, and some authorities believe that most are deserving of full generic rank. This means that in many cases we have a new "front name" for nearly every species. Most collectors prefer to use the name *Cypraea* for all of their species, and that system is followed here. However, for our few species the subgenus will be noted in parenthesis.

Genus *Cypraea* Linnaeus 1758

DEER COWRY *Cypraea (Trona) cervus* Linn.   **Pl. 48**
Range: S. Florida to West Indies.
Habitat: Shallow to moderately deep water.
Description: Our largest cowry, length up to 5 in. Once regarded as a subspecies of the Measled Cowry (below), but thinner and more inflated, and more heavily spotted with smaller whitish spots, which never are in rings.

GRAY COWRY *Cypraea (Luria) cinerea* Gmelin   **Pl. 48**
Range: S. Florida to West Indies.
Habitat: Shallow water.

Description: About 1½ in. long. A plump cowry, grayish brown on top, with lower portions shading to lilac. 2 paler bands commonly encircle the shell, and the sides are often decorated with black dots and sometimes streaks. Bottom creamy white, the apertural teeth rather small.

YELLOW COWRY Pl. 48
Cypraea (Erosaria) spurca acicularis Gmelin
Range: Florida to West Indies.
Habitat: Shallow water.
Description: About 1 in. long. Shell oval and solid; a tiny spire present in juveniles but concealed in adults. Aperture long and narrow, slightly curved and evenly toothed. Surface highly polished. Color yellowish tan, spotted with yellow; violet spots along sides.
Remarks: Typical *C. spurca* Linn. is a Mediterranean species.

MEASLED COWRY *Cypraea (Trona) zebra* Linn. **Pls. 5, 48**
Range: S. Florida to West Indies.
Habitat: Shallow water.
Description: About 3 to 4 in. long. Shell moderately inflated and oval, spire completely concealed by last whorl. Highly polished. Aperture narrow, notched at both ends. Lips toothed and dark brown. Color of shell purplish brown, with round whitish spots, often in rings. Young specimens are strongly banded with broad streaks of brown, and these streaks often persist in adult shells, occasional examples having their backs practically unspotted.
Remarks: Formerly listed as *C. exanthema* Linn.

Sea Buttons: Family Eratoidae

SHELLS somewhat like miniature cowries, many characterized by wrinkles or small ribs running around the shell from the narrow aperture to middle of the back. Chiefly confined to warm seas.

Genus *Erato* Risso 1826

MAUGER'S ERATO *Erato maugeriae* Gray Pl. 48
Range: N. Carolina to Florida, Gulf of Mexico, and West Indies.
Habitat: Moderately shallow water.
Description: About ¼ in. high. A tiny pear-shaped shell re-

sembling one of the *Marginella* shells (p. 231), but its rolled-in outer lip bears a row of distinct teeth. Rounded and wide at blunt apex and tapering to a bluntly rounded base. Aperture narrow, running full length of shell. Surface smooth and polished. Color pale tan, often with a rosy tint.

Genus *Trivia* Broderip 1837

ANTILLEAN TRIVIA   *Trivia antillarum* Schilder   **Pl. 48**
  Range: N. Carolina to West Indies.
  Habitat: Moderately shallow water.
  Description: About ¼ in. long. Oval, well arched. Narrow lines over upper surface, crossing weak median groove and continuing on base to curved aperture. Color dark purplish.
  Remarks: This is a relatively rare species.

WHITE GLOBE TRIVIA   *Trivia nix* Schilder   **Pl. 48**
  Range: S. Florida to West Indies.
  Habitat: Moderately shallow water.
  Description: About ½ in. long. Nearly globular. Marked medial groove. Sculpture of strong ridges that sometimes fork at the sides. Color white.

COFFEE BEAN TRIVIA   *Trivia pediculus* (Linn.)   **Pl. 48**
  Range: Florida to West Indies.
  Habitat: Shallow water.
  Description: ½ in. long. Shell solid, nearly spherical, the narrow aperture running full length of lower side, and strongly toothed within. Upper surface bears an impressed median furrow, with strong radiating ridges that extend around and into the aperture. Surface not polished. Color tan to violet-brown, with patches of chocolate.
  Remarks: Shells of this genus resemble those of *Cypraea,* but the animal is quite different. This species is the largest form on our coast.

FOUR-SPOTTED TRIVIA   **Pl. 48**
Trivia quadripunctata (Gray)
  Range: Florida to West Indies.
  Habitat: Shallow water.
  Description: About ¼ in. long. Shell nearly globular, with strong dorsal furrow. Fine but distinct ridges radiate from this to the elongate aperture. Color pale pink; upper surface bears 4 brownish spots, 2 on each side of the median furrow.
  Remarks: The body of this gastropod is vivid scarlet and surprisingly large when fully extended; one marvels that it can all be packed away in such a tiny shell.

PINK TRIVIA   *Trivia suffusa* (Gray)   **Pl. 48**
Range: S. Florida to West Indies.
Habitat: Shallow water.
Description: Usually less than ¼ in. long. Upper surface sculptured with numerous transverse lines divided along the median line by a longitudinal furrow. Lines weakly beaded, and continue around the shell and into the narrow aperture. Color rosy pink, sometimes a little darker at the side; generally there is a faint suggestion of spotting.

Simnias: Family Ovulidae

SHELLS usually long and slender, with a straight aperture notched at each end. Occurring in warm seas, the mollusks attach themselves to sea fans and various marine growths, and are usually colored to match their environment.

Genus *Primovula* Thiele 1925

DWARF RED OVULA   *Primovula carnea* (Poiret)   **Pl. 48**
Range: S. Florida to West Indies.
Habitat: Moderately deep water.
Description: About ½ in. long. A stout shell, with an arched aperture that is longer than body of shell and is notched at both ends. Outer lip rolled in and toothed. Surface bears fine spiral lines, but appears smooth. Color pinkish white or yellowish.

Genus *Neosimnia* Fischer 1884

COMMON WEST INDIAN SIMNIA   **Pl. 48**
Neosimnia acicularis (Lam.)
Range: N. Carolina to West Indies.
Habitat: Shallow water; on sea fans (*Gorgonia*).
Description: About ¾ in. long, a rather thin and elongate shell. Aperture extends entire length of shell, lip thin and sharp. Surface smooth and shiny. Columella bears a paler ridge. Color yellowish or purplish, according to shade of the sea fan on which it lives.

SINGLE-TOOTHED SIMNIA   **Pl. 48**
Neosimnia uniplicata (Sby.)
Range: N. Carolina to West Indies.
Habitat: Shallow water; on sea fans (*Gorgonia*).
Description: Length about ¾ in. Shell thin, elongate, somewhat cylindrical, the aperture a narrow slit running full

length of shell; ends bluntly pointed. Surface smooth and polished. Color usually some shade of pink or purple, although it may be white or yellowish.

Remarks: Like the Common West Indian Simnia, it lives attached to stems of some marine growth, usually sea fans, and is nearly always colored to harmonize with its surroundings.

Genus *Cyphoma* Röding 1798

FLAMINGO TONGUE *Cyphoma gibbosum* (Linn.) **Pl. 48**
Range: N. Carolina to West Indies.
Habitat: Shallow water; on sea fans (*Gorgonia*).
Description: About 1 in. long. Shell solid and durable, with a dorsal ridge, or hump, near center of shell and extending squarely across it; see McGinty's Flamingo Tongue for differentiation. Aperture long and narrow, running full length of lower side. Inner and outer lips without teeth. Color white or yellowish, highly polished.
Remarks: The humpbacked Flamingo Tongue is generally found living on the sea fan or some other branching aquatic growth, where it clings tightly to one of the stems. In life the shell is covered by the mantle, which is pale flesh color with squarish black rings.

McGINTY'S FLAMINGO TONGUE **Pl. 48**
Cyphoma macgintyi Pilsbry
Range: Florida Keys and Bahamas.
Habitat: Shallow water; on sea fans (*Gorgonia*).
Description: About 1 in. long. Shell very much like Flamingo Tongue, though slightly more elongate and with a somewhat narrower hump. The animal itself is colored quite differently. In Flamingo Tongue the mantle is flesh color decorated with squarish black rings; in McGinty's Flamingo Tongue the color is pinkish lavender, with marginal rows of round black dots.

Helmet Shells: Family Cassididae

THESE are mostly large and heavy shells, many of them used for cutting cameos. The shells are thick, and commonly 3-cornered when viewed from below. The aperture is long, terminating in front in a recurved canal. The outer lip is generally thickened and toothed within, and the inner lip commonly bears teeth, wrinklelike ridges, or pustules. Common on sandy bottoms in warm seas.

Genus *Sconsia* Gray 1847

ROYAL BONNET *Sconsia striata* (Lam.) **Pl. 49**
 Range: Bermuda, Florida, Bahamas, and south through West Indies to n. Brazil.
 Habitat: Deep water.
 Description: About 2 in. high. A solid shell of about 5 whorls. There is a short and pointed spire, but most of the shell is body whorl. Aperture elongate, outer lip thickened and toothed within. Inner lip broadly reflected and polished. Surface bears fine revolving lines, usually with 1 or 2 varices on last volution. Operculum horny. Color buffy gray with squarish spots of chestnut, the latter sometimes obscure.

Genus *Morum* Röding 1798

WOOD LOUSE *Morum oniscus* (Linn.) **Pl. 49**
 Range: Florida to West Indies.
 Habitat: Moderately shallow water.
 Description: About 1 in. high. Shell small but very solid, bluntly cone-shaped, with low spire. About 4 whorls. Surface covered with rows of warty tubercles, especially prominent on shoulders of last volution. Aperture elongate, outer lip considerably thickened, and toothed within. Inner lip with heavy white callus, ornamented with small pustules. Color soiled white, copiously mottled with brown and gray. During life there is a velvety gray periostracum.

Genus *Phalium* Link 1807

POLISHED SCOTCH BONNET **Pl. 8**
Phalium cicatricosum (Gmelin)
 Range: Bermuda; s. Florida to West Indies.
 Habitat: Moderately shallow water.
 Description: About 2 in. high. This shell is very similar to the Scotch Bonnet and is sometimes confused with it. The size and general form are the same, but this one lacks the deep revolving lines and its surface is smooth and shiny. The shell is not as solid, and the outer lip is usually not as thickened.

SCOTCH BONNET *Phalium granulatum* (Born) **Pl. 49**
 Range: N. Carolina south through West Indies to Brazil.
 Habitat: Moderately shallow water.
 Description: Height 2 to 4 in. Shell moderately solid, about

5 whorls, spire short, sutures indistinct. Surface bears deeply incised transverse lines. Aperture large, outer lip thickened and toothed within, canal strongly curved. 1 or more varices commonly present. Inner lip broadly reflected, pustulate on lower part. Color pale yellow or whitish, with squarish pale brown spots distributed over the shell with considerable regularity. See Polished Scotch Bonnet for similarity.

Remarks: Spots may be faded or obliterated over the shell in beach specimens. Formerly listed as *Semicassis inflata* (Shaw) and *S. abbreviata* (Lam.).

Genus *Cypraecassis* ·Stutchbury 1837

BABY BONNET *Cypraecassis testiculus* (Linn.) **Pl. 6**
Range: Bermuda, Florida, West Indies, and south to Brazil.
Habitat: Moderately shallow water.
Description: About 2 to 3 in. high. Shell solid, with very short spire. Some 5 whorls, the body whorl comprising most of shell. Aperture long and narrow, inner lip with a shield on body whorl and plicate (folded) for entire length along the aperture; outer lip rolled back and strongly toothed within. No operculum. Canal twisted to left and folded against shell. Sculpture of sharp vertical lines, crossed by strong revolving indentations. Color pinkish buff, mottled with orange and brown.

Genus *Cassis* Scopoli 1777

FLAME HELMET *Cassis flammea* (Linn.) **Pl. 49**
Range: Bahamas and south to Lesser Antilles.
Habitat: Shallow water.
Description: Height 3 to 6 in. About 6 whorls, small pointed spire. Body whorl large, ornamented with encircling knobs but lacking reticulate sculpture of King Helmet (below). Aperture elongate, outer lip strongly toothed within. Inner lip reflected on last volution and decorated with distinct wrinkles. Canal strongly recurved. Surface generally polished, color yellowish, with brown streaks and splashes. Outer lip with no brown spots between the teeth but with heavy brown bars on outer margin.

CLENCH'S HELMET **Pl. 49**
Cassis madagascariensis spinella Clench
Range: S. Florida to West Indies.
Habitat: Shallow water.
Description: Our largest helmet shell attains a height of 14 in. Heavy and ponderous, practically no spire, the greatly

enlarged body whorl constituting most of the shell. This body whorl bears 3 spiral ridges with blunt knobs. The thickened outer lip has a few well-separated large teeth and the inner lip bears many riblike smaller teeth, or plications. These are pale buff, the area between them dark brown. Front of aperture recurved and folded back on shell. Color grayish yellow, clouded with brown markings.

Remarks: The Emperor Helmet, *C. madagascariensis,* shown on Plate 49, is smaller, proportionately heavier and has one or more knobs larger than the others.

KING HELMET *Cassis tuberosa* (Linn.) **Pl. 49**
Range: S. Florida to West Indies and south to Brazil.
Habitat: Shallow water.
Description: This shell is much like the Flame Helmet (above), but more triangular when viewed from below. Color buffy or rufous yellow, mottled and blotched with various shades of brown. In addition to the brown stain between the folds of the inner lip, and the strong teeth lining the outer lip, there is a conspicuous patch of bright chestnut toward the posterior end of the aperture. Outer lip rolled over and strongly marked with brown patches.
Remarks: This species is preferred for cameo cutting: there is a very dark coat beneath the outer layer, so the figure stands out well against the "onyx" background. The shell is sometimes known as the Sardonyx Helmet. Most of the helmet shells yield a cameo with a reddish-orange or pink background. This species is sometimes confused with the Flame Helmet, *C. flammea,* which does not occur on our shores and has a smooth, almost polished surface; *C. tuberosa* is marked with fine but sharp longitudinal lines.

Tritons: Family Cymatiidae

THESE are rather decorative shells, rugged and strong, with no more than 2 varices to a volution. The closely related *Murex* shells (p. 189) have 3. The canal is prominent, and teeth are usually present on the lips. They are distributed in warm and temperate seas.

Genus *Cymatium* Röding 1798

DOG-HEAD TRITON **Pl. 50**
Cymatium caribbaeum Clench & Turner
Range: Florida to West Indies; Texas.
Habitat: Moderately shallow water.

Description: Height 2 to 3 in. About 5 whorls, a low spire, and a long, somewhat curved canal nearly closed at its upper end. Surface bears very heavy revolving ribs, and there are 2 or 3 robust varices. Aperture large, inner lip reflected on last whorl; outer lip strongly crenulate within. Color pale yellow, variously clouded with gray and white.
Remarks: Formerly listed as *C. cynocephalum* (Lam.).

ANGULAR TRITON *Cymatium femorale* (Linn.) **Pl. 7**
Range: S. Florida, West Indies, and south to Brazil.
Habitat: Shallow water.
Description: A large and spectacular species, to 7 in. high. 7 or 8 well-shouldered whorls. Sculpture consists of revolving ribs of 2 sizes, the larger ones studded with coarse knobs at regular intervals; 2 prominent varices stand out on each volution. Outer lip rolled inward at margin and decorated with large nodules at ends of ribs, the top one curving upward in direction of the spire. Canal moderately long, partly reflected, and nearly closed. Pale yellowish brown, with bands of darker shade. Aperture white.
Remarks: A common West Indies shell, infrequently taken in s. Florida.

LIP TRITON *Cymatium labiosum* (Wood) **Pl. 50**
Range: Florida Keys to West Indies.
Habitat: Moderately shallow water.
Description: Height ¾ in. A small squat shell of 4 or 5 stout whorls, each decorated with coarse vertical and revolving ribs. Aperture small and notched above, canal short, outer lip greatly thickened and ornamented with sharp transverse lines. Color yellowish white.

WHITE-MOUTHED TRITON **Pl. 50**
Cymatium muricinum (Röding)
Range: Bermuda; s. Florida to West Indies.
Habitat: Shallow water.
Description: About 2 in. high. Rugged and strong, with 5 whorls and a sharp spire. Surface sculptured with revolving nodular ribs, and usually with 2 varices prominently displayed on each volution. Aperture large, notched at upper end, canal moderately long and open. Inner lip reflected on body whorl. Color gray, marked and sometimes mottled with brown. Aperture white.
Remarks: Formerly listed as *C. tuberosum* (Lam.).

GOLD-MOUTHED TRITON **Pl. 50**
Cymatium nicobaricum (Röding)
Range: Bermuda; s. Florida to West Indies.

Habitat: Moderately shallow water.
Description: Usually 2 to 3 in. high. Rugged and solid, about 5 whorls, a short spire, and 2 stout varices on each volution. Surface divided into squares by the crossing of horizontal and vertical ribs. Outer lip thick and heavy, with double row of teeth inside. Canal short and curved. Color gray or whitish, clouded and mottled with brown. Inside of aperture usually orange or bright yellow, the white teeth standing out vividly.
Remarks: Formerly listed as *C. chlorostomum* (Lam.).

NEAPOLITAN TRITON   **Pl. 50**
Cymatium parthenopeum (Salis)
  Range: Bermuda; Florida to West Indies.
  Habitat: Moderately shallow water.
  Description: Height 3 to 4 in. About 5 whorls, sutures indistinct. Sculpture consists of prominent revolving ribs, those on shoulders somewhat wavy. Aperture moderately large, outer lip thick and made knobby by the terminating ribs. Color yellowish brown, somewhat mottled with darker hues. In life there is a hairy periostracum.
  Remarks: Formerly listed as *C. costatum* (Born).

HAIRY TRITON   *Cymatium pileare* (Linn.)   **Pl. 7**
  Range: S. Florida, West Indies, and south to Brazil.
  Habitat: Moderately shallow water.
  Description: About 3 to 5 in. high. Strong and solid, 5 or 6 whorls, rather pointed spire. Surface crosshatched with lines running in 2 directions, the shoulders somewhat nodulous, and there are 2 robust varices on each volution. Aperture large, both lips wrinkled into small teeth. Color pale brown, banded with gray and white.
  Remarks: Formerly listed as *C. aquitila* (Reeve) and *C. martinianum* (Orb.).

POULSEN'S TRITON   *Cymatium poulseni* (Mörch)   **Pl. 50**
  Range: Florida and Texas south through West Indies to Venezuela.
  Habitat: Moderately deep water.
  Description: About 2½ in. high. 4 or 5 whorls, a moderately short acute spire, a moderate canal, and a wide aperture. Whorls well shouldered, decorated with revolving ribs that alternate in size. Color yellowish white.

DWARF HAIRY TRITON   **Pl. 50**
Cymatium vespaceum (Lam.)
  Range: S. Florida to West Indies.

Habitat: Moderately deep water.
Description: About 1½ in. high. 5 or 6 shouldered whorls, pointed apex, sutures distinct. Sculpture of strong revolving lines, prominent varices. Aperture small, toothed within on both lips, canal long and nearly closed. Color yellowish white, varices brown and white.

Genus *Distorsio* Röding 1798

WRITHING SHELL   *Distorsio clathrata* (Lam.)   **Pl. 50**
Range: Florida to West Indies.
Habitat: Moderately deep water.
Description: About 1 to 2 in. high. 6 or 7 whorls, sculptured by strong revolving and vertical ribs, giving the surface a checkered appearance. Aperture small, constricted, with deep notch on inner lip. Both lips toothed. Columella strongly reflected on body whorl. Canal open and moderately long. Color yellowish white.

McGINTY'S WRITHING SHELL   **Pl. 50**
Distorsio macgintyi Emerson & Puffer
Range: Florida to West Indies.
Habitat: Deep water.
Description: Height 1 to 1½ in. About 6 whorls, the last bulges noticeably. Sculpture of revolving cords and vertical ridges. Aperture small and constricted, outer lip strongly toothed, and columella with distinct gouge. Canal short. Color yellowish white.

Genus *Charonia* Gistel 1848

TRUMPET SHELL   *Charonia variegata* (Lam.)   **Pl. 51**
Range: S. Florida to West Indies.
Habitat: Moderately shallow water.
Description: Height 10 to 15 in. Shell strong and solid, gracefully elongate, apex bluntly pointed. 8 or 9 whorls, sutures plainly marked; body whorl shows distinct shoulders. Widely separated round ribs encircle the shell, and each volution bears 2 rather obscure varices. Inner lip reflected, and stained a dark purplish brown, crossed by whitish wrinkles. Shell richly variegated with buff, brown, purple, and red, in crescentic patterns suggestive of pheasant plumage. Aperture pale orange.

Remarks: Formerly considered a subspecies of the Pacific Trumpet Shell, and listed as *C. tritonis nobilis* (Conrad).

Frog Shells: Family Bursidae

SHELLS ovate or oblong, somewhat compressed laterally, with 2 varices of continuous nature, 1 on each side. Deep-water forms have the varices thin and bladelike, whereas those living among the rocks and on coral reefs closer to shore are more nodular. They inhabit warm seas.

Genus *Bursa* Röding 1798

CORRUGATED FROG SHELL   Pl. 51
Bursa corrugata (Perry)
   Range: S. Florida to West Indies.
   Habitat: Moderately shallow water.
   Description: Height 2 to 3 in. About 5 whorls, somewhat flattened laterally. Each volution bears a series of encircling knobs and a pair of knobby varices. Outer lip broad and toothed within, inner lip strongly reflected. Aperture with deep notch at upper angle. Canal short and nearly closed. Color yellowish brown, more or less clouded with darker hues.

CUBAN FROG SHELL   *Bursa cubaniana* (Orb.)   Pl. 51
   Range: S. Florida to West Indies.
   Habitat: Moderately deep water.
   Description: Height 2 in. About 7 whorls, a moderately tall spire with pointed apex. Whorls well inflated but somewhat compressed laterally. 2 prominent varices on each whorl; they are directly opposite each other, so there is a nearly continuous ridge running up each side of shell. Surface decorated with revolving lines, some of them beaded; a row of larger beads, or nodes, on the shoulders. Aperture oval, outer lip thickened and toothed within, inner lip plicate. Short anterior canal, strong notch at upper angle of aperture. Color reddish brown, blotched with white.
   Remarks: Formerly listed as *B. granularis* (Röding), which is an Indo-Pacific species.

CHESTNUT FROG SHELL   Pl. 51
Bursa spadicea (Montfort)
   Range: S. Florida to West Indies.

Habitat: Moderately deep water.
Description: About 2 in. high. 5 whorls, sutures indistinct. A rather flat shell with prominent varix on each side. Aperture with strong notch above, short canal below, toothed on both lips. Sculpture of 2 revolving rows of beads on upperpart of each volution, with numerous rows of tiny beads below. Color yellowish brown, banded with orange-brown.

ST. THOMAS FROG SHELL *Bursa thomae* (Orb.)   **Pl. 51**
Range: S. Florida, West Indies, and south to Brazil.
Habitat: Moderately shallow water.
Description: About 1 in. high. A rather squat shell of 4 or 5 whorls. Varices along each side knobby and prominent, and the volutions themselves bear many rounded tubercles and beadlike knobs. Aperture moderate in size, strongly notched above. Color yellowish white, sometimes spotted with reddish.
Remarks: Formerly listed as *B. cruentatum* (Reeve).

Tun Shells: Family Tonnidae

A SMALL group of large or medium-sized gastropods, chiefly of the tropics. The shell is thin and nearly spherical (subglobular) usually with a greatly swollen body whorl. They are sometimes called cask shells or wine jars.

Genus *Tonna* Brünnich 1772

GIANT TUN SHELL   *Tonna galea* (Linn.)   **Pl. 50**
Range: N. Carolina to West Indies; Texas.
Habitat: Moderately deep water.
Description: Averages 6 in. high, but 10-in. specimens have been recorded. Shell thin but quite sturdy and greatly inflated, being almost as broad as tall. About 5 whorls, sutures somewhat sunken. Very low spire. Surface with widely spaced encircling grooves. Aperture very large, outer lip thickened in mature examples, thin and sharp in partly grown individuals. Columella strongly twisted. Juvenile specimens possess a small horny operculum, but this structure is lacking in adults. Color yellowish white, blotched or banded with brown.
Remarks: Very similar shells, now believed by some authorities to belong to this species, are found in the e. Atlantic along the African coast, and in both the Pacific and Indian Oceans.

PARTRIDGE SHELL *Tonna maculosa* (Dill.) **Pl. 50**
Range: S. Florida to West Indies.
Habitat: Moderately deep water.
Description: About 5 or 6 in. high. 4 or 5 whorls, sutures impressed. Shell thin and inflated, spire short and with pointed apex. Aperture large, outer lip thin and sharp, slightly crenulate at edge. Sculpture of widely spaced encircling grooves. Color pale to rich brown, heavily mottled with darker shades and bearing crescent-shaped patches of white, so the surface does indeed remind one of the plumage of a partridge.
Remarks: This shell has long been considered a subspecies of the Indo-Pacific *Tonna perdix* (Linn.) and has been listed as *T. perdix pennata* (Mörch), but it is now recognized as a distinct species, entitled to full specific rank. It has been given the earliest available name, *maculosa,* proposed by Lewis Dillwyn in 1817.

Genus *Eudolium* Dall 1889

CROSSE'S TUN **Pl. 49**
Eudolium crosseanum (Monterosato)
Range: New Jersey to West Indies.
Habitat: Deep water.
Description: About 3 in. high. Approximately 6 rounded whorls, a short spire, and a rather blunt apex. Shell substance thin, shape moderately stout. Decorated with low and widely spaced encircling ribs. Aperture large, outer lip slightly rolled at edge in mature shells. Inner lip reflected on body whorl. Color buffy white.

Fig Shells: Family Ficidae

THIN and light in substance but surprisingly strong. In life, when the mollusk is crawling the shell appears to be almost buried in the mantle. Occur in warm seas around the world.

Genus *Ficus* Röding 1798

PAPER FIG SHELL *Ficus communis* Röding **Pl. 49**
Range: N. Carolina to Gulf of Mexico.
Habitat: Moderately deep water.

Description: About 3 to 4 in. high. Shell thin and pearshaped, flat on top, no spire. About 4 whorls, the body whorl enlarged above and narrowing to a long, straight canal. Surface sculptured with fine growth lines crossed by small revolving, cordlike ribs. Aperture large, lip thin and sharp. No operculum. Color pinkish gray, with widely spaced pale brown dots that are sometimes scarcely discernible. Interior polished orange-brown.

Remarks: This snail is uncommon on the Atlantic Coast but quite abundant on the western coast of Florida. It was formerly listed as *Ficus* (or *Pyrula*) *papyratia* (Say).

Rock or Dye Shells: Family Muricidae

SHELLS thick and solid, generally more or less spiny. There are usually 3 varices to a volution. These are active, carnivorous snails, preferring a rocky or gravelly bottom and moderately shallow water. They are found in all seas, but are most abundant in the tropics. The famous Tyrian purple of the ancients was a dye obtained by crushing the bodies of certain Mediterranean species of this group and skimming the dye off the water surface.

Genus *Murex* Linnaeus 1758

AGUAYO'S MUREX *Murex aguayoi* Clench & Far. **Pl. 52**
Range: West Indies.
Habitat: Deep water.
Description: Height 1½ in. 8 or 9 well-rounded whorls, sutures well impressed. 3 prominent varices, each with several spines, those at the shoulders commonly larger. Sculpture of nodulous revolving ridges. Aperture small, parietal wall reflected. Canal long, closed, and usually curving. Color yellowish white.

ANTILLEAN MUREX *Murex antillarum* Hinds **Pl. 8**
Range: S. Florida to West Indies.
Habitat: Moderately deep water.
Description: Nearly 4 in. high. 8 or 9 convex whorls, a moderate spire, and a long, partially closed canal. 3 equidistant varices, each lined with spines of varying sizes, but with 1 long spine at the shoulder. These spines continue partway

down the canal. Between the varices the shell is sculptured with revolving ribs, some of them moderately beaded. Color creamy white, varying to brownish purple; aperture commonly purplish within.

BEAU'S MUREX *Murex beaui* Fischer & Bern. **Pl. 52**
Range: S. Florida to West Indies.
Habitat: Deep water.
Description: Height 3 to 4 in. 6 or 7 rounded whorls, a rather tall spire, and a long canal. The 3 varices are armed with spines, and in addition they often bear thin bladelike webs. Surface of shell decorated with several rows of encircling beads. Color creamy yellowish or brownish.
Remarks: One of the handsomest members of its group in our waters, and one of the rarest. Considered a collector's item.

BEQUAERT'S MUREX **Pl. 52**
Murex bequaerti Clench & Far.
Range: N. Carolina to Florida Keys.
Habitat: Deep water.
Description: About 1½ in. high. 7 whorls, sutures conspicuous, apex sharp. 3 thin bladelike or weblike varices, surface in between with revolving lines, and a single vertical ridge. Aperture small, canal short and closed. Color white.

SHORT-FROND MUREX *Murex brevifrons* Lam. **Pl. 51**
Range: S. Florida to West Indies.
Habitat: Moderately shallow water.
Description: Height 5 in. Shell rough and strong, with about 5 whorls having rather indistinct sutures. Surface roughened by many transverse wrinkles. 3 varices, decorated with spines and hollow fronds, those near the shoulders curving upward. Aperture oval, canal fairly long and partially closed. Color grayish, mottled with white and brown.

CABRIT'S MUREX *Murex cabriti* Bern. **Pl. 51**
Range: Florida to West Indies.
Habitat: Moderately deep water.
Description: Height to 3 in. Shell quite sturdy, generally spiny, with about 6 whorls, sutures indistinct. Surface bears revolving ridges. Aperture small, canal narrow and much elongated, being fully twice as long as rest of shell. 3 varices on each volution, decorated with slender spines that continue on down the canal. Color pinkish buff.
Remarks: In number and length of spines there is considerable variation, and in some individuals they may be completely lacking.

CAILLET'S MUREX *Murex cailleti* Petit **Pl. 52**
Range: S. Florida to West Indies.
Habitat: Deep water.
Description: About 1½ in. high. 8 or 9 whorls, showing slight shoulders. 3 main varices, with rounded vertical ridges in between. Varices usually bear short spines. Aperture moderate, bordered by thickened varix, parietal wall reflected. Canal rather long, closed, and curved. Color yellowish white, with brownish bands.

PITTED MUREX *Murex cellulosus* Conrad **Pl. 51**
Range: N. Carolina to West Indies.
Habitat: Shallow water.
Description: About ¾ in. high. A solid and rough shell of about 6 whorls. 5 or 6 poorly defined varices with strong wrinkled lines between them. Aperture round, canal moderate, closed, and termination of the varices at that point produces a forked appearance. Color grayish; aperture purplish within.
Remarks: Formerly listed as *Tritonalia cellulosa* (Conrad).

CIBONEY MUREX *Murex ciboney* Clench & Far. **Pl. 52**
Range: West Indies.
Habitat: Deep water.
Description: About 2 in. high. 9 rounded whorls, sutures impressed. Sculpture of 3 varices, armed with several short spines, and 3 vertical ridges in between, in addition to strong revolving lines. Aperture moderate, canal nearly closed. Color white, sometimes with well-spaced thin encircling lines of brown.

LACE MUREX *Murex florifer* Reeve **Pl. 51**
Range: Florida to West Indies.
Habitat: Shallow water.
Description: Height 2 to 3 in. Shell rough, sturdy, and ornate, with about 7 whorls. 3 prominent varices, decorated with frondlike spines. Sculpture of revolving cords. Canal curved backward, nearly closed, and flattened below. Operculum horny. Color deep brownish black, the aperture and apex pink. Young specimens are commonly pink all over.
Remarks: Formerly listed as *M. rufus* Lam.

TAWNY MUREX *Murex fulvescens* Sby. **Pl. 51**
Range: N. Carolina to Florida; Texas.
Habitat: Moderately shallow water.
Description: To 7 in. high, the largest member of this genus on our shores. Body whorl quite large, giving the shell a somewhat globular look. 7 or 8 whorls, 3 prominent varices

with what might be termed subvarices between them. Apex sharp, canal broadly thickened and nearly closed. Outer lip and canal supplied with spines; shorter spines are usually present on varices as well. Color buffy white.

Remarks: Formerly listed as *M. spinicostatus* Val.

McGINTY'S MUREX *Murex macgintyi* M. Smith **Pl. 52**
Range: Florida Keys and Bahamas.
Habitat: Deep water.
Description: Height 1 in. 6 whorls, with strong shoulders. Surface bears robust well-spaced revolving ridges that cross over the 6 varices. Canal short, broad, and partially closed. Color grayish white, sometimes with brownish marks on varices.
Remarks: This species was named in 1939 by Maxwell Smith from a fossil found at Clewiston, Florida.

PAZ'S MUREX *Murex pazi* Crosse **Pl. 52**
Range: Bahamas to West Indies.
Habitat: Deep water.
Description: About 2 in. high. 7 or 8 rounded whorls, sutures impressed, apex sharp. Sculpture of 4 or 5 varices, each with a long curved spine at shoulder and smaller spines below. Aperture small, canal short, closed, and slightly curved. Color white.

APPLE MUREX *Murex pomum* Gmelin **Pls. 4, 51**
Range: N. Carolina to West Indies and south to Brazil.
Habitat: Shallow water.
Description: Height 2 to 3 in. Shell rough and heavy, spire well developed, body whorl large. 5 or 6 whorls, sutures indistinct. Surface nodular and sculptured with revolving ridges and cordlike ribs. 3 prominent varices on each volution. Aperture large and round, outer lip thickened, columellar callus plastered on body whorl. Canal short, nearly closed, and curving backward. Operculum horny. Color yellowish brown, with broad stripes and mottlings of darker shades. Aperture tinged with rose, and marked with brownish on inner lip.
Remarks: In Jamaica this species grows to a height of nearly 5 in.

HANDSOME MUREX *Murex pulcher* A. Adams **Pl. 52**
Range: West Indies.
Habitat: Deep water.
Description: About 2 in. high. 8 or 9 rounded whorls, apex short but sharp. 3 varices, with 2 vertical ridges in between. A few short spines at shoulders and on long curved canal that is partially closed. Color yellowish white, with suggested banding of reddish brown, especially on varices.

RED MUREX *Murex recurvirostris rubidus* Baker   **Pl. 51**
Range: S. Florida to West Indies.
Habitat: Shallow water.
Description: Height 1½ in. Small but stocky, has 5 whorls sculptured with revolving ribs. 3 rounded varices, with 2 or 3 vertical ridges between them. Varices may bear short spines or they may be spineless. Aperture round, canal long and nearly closed. Mottled gray, brown, pink, or reddish, sometimes showing weak bands. Horny operculum pale yellow.
Remarks: Typical *M. recurvirostris* Brod. is a Pacific shell.

TRYON'S MUREX *Murex tryoni* Hidalgo   **Pl. 52**
Range: Florida to West Indies.
Habitat: Deep water.
Description: Slightly more than 1 in. high. 6 or 7 whorls, short spire, and long canal. 3 varices, each with alternate long and short nublike spines. Surface between varices with bumpy revolving cords. Canal closed, spineless except near aperture. Color yellowish or creamy white.

WOODRING'S MUREX   **Pl. 52**
Murex woodringi Clench & Far.
Range: Jamaica south.
Habitat: Deep water.
Description: Nearly 3 in. high. 8 or 9 well-rounded whorls, pointed apex and long slender canal. 3 varices with short spines; surface in between bears encircling ridges that are nodulous and produce some 3 or 4 vertical rows of bumps. Aperture moderate, parietal wall reflected, canal partially closed. Color yellowish white, sometimes with faint darker markings.

Genus *Muricopsis*
Bucquoy, Dautzenberg, & Dollfus 1882

MAUVE-MOUTHED MUREX   **Pl. 52**
Muricopsis ostrearum (Conrad)
Range: W. Florida.
Habitat: Shallow water.
Description: About 1 in. high. A spindle-shaped shell of 6 whorls, sutures rather indistinct. Spire well elevated, canal quite long and open. Sculpture consists of strong revolving riblets crossed by sharp vertical lines, and there is a series of nodes at the shoulders. Color grayish; aperture commonly mauve.

HEXAGONAL MUREX   **Pl. 52**
Muricopsis oxytatus M. Smith
Range: S. Florida to West Indies.

Habitat: Moderately shallow water.
Description: About 1¼ in. high. Some 6 whorls, rather indistinct sutures. Well-developed spire, short open canal. Surface somewhat spiny, with about 5 rows of short spines on body whorl. Outer lip margined with frondlike protuberances. Color grayish white, tinged with reddish, especially on apex.
Remarks: Formerly listed as *Muricidea hexagona* (Lam).

Genus *Aspella* Mörch 1877

WRETCHED ASPELLA   **Pl. 52**
Aspella paupercula (C. B. Adams)
   Range: S. Florida to West Indies.
   Habitat: Moderately deep water.
   Description: Height ½ in. 7 or 8 whorls, sutures fairly impressed. Shell decorated with thin varices and vertical knobs. Aperture small, canal short and closed. Color grayish, clouded with brown or black.

Genus *Pseudoneptunea* Kobelt 1882

FALSE DRILL   *Pseudoneptunea multangula* (Phil.)   **Pl. 52**
   Range: N. Carolina to West Indies; Texas.
   Habitat: Moderately shallow water.
   Description: About 1 in. high. 6 whorls, body whorl large, sutures distinct. Sculpture of 7 prominent vertical ribs interrupted at the sutures and crossed by numerous revolving threadlike lines. Aperture oval, lip thin, canal short, operculum horny. Color variable, usually creamy white flecked with brown, but occasional specimens are solid brown or orange. Aperture commonly rosy pink.

Genus *Urosalpinx* Stimpson 1865

OYSTER DRILL   *Urosalpinx cinerea* (Say)   **Pl. 52**
   Range: Nova Scotia to Florida.
   Habitat: Shallow water.
   Description: About 1¼ in. high. A rugged shell of approximately 5 whorls that are rather convex. Shell coarse and quite solid, surface sculptured with raised revolving lines, made wavy by rounded vertical folds. Aperture oval, lip thin and sharp, canal short. Operculum horny. Color dingy gray; aperture dark purple. See Gulf Oyster Drill, which is similar.
   Remarks: Next to the starfish, this snail is the worst enemy

the oystermen have to contend with. In some localities, notably the Chesapeake Bay area, their depredations sometimes exceed those of the starfish. Settling upon a young bivalve, the Oyster Drill quickly bores a neat round hole through a valve, making expert use of its sandpaperlike radula. Through this perforation the Drill is able to insert its long proboscis and consume the soft parts of the oyster.

GULF OYSTER DRILL   **Pl. 52**
Urosalpinx perrugata (Conrad)
   Range: W. Florida.
   Habitat: Shallow water.
   Description: Height about 1 in. A sturdy shell much like the Oyster Drill, but with vertical folds more prominent and with a coarser sculpture of revolving lines; also a somewhat slimmer shell. Outer lip thickened and weakly toothed within. Operculum horny. Color yellowish gray.

TAMPA OYSTER DRILL   **Pl. 52**
Urosalpinx tampaensis (Conrad)
   Range: W. Florida.
   Habitat: Mudflats.
   Description: Height 1 in. Shell rugged, 5 or 6 broadly rounded, well-shouldered whorls with distinct sutures. About 10 vertical ridges to a volution, cut by strongly sculptured revolving cords. Ridges generally paler-colored than spaces in between. Canal short and open, slightly curved. Operculum horny and yellow. Shell grayish brown, rather mottled with white.

Genus *Eupleura* H. & A. Adams 1853

THICK-LIPPED OYSTER DRILL   **Pl. 52**
Eupleura caudata (Say)
   Range: Massachusetts to Florida.
   Habitat: Shallow water.
   Description: Height ¾ in. Shell solid, with about 5 distinctly shouldered whorls. Surface bears about 11 stout vertical ribs, the one bordering the aperture and one directly opposite (on left side of body whorl) enlarged into stout knobby ridges. Numerous revolving lines. Outer lip thick and bordered within by raised granules. Canal short and nearly closed, operculum horny. Color varies from reddish brown to bluish white.
   Remarks: This little snail also drills into bivalve shells, but it is not numerous enough to be the pest that the common Oyster Drill, *Urosalpinx cinerea* (above) is.

Genus *Boreotrophon* Fischer 1884

CLATHRATE TROPHON Pl. 52
Boreotrophon clathratum (Linn.)
   Range: Arctic Ocean to Maine.
   Habitat: Moderately deep water.
   Description: Height 1 to 2 in. A sturdy shell of 5 or 6 whorls, sutures distinct and shoulders angled. Sculpture of sharp vertical ridges. Aperture oval, canal open and moderately long. Color chalky white.
   Remarks: A variable shell, and several forms have been described. Also shown on Plate 52 is a form with a long and somewhat curved canal that was named *B. clathratum rostratum* (Verrill), Curved Trophon.

LATTICED TROPHON Pl. 52
Boreotrophon craticulatum (Fabr.)
   Range: Greenland to Nova Scotia.
   Habitat: Deep water.
   Description: Height 1¼ in. Shell rather light in substance, with 6 fairly well shouldered whorls. Spire elevated, apex sharp. Sculptured with rather widely spaced thin and sharp varices. Color grayish white.
   Remarks: Formerly listed as *Trophon fabricii* Möller.

GUNNER'S TROPHON Pl. 52
Boreotrophon gunneri (Lovén)
   Range: Greenland to Massachusetts.
   Habitat: Shallow to deep water.
   Description: About 1 in. high. 5 or 6 squarely shouldered whorls, decorated with sharp bladelike varices. Apex sharply pointed. Aperture fairly large, canal long and open. Color chalky white.

STAIRCASE TROPHON Pl. 52
Boreotrophon scalariformis (Gould)
   Range: Iceland to Massachusetts.
   Habitat: Deep water.
   Description: Height 1 to 2 in. Shell quite solid, consisting of 7 or 8 convex whorls, moderately shouldered. Surface marked at close intervals with compressed whitish ribs, the edges sharp when young but smooth in old specimens. Faint revolving lines between ribs. Aperture oval, canal fairly long, open, and slightly curved. Color grayish brown.
   Remarks: Probably the largest member of this genus (formerly *Trophon* Montfort 1810) living on our coast. Some au-

thorities regard it as a subspecies of Clathrate Trophon, *B. clathratum* (above).

Coral Snails: Family Magilidae

RATHER small but thick and solid shells, living on corals and on stony bottoms. Shells may be spiny or relatively smooth, with close lines of growth. The aperture is generally pinkish or purplish.

Genus *Coralliophila* H. & A. Adams 1853

CORAL SNAIL   *Coralliophila abbreviata* (Lam.)   **Pl. 52**
 Range: S. Florida to West Indies and south to Brazil.
 Habitat: At bases of sea fans.
 Description: About 1 in. high, sometimes a little more. A solid shell of 5 or 6 shouldered whorls, the last making up most of shell. Sculpture of weak vertical folds and closely spaced encircling threadlike lines. Aperture flaring, outer lip simple, and inner lip rolled over to form a distinct umbilicus. Color grayish white; aperture pinkish or violet.

GLOBULAR CORAL SNAIL   **Pl. 52**
Coralliophila aberrans (C. B. Adams)
 Range: West Indies.
 Habitat: Shallow water.
 Description: Height ½ in. About 4 whorls, slightly shouldered. Moderate spire. Strong nodes at shoulders. Sculpture of revolving ridges. Aperture pear-shaped. Color white; aperture purple.

CARIBBEAN CORAL SNAIL   **Pl. 52**
Coralliophila caribaea Abbott
 Range: S. Florida to West Indies.
 Habitat: Shallow water.
 Description: Height ¾ in. 5 whorls, angled at shoulder, so spire appears turreted. Apex pointed. Sculpture of slanting vertical folds and weak revolving lines. Color whitish; aperture purplish.

Rock Shells or Dogwinkles: Family Thaididae

STOUT shells with an enlarged body whorl, wide aperture, and usually a short spire. These are predatory mollusks living close to shore in rocky situations. The popular name dye shells derives from the fact that the body secretes a colored fluid that may be green, scarlet, or purple.

Genus *Drupa* Röding 1798

BLACKBERRY SNAIL Pl. 53
Drupa nodulosa (C. B. Adams)
  Range: S. Florida to West Indies.
  Habitat: Shallow water; under stones.
  Description: A bumpy little shell about ½ in. high. Shell strong and solid, moderately high spired, with 5 or 6 whorls. Surface sculpture with spiral rows of prominent nodules, about 5 rows on body whorl and 3 on each of the early volutions. Aperture long and rather narrow, rather constricted by teeth on both lips. Color grayish green, the nodules shiny black.

Genus *Purpura* Bruguière 1789

WIDE-MOUTHED ROCK SHELL Pls. 7, 53
Purpura patula (Linn.)
  Range: Florida to West Indies.
  Habitat: Intertidal.
  Description: Attains height of about 3 in. Rough and solid, body whorl greatly enlarged and makes up most of shell. Surface bears revolving lines and numerous nodules, very pronounced in partly grown specimens but often worn and indistinct in old individuals. Aperture very large, outer lip thin and sharp, a broad polished area on inner lip. Small horny operculum, too small to close the aperture. Color grayish green or brown; interior often salmon-pink.

Genus *Thais* Röding 1798

DELTOID ROCK SHELL *Thais deltoidea* (Lam.) Pl. 53
  Range: Bermuda; Florida to West Indies.
  Habitat: Intertidal.
  Description: About 1½ in. high. A squat shell, 3 or 4 whorls,

sutures quite indistinct. 3 series of nodules on body whorl, those on shoulders forming pronounced knobs. Aperture large, lip simple and thin. Columella with fold at lower part. Grayish or pinkish white, blotched with brown and purple, typically in form of broad bands encircling shell.

Remarks: Usually common on rocks at low tide, although its shell is often so covered with encrustations that it matches the rock and therefore is not easily seen.

FLORIDA ROCK SHELL Pl. 53
Thais haemastoma floridana (Conrad)

Range: N. Carolina to West Indies, Gulf of Mexico, and south to Brazil.

Habitat: Intertidal.

Description: About 2 in. high. Strong and solid, large body whorl, pointed apex. About 5 whorls, sutures indistinct. Shoulders sloping, surface decorated with sharp revolving lines and a double row of tubercles on last volution. Outer lip thick, crenulate within, short canal. Columella folded at base to form polished area. Color grayish, somewhat blotched with brown; aperture pinkish flesh color.

Remarks: Typical *T. haemastoma* (Lam.) is a Mediterranean snail.

HAYS' ROCK SHELL Pl. 53
Thais haemastoma haysae Clench

Range: W. Florida to Texas.

Habitat: Shallow water.

Description: Nearly 5 in. high. This snail is considerably larger than Florida Rock Shell, with a taller spire and sharper revolving lines. Sutures rather deeply channeled. Color grayish, clouded with brown; aperture pinkish flesh color.

Remarks: This subspecies is very destructive to oyster beds in the Gulf of Mexico.

DOGWINKLE *Thais lapillus* (Linn.) Pls. 1, 53

Range: Labrador to Rhode Island; Europe.

Habitat: Intertidal.

Description: Height 1 to 1½ in. Shell thick and solid, with 5 whorls, a short spire, and bluntly pointed apex. Surface sculptured with deep revolving ridges and furrows, as well as many transverse wrinkles. Aperture oval, lip arched, operculum horny. Very short, open canal. Variable color ranges from white to lemon-yellow and purplish brown. Some individuals are banded with white or yellow.

Remarks: There is considerable variation in the shells aside from color: some are more angular than others, some larger, and some stouter. In a series of several-dozen shells it is often possible to find extremes that if studied alone would almost seem to represent different species.

RUSTIC ROCK SHELL *Thais rustica* (Lam.) **Pl. 53**
Range: Bermuda; Florida to West Indies.
Habitat: Intertidal.
Description: Height 1½ in. About 5 whorls with sloping shoulders, a sharply pointed apex, and a long and wide aperture. Sculpture of revolving riblets and vertical lines, plus a row of nodes at shoulders and a 2nd row of weaker nodes midway on body whorl. Color gray or brown, sometimes banded with white; aperture whitish.
Remarks: Formerly listed as *T. undata* (Lam.) but that name rightfully belongs to an Indo-Pacific species.

Genus *Ocenebra* Gray 1847

FRILLY DWARF ROCK SHELL **Pl. 52**
Ocenebra intermedia (C. B. Adams)
Range: West Indies.
Habitat: Moderately shallow water.
Description: Height 1 in. 5 or 6 whorls, elongate spire. Sculpture of robust revolving ridges, and strong frilly varices. Aperture small, outer lip thickened. Canal short and closed. Color white, occasionally banded with pale brown.

Dove Shells:
Family Columbellidae

SMALL fusiform shells, with an outer lip commonly thickened in the middle. The inner lip has a small tubercle at its lower end. The shells are sometimes quite colorful, and often shiny. Chiefly inhabitants of warm seas.

Genus *Columbella* Lamarck 1799

MOTTLED DOVE SHELL **Pl. 54**
Columbella mercatoria (Linn.)
Range: Florida to West Indies.
Habitat: Shallow water.
Description: Height ½ in., sometimes slightly more. Sturdy, solid, 5 or 6 whorls, short and blunt spire. Surface bears numerous revolving grooves. Aperture long and narrow, outer lip thickened, particularly in the middle, and strongly toothed

within. Inner lip with series of fine teeth on lower part. Tiny horny operculum. Color very variable, usually gray, clouded and spotted with purplish brown, rusty red, and white, often in the form of encircling bars.

SPOTTED DOVE SHELL Pl. 54
Columbella rusticoides (Heilprin)
Range: W. Florida.
Habitat: Shallow water.
Description: Height about ¾ in. Shell oval, 6 or 7 whorls, well-elevated spire with pointed apex. Surface smooth, often polished. Aperture long and narrow, outer lip thickened at center, toothed within. But few teeth on lower portion of inner lip. Color yellowish or whitish, heavily marked with chestnut-brown.

Genus *Pyrene* Röding 1798

OVATE DOVE SHELL *Pyrene ovulata* Lam. Pl. 54
Range: West Indies.
Habitat: Moderately shallow water.
Description: Slightly more than ½ in. high, shape oval, 5 or 6 whorls, blunt apex. Surface smooth, usually shiny. Aperture long and narrow, feebly toothed within. A few weak spiral lines at base of shell. Color reddish brown, with whitish scrawls.

Genus *Anachis* H. & A. Adams 1853

GREEDY DOVE SHELL *Anachis avara* (Say) Pl. 54
Range: New Jersey to Florida.
Habitat: Shallow water.
Description: Less than ½ in. high. Shell thick and strong, spire elevated and acute. 6 or 7 whorls, sculptured with impressed spiral lines and prominent vertical folds. The folds usually do not extend beyond middle of last whorl, leaving only revolving lines on that portion of shell. Outer lip feebly toothed within. Color brownish yellow.
Remarks: See Well-ribbed Dove Shell (below) for similarity.

CHAIN DOVE SHELL *Anachis catenata* (Sby.) Pl. 54
Range: Bermuda, West Indies.
Habitat: Shallow water.
Description: Height ¼ in. About 5 whorls, sutures plainly marked. Apex rather blunt. Sculpture of well-developed vertical ribs with faint encircling lines. Color yellowish white, with scattered brown dots.

HALIAECT'S DOVE SHELL   Pl. 54
Anachis haliaecti (Jeffreys)
  Range: Maine to N. Carolina.
  Habitat: Moderately deep water.
  Description: About ½ in. high. Moderately stout, with 5 or 6 whorls, sutures well defined. Sculpture consists of rather prominent and regularly spaced vertical folds. Aperture long, somewhat narrow. Color yellowish brown.
  Remarks: Formerly listed as *A. costulata* (Sars).

FAT DOVE SHELL   *Anachis obesa* (C. B. Adams)   Pl. 54
  Range: Virginia to West Indies.
  Habitat: Shallow water.
  Description: Height about ¼ in. A stocky shell of 5 whorls, sutures not very distinct. Numerous vertical folds and fine revolving lines. Outer lip with a few small teeth on inner margin. Color yellowish gray.

BEAUTIFUL DOVE SHELL   Pl. 54
Anachis pulchella (Sby.)
  Range: S. Florida to West Indies.
  Habitat: Shallow water.
  Description: Nearly ½ in. high. 6 whorls, sutures indistinct. Fusiform (spindle-shaped) in outline, pointed apex. Sculpture of weak vertical ribs and revolving lines. Aperture elongate. Color yellowish white, with reddish-brown scrawls.

SPRINKLED DOVE SHELL   Pl. 54
Anachis sparsa (Reeve)
  Range: West Indies.
  Habitat: Shallow water.
  Description: Nearly ½ in. high. A fusiform shell of about 6 whorls. Sculpture of relatively large vertical ribs. Sutures indistinct. Aperture small, outer lip toothed within. Color yellowish white, with vertical splashes of orange-brown. Surface glossy.

WELL-RIBBED DOVE SHELL   Pl. 54
Anachis translirata (Rav.)
  Range: Massachusetts to Florida.
  Habitat: Shallow water.
  Description: About ½ in. high. This species closely resembles Greedy Dove Shell (above), and was once considered a subspecies. It is slightly stouter, and the vertical folds are more numerous, occasionally running the full height of each volution. Color brown, sometimes weakly banded with darker.

Genus *Nitidella* Swainson 1840

TWO-COLORED DOVE SHELL   Pl. 54
Nitidella dichroa (Sby.)
  Range: West Indies.
  Habitat: Shallow water; under stones.
  Description: Height ¼ in. About 6 whorls, sutures plainly
  marked. Apex pointed. Aperture elongate, lip thickened.
  Surface smooth and shiny. Color whitish, with vertical
  streaks of brown.

SMOOTH DOVE SHELL   Pl. 54
Nitidella laevigata (Linn.)
  Range: Florida Keys to West Indies.
  Habitat: Shallow water.
  Description: About ½ in. high. 4 or 5 whorls, sutures well
  impressed. Apex pointed, aperture rather wide. Surface
  smooth and glossy as a rule. Color variable; usually yellowish
  orange, with vertical white streaks and a broken band of
  brown at middle of body whorl.

GLOSSY DOVE SHELL   *Nitidella nitida* (Lam.)   Pl. 54
  Range: S. Florida to West Indies.
  Habitat: Shallow water.
  Description: Height ½ in. A fusiform (spindle-shaped) shell
  of about 5 whorls, sutures fairly distinct. Aperture long and
  narrow, lip smooth. Surface of shell polished. Color pale yel-
  low, heavily mottled with chestnut-brown.
  Remarks: Formerly listed as *N. nitidula* (Sby.).

WHITE-SPOTTED DOVE SHELL   Pl. 54
Nitidella ocellata (Gmelin)
  Range: Bermuda; s. Florida to West Indies.
  Habitat: Shallow water.
  Description: Nearly ½ in. high. A spindle-shaped shell of
  about 7 sloping whorls, no shoulders. Aperture long and narrow, outer lip thickened and feebly toothed within. Surface
  smooth and shining. Color yellowish tan to nearly black,
  round white spots variable. In life there is a dark brown periostracum.
  Remarks: Formerly listed as *N. cribraria* (Lam.).

Genus *Mitrella* Risso 1826

LITTLE WHITE MITRELLA   Pl. 54
Mitrella albella (C. B. Adams)

Range: S. Florida.
Habitat: Moderately shallow water.
Description: Height ⅕ in. A small glossy shell of 5 or 6 whorls, sutures quite distinct. Spire elevated, apex pointed. Aperture small and narrow, lip thin and sharp. Sculpture of obscure vertical folds. Color whitish, with darker markings.

CRESCENT MITRELLA *Mitrella lunata* (Say) **Pl. 54**
Range: Massachusetts to Florida and Texas; West Indies to Brazil.
Habitat: Shallow water.
Description: Height ⅕ in. Shell rather stoutly fusiform, 6 whorls separated by shallow sutures. Surface smooth except for a single revolving line below the sutures and a few at base. Aperture oval and narrow, outer lip simple, a few teeth on inner margin. Color yellowish tan, with reddish-brown crescent-shaped markings.

BANDED MITRELLA *Mitrella zonalis* (Gould) **Pl. 54**
Range: Maine to Connecticut.
Habitat: Moderately shallow water.
Description: Height ⅕ in. This snail is about the same size as *M. lunata,* and also about the same shape. Color reddish brown, with no evidence of spotting; surface rather well polished.

Genus *Parastola* Rehder 1943

MANY-SPOTTED DOVE SHELL **Pl. 54**
Parastola monilifera (Sby.)
Range: Florida to West Indies.
Habitat: Shallow water.
Description: About ¼ in. high. About 6 whorls, sutures faint. Shell elongate, apex rather blunt. Aperture small, outer lip toothed within. Sculpture of strong revolving ridges and vertical folds. Color whitish, with broken bands of brown.

Dog Whelks: Family Nassariidae

SMALL carnivorous snails, present in all seas. The shells are rather stocky, with pointed spires. The aperture is strongly notched at both ends, and commonly there is a marked columellar callus (calcareous deposit). Mostly inhabitants of shallow water.

Genus *Nassarius* Duméril 1806

VARIABLE DOG WHELK *Nassarius albus* (Say) **Pl. 54**
Range: N. Carolina to West Indies.
Habitat: Shallow water.
Description: Height ½ in. Shell stout and well elevated, apex sharp. About 5 whorls, well shouldered and decorated with vertical ribs (suture to suture) crossed by ridges that vary in size. Aperture small, outer lip thickened, inner lip twisted at base. Variable in color but predominantly white, often spotted with brown.
Remarks: Formerly listed as *N. ambiguus* (Pulteney).

MUD DOG WHELK *Nassarius obsoletus* (Say) **Pl. 54**
Range: Gulf of St. Lawrence to Florida.
Habitat: Mudflats.
Description: Nearly 1 in. high, with 6 whorls and a moderately elevated spire. Apex blunt. Surface marked with numerous unequal revolving lines crossed by minute growth lines and rather oblique folds, especially on early volutions. Aperture oval, outer lip thin and sharp, inner lip strongly arched, with fold at front. Canal a mere notch. Color dark reddish purple to almost black.
Remarks: This dark and unattractive mollusk is one of the most abundant univalves to be found along the Atlantic Coast. It is a scavenger, and a dead fish or a crushed clam thrown into shallow water will quickly attract hundreds of individuals. Adult specimens very commonly have the apex somewhat corroded.

NEW ENGLAND DOG WHELK **Pl. 54**
Nassarius trivittatus (Say)
Range: Nova Scotia to S. Carolina.
Habitat: Shallow water.
Description: About ¾ in. high. A robust shell with an acute apex. 6 or 7 whorls, each slightly flattened at the shoulder. A series of spiral grooves cuts across a series of beaded lines, giving surface a pimpled appearance. Outer lip sharp and scalloped by the revolving lines. Inner lip strongly arched, aperture notched at both ends. Color white or yellowish white, sometimes banded with brown.

MOTTLED DOG WHELK *Nassarius vibex* (Say) **Pl. 54**
Range: Massachusetts to West Indies.
Habitat: Shallow water.
Description: About ½ in. high. Shell short and heavy, with about 5 whorls. Sutures well defined, apex pointed. Surface

bears strong vertical folds crossed by indistinct revolving lines. Aperture notched at both ends, outer lip thick, and there is a heavy patch of enamel on inner lip. Color white, variously mottled and marked with brown and gray.

Remarks: This is a variable species, and individuals from different beaches are apt to show different color patterns. Very abundant in the south, uncommon north of Cape Hatteras.

Whelks: Family Buccinidae

SHELLS generally large, with convex whorls. The aperture is large and usually notched below. These carnivorous snails occur in northern seas all around the world. Certain species are used for food in some countries.

Genus *Bailya* M. Smith 1944

INTRICATE BAILY SHELL   **Pl. 55**
Bailya intricata (Dall)
    Range: S. Florida to West Indies.
    Habitat: Intertidal; under stones.
    Description: Height ½ in. About 6 shouldered whorls, pointed apex, elongate aperture, weakly notched above. Outer lip thickened. Sculpture of strong revolving lines and weaker vertical ribs. Color grayish white.

LITTLE BAILY SHELL   **Pl. 55**
Bailya parva (C. B. Adams)
    Range: West Indies.
    Habitat: Shallow water.
    Description: Height ¾ in. About 6 whorls, no shoulders. Sutures well indented. Sculpture of strong vertical ribs and weaker revolving lines. Aperture notched above. Color yellowish white, with orange-brown bands.

Genus *Buccinum* Linnaeus 1767

DEEP-SEA WHELK   *Buccinum abyssorum* Verrill   **Pl. 55**
    Range: Off the Carolinas.
    Habitat: Deep water.
    Description: About 2 in. high. Shell moderately thin, with acute spire and 7 or 8 strongly carinated whorls. Volutions angulated by sharp revolving carinae (ridges), of which 3 usually prominent on whorls of the spire; the upper ridge forms a pronounced shoulder. Aperture ovate, outer lip thin, canal

short and nearly straight. Inner lip bears small patch of enamel. Dull yellowish white; periostracum inconspicuous or lacking.

BLUISH WHELK *Buccinum cyaneum* Brug. **Pl. 55**
Range: Labrador to Massachusetts.
Habitat: Moderately deep water.
Description: About 2½ in. high. 6 rounded whorls, sutures well indented. Aperture moderately large, apex sharply pointed. Inner lip reflected on body whorl, canal notched. Surface quite smooth, with no vertical folds and only weak spiral lines. Dull bluish white.
Remarks: Formerly listed as *B. groenlandicum* Hancock.

DONOVAN'S WHELK *Buccinum donovani* Gray **Pl. 55**
Range: Arctic Ocean to Nova Scotia.
Habitat: Moderately deep water.
Description: About 2½ in. high. 7 rounded whorls, sutures well marked. Spire fairly tall, apex pointed. Early volutions show weak vertical folds, but they fade rapidly, and most of the shell is marked only by revolving lines. Aperture nearly round, lip moderately thickened. Color yellowish brown.

SILKY WHELK *Buccinum tenue* Gray **Pl. 55**
Range: Labrador to Maine.
Habitat: Deep water.
Description: About 2½ in. high. 6 or 7 convex whorls, sutures deep. Surface decorated with numerous vertical folds, with very fine, slightly beaded revolving lines that impart a silky appearance. Aperture nearly round, operculum horny. Outer lip simple, canal little more than a notch. Color yellowish brown.

TOTTEN'S WHELK *Buccinum totteni* Stimpson **Pl. 55**
Range: Labrador to Maine.
Habitat: Moderately deep water.
Description: About 2 in. high. A rather strong shell with 6 rounded whorls, sutures distinct. Sculptured with numerous closely spaced striations that are deeply cut, but the surface appears relatively smooth. Color yellowish brown, with thin yellowish periostracum. Aperture straw color within.

WAVED WHELK *Buccinum undatum* Linn. **Pl. 55**
Range: Arctic Ocean to New Jersey; Europe.
Habitat: Moderately shallow water.
Description: Height 3 to 4 in. Moderately solid, regularly convex, sharp apex. About 6 whorls, decorated with distinct revolving lines, and wavy vertical folds. Aperture large, lip sharp, canal notched. Operculum horny. Color pale reddish

or yellowish brown. Aperture yellow in fresh specimens.
Remarks: This is the edible whelk of Scotland and Ireland.

Genus *Colus** Röding 1798

CARVED WHELK *Colus caelatula* (Verrill) **Pl. 56**
Range: Labrador to N. Carolina.
Habitat: Deep water.
Description: Height 1½ in. Fusiform, sturdy, about 7 whorls, blunt apex. Well impressed sutures. Surface with vertical ribs, about 12 to a volution. Outer lip thin and sharp, canal moderate and turned slightly to left. Inner lip strongly bent and spirally twisted. Color pinkish gray.

SPITEFUL WHELK *Colus livida* (Mörch) **Pl. 56**
Range: Labrador to Nova Scotia.
Habitat: Deep water.
Description: Height 2 to 4 in. Shell gracefully elongate, composed of 5 or 6 whorls. Sculpture of revolving ribs, unequal in size. Aperture long and oval, and, with the canal, equals about ½ of the entire shell. Outer lip thin and somewhat crenulate, canal open and quite long. Color yellowish gray or pale brown.

OBESE WHELK *Colus obesa* (Verrill) **Pl. 56**
Range: Nova Scotia to S. Carolina.
Habitat: Moderately deep water.
Description: About 1 in. high. Obesely fusiform, 4 or 5 whorls, sculptured with numerous spiral lines and a few strong vertical ribs. Outer lip thin, inner lip excavated at center. Canal moderate, open. Color pinkish gray, with grayish periostracum that bears slender hairs along the spiral lines.

HAIRY WHELK *Colus pubescens* (Verrill) **Pl. 56**
Range: Newfoundland to Cape Hatteras.
Habitat: Moderately deep water.
Description: About 2½ in. high. 6 rounded whorls, sutures distinct. Aperture oval, canal moderate and open. Outer lip thin and sharp, inner lip arched at center. Obscure revolving lines, under velvety periostracum that is sometimes hairy during life. Color yellowish brown.

PYGMY WHELK *Colus pygmaea* (Gould) **Pl. 56**
Range: Gulf of St. Lawrence to N. Carolina.
Habitat: Moderately deep water.

*Generally *Colus* has been considered masculine in conchology, evidently in error, according to Cassell's and Harper's Latin dictionaries; it is a sixth-declension noun and therefore feminine.

Description: About 1 in. high. Shell fusiform (spindle-shaped), with 7 or 8 whorls. Surface marked with numerous revolving lines, but these can be seen only after the periostracum has been removed. Aperture oval, canal rather short. Operculum horny. Color grayish; periostracum greenish gray.

Remarks: This snail occurs with Stimpson's Whelk (below), but it may be distinguished from young specimens of the latter by the number of volutions when compared with a shell of the same size.

SPITSBERGEN WHELK   Pl. 56
Colus spitzbergensis (Reeve)
   Range: Arctic Ocean to Nova Scotia; Europe.
   Habitat: Deep water.
   Description: Height nearly 4 in. 7 or 8 well-rounded whorls, sutures deep. Apex sharp, canal short. Sculpture of prominent revolving ribs, about 20 to a volution. Aperture roundish, lip thin and sharp. Color yellowish gray.

STIMPSON'S WHELK   *Colus stimpsoni* (Mörch)   Pl. 56
   Range: Labrador to Cape Hatteras.
   Habitat: Deep water.
   Description: 3 to 4 in. high. Spindle-shaped (fusiform) and elongate, with 7 or 8 rather flat whorls, sutures indistinct. Surface bears weak revolving lines beneath the tough periostracum. Aperture oval, white within, canal moderately long, open, and inclined to turn backward. Color bluish white; velvety periostracum dark greenish brown.
   Remarks: See Remarks under Pygmy Whelk (above).

STOUT WHELK   *Colus ventricosa* (Gray)   Pl. 56
   Range: Labrador to Maine.
   Habitat: Deep water.
   Description: Height 2 in. Stout, well inflated, relatively short spire. About 5 swollen whorls, sutures distinct. Aperture wide, terminating in short open canal. Operculum horny. Faint revolving lines decorate the surface, with polished white area on reflected inner lip. Color greenish yellow, with velvety brownish periostracum.

Genus *Beringius* Dall 1879

OSSIAN'S WHELK   *Beringius ossiani* (Friele)   Pl. 57
   Range: Off Nova Scotia; Europe.
   Habitat: Deep water.
   Description: Height 3 in. A fine large shell of 6 or 7 well-

rounded whorls. Aperture large and oval, canal short. Inner lip reflected on body whorl, and polished white. Surface with revolving riblets, numerous and closely spaced. Color yellowish brown, with thin brownish periostracum.

Remarks: This gastropod will be found in some books under the generic name of *Jumula* Friele 1882. However, "Jumula" is the Finnish word for God, and the International Committee on Zoological Nomenclature in recent years bars names that might give offense on religious or political grounds, so it seems best to adopt Dall's name, *Beringius,* for this group.

Genus *Volutopsius* Mörch 1857

LARGILLIERT'S WHELK   Pl. 57
Volutopsius largillierti (Petit)
   Range: Labrador to Massachusetts.
   Habitat: Deep water.
   Description: Height 4 in. About 6 convex whorls, well-indented sutures. Surface bears weak revolving lines. Aperture large, outer lip somewhat flaring, inner lip reflected back against body whorl. Operculum horny. Color yellowish white, with heavy greenish-brown periostracum. Aperture white.

Genus *Plicifusus* Dall 1902

KROYER'S WHELK   *Plicifusus kroyeri* (Möller)   Pl. 57
   Range: Arctic Ocean to Newfoundland; Europe.
   Habitat: Deep water.
   Description: Height 2 in. 6 whorls, sutures distinct. Early volutions sculptured with vertical ribs, the major portion of shell smooth. Aperture large, outer lip thin and sharp. Inner lip partially reflected, canal open and short. Operculum horny. Color yellowish white, periostracum brownish.

Genus *Neptunea* Röding 1798

TEN-RIDGED WHELK   Pl. 1
Neptunea decemcostata (Say)
   Range: Nova Scotia to Massachusetts.
   Habitat: Moderately shallow water.
   Description: About 3 to 4 in. high. Shell large and stout, composed of 6 or 7 whorls, spirally ribbed with raised keels, or ridges. 10 of these on body whorl, 3 on upper whorls. Short open canal, operculum horny. Color yellowish gray,

ridges reddish brown. Aperture white within, the darker ridges faintly showing through.

Remarks: Quite colorful for a mollusk of northern seas. Fishermen often bring them up in their nets and the snail frequently manages to get into lobster traps, but good specimens are not commonly found on the shore.

UNFINISHED WHELK Pl. 57
Neptunea despecta tornata (Gould)
Range: Labrador to Massachusetts.
Habitat: Moderately deep water.
Description: Height 4 in. A sturdy shell of about 8 convex whorls with sloping shoulders, producing a rather high, turreted spire. Volutions encircled with ridges of 2 sizes. On body whorl there are 5 large and 3 small, uncolored ridges; these ridges not nearly so robust as those of Ten-ridged Whelk. Aperture oval, inner lip reflected on body whorl, canal open and short, and strongly curved. Color buffy yellow, aperture pale yellow.
Remarks: Typical *N. despecta* (Linn.) is a European species.

Genus *Antillophos* Woodring 1928

CANDÉ'S PHOS *Antillophos candei* (Orb.) Pl. 57
Range: N. Carolina to West Indies.
Habitat: Moderately deep water.
Description: About 1 in. high. 8 or 9 whorls, an elevated spire, and a sharply pointed apex. Sculpture of sharp vertical ribs, about 15 to a volution, and these are crossed by distinct revolving lines, so the surface has a beaded appearance. Aperture oval and quite long, outer lip somewhat thickened, inner lip with a few folds at lower end. Operculum small and clawlike, horny. Color pale brownish yellow, sometimes almost white; occasionally lightly banded with pale orange.
Remarks: Formerly in the genus *Phos* Montfort 1810.

Genus *Engoniophos* Woodring 1928

GUADELOUPE PHOS Pl. 57
Engoniophos guadelupensis (Petit)
Range: West Indies.
Habitat: Shallow water.
Description: Height 1 in. About 6 shouldered whorls, spire moderately turreted, apex sharp. Sculpture of strong vertical ribs, about 10 on last volution, and distinct revolving lines. Color whitish.
Remarks: Formerly in the genus *Phos* Montfort 1832.

Genus *Engina* Gray 1839

WHITE-SPOTTED ENGINA   **Pl. 57**
Engina turbinella (Kiener)
  Range: S. Florida to West Indies.
  Habitat: Moderately shallow water; reefs.
  Description: About ½ in. high. A stoutly fusiform (spindle-shaped) shell of 6 whorls. Aperture elongate, lip somewhat thickened and toothed within. Columella with slight twist. A number of white knobs encircle each volution; rest of shell is purplish brown.

Genus *Pisania* Bivona 1832

PISA SHELL *Pisania pusio* (Linn.)   **Pl. 57**
  Range: S. Florida to West Indies.
  Habitat: Moderately shallow water; reefs.
  Description: About 1½ in. high. 5 or 6 sturdy and strong whorls, sutures indistinct. Well-developed spire, sharp apex. Aperture oval, outer lip toothed within, inner lip with prominent tooth at upper angle. Canal short and straight. Surface smooth and polished, purplish brown with bands of irregular dark and light spots often resembling chevrons.

Genus *Colubraria* Schumacher 1817

ARROW TRITON *Colubraria lanceolata* Menke   **Pl. 57**
  Range: N. Carolina to West Indies.
  Habitat: Offshore rocky bottoms.
  Description: About 1 in. high. A slender, elongate shell of 5 or 6 whorls, each bearing many finely cut vertical lines, and 2 distinct riblike varices. Aperture moderately small and narrow, canal short. Inner lip sometimes forms bladelike ridge. Color pale brown or yellowish buff, with scattered spots of orange-brown.

LEANING DWARF TRITON   **Pl. 57**
Colubraria obscura Reeve
  Range: Florida Keys to West Indies.
  Habitat: Moderately shallow water.
  Description: Height 1½ in. About 7 whorls, sutures faint. Aperture narrow, parietal wall reflected. Sculpture of weak revolving and vertical lines. Outer lip thickened; previous lips leave prominent varices at irregular intervals. Color pale brown, with marks of darker orange-brown, the varices decorated with vivid brown and white.

SWIFT'S DWARF TRITON Pl. 57
Colubraria swifti Tryon
Range: Bermuda, West Indies.
Habitat: Shallow water.
Description: Height ¾ in. A slender shell of 6 or 7 whorls, sutures faintly impressed. No varices on spire. Sculpture of distinct revolving lines and weak vertical ridges. Parietal wall reflected. Color yellowish gray, blotched with brown.

Genus *Cantharus* Röding 1798

GAUDY SPINDLE *Cantharus auritula* (Link) Pl. 57
Range: S. Florida to West Indies.
Habitat: Shallow water.
Description: About 1 in. high, stout and solid. 5 or 6 whorls decorated with strong vertical folds and weaker revolving lines. Whorls somewhat shouldered and knobby. Outer lip thickened, canal short. 2 teeth at upper angle of aperture. Color mottled gray, brown, and black.

CROSS-BARRED SPINDLE Pl. 57
Cantharus cancellaria (Conrad)
Range: Florida to Texas.
Habitat: Shallow water.
Description: Height 1¼ in. Shell spindle-shaped, elongate, with 5 or 6 whorls. Sculpture of rather indistinct vertical folds crossed by wavy encircling lines. Aperture elongate-oval, outer lip thickened, inner lip with tooth at upper angle, pleat at bottom of columella. Color reddish brown, somewhat mottled with white.

MOTTLED SPINDLE *Cantharus tinctus* (Conrad) Pl. 57
Range: N. Carolina to West Indies.
Habitat: Shallow water.
Description: About 1 in. high. Shell solid and fusiform, with 5 or 6 whorls. Shoulders slightly constricted to form small nodules. Ornamentation of low vertical folds crossed by revolving ridges. Aperture oval, outer lip thickened, inner lip with prominent tooth above. Canal very short. Color reddish brown, mottled with white.

Crown Conchs: Family Melongenidae

MODERATELY large and solid shells, living from shoreline to moderate depths. They are carnivorous and predacious, and occur in tropical and temperate seas.

Genus *Melongena* Schumacher 1817

CROWN CONCH *Melongena corona* (Gmelin) **Pls. 5, 58**
Range: Florida to Mexico.
Habitat: Shallow water.
Description: About 2 to 5 in. high. Shell roughly pear-shaped, with short spire and large inflated body whorl. 5 or 6 whorls, the last few bearing a single or double row of short but sharp spines. An additional row or two of blunter spines encircles base of shell, or may be lacking. Aperture oval and wide, outer lip simple, with notch at base, columella twisted. Operculum clawlike, horny. Color dark brown to black, spirally banded with bluish white or yellow, in various shades and irregular arrangements.
Remarks: Several subspecies have been named, such as the High-spired Crown Conch, *M. c. altispira* Pilsbry & Vanatta, shown on Plate 58. The separations are based on the shell's differences in height, spine size, and general sturdiness, but since it is possible to find all gradations, most authorities do not believe that the rank of subspecies is valid and prefer to call them forms or varieties.

BROWN CROWN CONCH **Pl. 58**
Melongena melongena (Linn.)
Range: West Indies.
Habitat: Shallow water.
Description: Height 4 or 5 in. About 6 whorls; body whorl makes up of shell and folds over the short spire, forming a groove at top of shoulders. Vertical ribs on spire but not on last whorl. There may be 2 or 3 rows of sharp spines on body whorl, or it may be smooth. Aperture oval and wide. Color rich brown, with yellowish or white bands.
Remarks: This is a West Indian shell but it has been reported from the Florida Keys in error.

Genus *Busycon* Röding 1798

CHANNELED WHELK **Pl. 58**
Busycon canaliculatum (Linn.)
Range: Massachusetts to e. Florida.
Habitat: Shallow water.
Description: About 4 to 7 in. high. A large pear-shaped shell with 5 or 6 turreted whorls. Body whorl very large above, gradually diminishing downward and terminating in a long, nearly straight canal. A broad and deep channel at the suture forms a winding terrace up the spire. Shoulders bear

weak knobs. Sculpture of fine revolving lines. Color buffy gray; aperture yellowish. In life there is a yellowish-brown periostracum bristling with stiff hairs.
Remarks: Indian wampum was made from the twisted columellas of this and the next species, cut into elongate beads.

KNOBBED WHELK *Busycon carica* (Gmelin) **Pl. 58**
Range: Massachusetts to Florida.
Habitat: Shallow water.
Description: Height 4 to 9 in. Shell thick and solid, pear-shaped (see Lightning Whelk, below). About 6 whorls, body whorl large and broad, crowned by a series of blunt spines or nodes. Spire low, with series of nodules encircling the shoulders of each volution. Aperture long and oval, canal long and open. Operculum horny. Color yellowish gray; interior orange-red. Young specimens often streaked with violet.
Remarks: The largest gastropod found north of Cape Hatteras. Most visitors at the seashore have noted the egg ribbons of the mollusks of this genus. The strings of curiously flattened, disklike capsules (see Plate 58) may be picked up alongshore during the summer months and are popularly called Venus necklaces. See Remarks under preceding species.

TURNIP WHELK *Busycon coarctatum* (Sby). **Pl. 6**
Range: Mexico.
Habitat: Moderately shallow water.
Description: About 6 in. high. Rather solid, a short sloping spire, a somewhat bulbous body whorl, and a long open canal. Whorl count about 5. A few short spines may be present on shoulders. Buffy gray, decorated with vertical streaks of purplish brown. Aperture orange-yellow, ribbed within.
Remarks: This species was first described by Sowerby in 1825. A very similar shell, *B. rapum* (Heilprin), occurs as a fossil in the Pliocene of Florida, and *B. coarctatum* is believed to be a modern descendant of that ancient gastropod. Very few examples were known, none had been collected for about a century, and the species was generally regarded as being extinct. Since 1950 the shrimp fishermen, dredging in Campeche Bay, Mexico, have been bringing them up occasionally, but it is still considered a rare and coveted shell.

LIGHTNING WHELK **Pl. 5**
Busycon contrarium (Conrad)
Range: S. Carolina to Florida.
Habitat: Shallow water.
Description: Height 7 to 10 in., but 15-in. specimens are recorded. Shell thick and strong, shaped very much like the

Knobbed Whelk (above), except that the spiral turns to the left. In other words, this is a left-handed (sinistral) snail and the Knobbed Whelk is right-handed (dextral). Spire rather flatter, and whole shell more graceful. Fawn color, vertically streaked with violet-brown. Young specimens are brightly colored, but after the shell grows to 8 or 9 in. the colors fade, and the shell becomes dull grayish white.

PERVERSE WHELK *Busycon perversum* (Linn.)   **Pl. 58**
Range: N. Carolina to Florida.
Habitat: Moderately shallow water.
Description: Height 6 to 8 in. A heavy and rugged shell of 5 or 6 whorls. Body whorl very thick and swollen, particularly in region of canal. May be either left-handed or right-handed. Color grayish, somewhat streaked with violet-brown.
Remarks: The right-handed specimens were formerly regarded as a subspecies of Knobbed Whelk, *B. carica,* and were listed as *B. carica eliceans* (Montfort). The name *B. perversum* was applied to all the left-handed shells, and this thick and swollen form was listed as *B. perversum kieneri* (Phil.). Authorities now agree that the common left-handed shell is the Lightning Whelk, *B. contrarium,* and that *B. perversum* is the swollen shell that may be either dextral or sinistral.

FIG WHELK *Busycon spiratum* (Lam.)   **Pl. 58**
Range: N. Carolina to Florida and Gulf states.
Habitat: Shallow water.
Description: Height 3 to 5 in. 4 or 5 whorls, body whorl very large. Sutures wide and deeply channeled. Shell thin but sturdy, spire very short. Aperture wide and prolonged into a straight, open canal. Operculum horny. Surface sculptured with weak revolving lines. Flesh-colored, with reddish-brown streaks.
Remarks: Formerly listed as *B. pyrum* (Dill.).

Tulip Shells:
Family Fasciolariidae

LARGE snails, with strong, thick, fusiform (spindle-shaped) shells. The spire is elevated and sharply pointed, and there is no umbilicus. Inner lip usually decorated with a few oblique folds. These are predatory mollusks, slow and deliberate in their movements. Generally distributed in warm seas.

Genus *Leucozonia* Gray 1847

CHESTNUT LATIRUS *Leucozonia nassa* (Gmelin) **Pl. 60**
   Range: Florida to West Indies; Texas.
   Habitat: Shallow water.
   Description: About 2 in. high. Shell solid and fusiform, about 7 whorls, sutures quite distinct. Strong tubercles on body whorl form shoulders. Sculpture of revolving lines. Aperture oval, outer lip grooved within, operculum horny. Short open canal. Color varies from chestnut-brown to almost black, usually with a paler band near base.
   Remarks: Formerly listed as *L. cingulifera* (Lam.).

WHITE-SPOTTED LATIRUS **Pl. 60**
Leucozonia ocellata (Gmelin)
   Range: W. Florida to West Indies.
   Habitat: Shallow water.
   Description: About 1 in. high. A stoutly fusiform shell of some 5 whorls. Sculpture of revolving lines, with elongate knobs at shoulders. Canal short. Color dark brown, with a circle of whitish knobs at periphery and smaller white spots at base of shell. Nearly all mature shells have the apex white.

Genus *Latirus* Montfort 1810

SHORT-TAILED LATIRUS **Pl. 60**
Latirus brevicaudatus Reeve
   Range: Florida Keys to West Indies.
   Habitat: Shallow water.
   Description: About 1 to 2 in. high. 8 knobby whorls, sutures well defined. Apex pointed; aperture moderate, notched above, short open canal below. Base of columella with weak plications. Sculpture of rounded vertical knobs. Color light brown, with numerous dark brown encircling lines.

KEY WEST LATIRUS *Latirus cayohuesonicus* Sby. **Pl. 60**
   Range: Florida Keys and West Indies.
   Habitat: Moderately shallow water.
   Description: Height 1 in. A nondescript shell with 5 or 6 whorls, sutures indistinct. Sculptured with revolving ribs made somewhat wavy by a series of weak vertical folds. Color dull gray.
   Remarks: This species is now placed in the genus *Fusilatirus* McGinty 1955 by some authors.

DISTINCT LATIRUS   *Latirus distinctus* (Kobelt)   **Pl. 60**
Range: Florida Keys and West Indies.
Habitat: Moderately shallow water.
Description: Height 1½ in. A rugged shell of 6 whorls, sutures indistinct. Surface bears strong revolving ribs, which at intervals are raised into rather sharp knobs. Operculum horny. Inner lip rolled back over body whorl. Color brownish gray.

BROWN-LINED LATIRUS   **Pl. 6**
Latirus infundibulum (Gmelin)
Range: S. Florida to West Indies.
Habitat: Moderately shallow water.
Description: About 2 in. high. Shell solid and elongate, with about 7 whorls, a tall spire with blunt apex, sutures strongly marked. Each volution bears 6 vertical folds, and the surface is further decorated with prominent spiral ridges. The crests of the folds are commonly worn and shiny, obliterating the ridges at that point. Canal quite long and narrow, with 2 or 3 pleats on inner lip. Color grayish, the spiral ridges reddish.

McGINTY'S LATIRUS   *Latirus macgintyi* Pilsbry   **Pl. 60**
Range: S. Florida.
Habitat: Deep water.
Description: About 2½ in. high. Heavy and solid, with 8 or 9 whorls. Apex pointed, aperture narrow, funnel-shaped umbilicus. Sculpture of strongly rounded vertical ribs and coarse revolving ridges. Color yellowish, with dark brown stains between the ribs.

VIRGIN ISLAND LATIRUS   **Pl. 60**
Latirus virginensis Abbott
Range: Greater Antilles.
Habitat: Moderately shallow water.
Description: About 1 in. high. A solid shell of 6 or 7 whorls, sutures moderately impressed. Canal short and open, aperture notched above. Sculpture of rounded vertical knobs and fine revolving lines. Color yellowish brown, the knobs generally whitish.

Genus *Fasciolaria* Lamarck 1799

BANDED TULIP SHELL   **Pl. 5**
Fasciolaria hunteria Perry
Range: N. Carolina to Florida and Gulf states.
Habitat: Shallow water.
Description: Height about 3 in. Gracefully spindle-shaped, 7

or 8 whorls, sutures distinct. Surface smooth, often shiny, with only a few wrinkles at base. Aperture long and oval, canal short, operculum horny. Color bluish gray, with vertical cloudings of white and revolving lines of rich dark brown. **Remarks:** This species has long been listed as *F. distans* Lam. In fairly recent years a larger and more colorful (often orange) form with stronger ridges on the base was discovered at Campeche Bay in Mexico. A new name was proposed for this snail, *F. branhamae* Rehder & Abbott, but a restudy of Lamarck's types, still preserved in the Natural History Museum of Geneva, Switzerland, revealed the fact that his specimens actually came from Campeche Bay and clearly represented this larger form and not the common Florida shell at all. The name *distans,* then, could only be used for this Mexican snail, and our familiar Florida mollusk was left without a name. The earliest one available — *hunteria* — was published by Perry in 1811.

TULIP SHELL *Fasciolaria tulipa* (Linn.) **Pl. 60**
Range: N. Carolina to West Indies.
Habitat: Shallow water.
Description: Height 4 to 6 in. Shell spindle-shaped and moderately high-spired, with 8 or 9 convex whorls and distinct sutures. Surface relatively smooth, with weak revolving lines and a few strong wrinkles just below each suture. Aperture long and oval, operculum horny, canal moderately short and open. Color pinkish gray to reddish orange, with interrupted spiral bands of dark brown and many streaks and blotches of reddish brown and amber.

Genus *Pleuroploca* Fischer 1884

HORSE CONCH *Pleuroploca gigantea* (Kiener) **Pl. 59**
Range: N. Carolina to Florida.
Habitat: Shallow water.
Description: Height reaches 2 ft. Shell ponderous in size and weight. About 10 whorls, the shoulders bearing large but low nodules. Spire high and somewhat turreted. Sculpture of revolving ridges. Aperture wide and oval, with lengthy canal. Columella bears 3 pleats, operculum leathery. Color brown, aperture orange-red; animal is brick-red. Young shells are orange.
Remarks: Easily the largest snail to be found in American waters, it shares with one other species, *Megalotractus auruanus* (Linn.) of Australia, the honor of being the largest univalve in the world. It used to be in the genus *Fasciolaria*.

Genus *Ptychatractus* Stimpson 1865

RIDGED TULIP SHELL   **Pl. 60**
Ptychatractus ligatus (Mighels & Adams)
  Range: Gulf of St. Lawrence to Connecticut (?).
  Habitat: Moderately deep water.
  Description: About ¾ in. high. A fusiform (spindle-shaped)
  shell of 5 rounded whorls, the apex bluntly pointed. Canal
  open and moderately long. 2 weak folds on columella.
  Sculpture of numerous distinct revolving lines. Color grayish
  brown.
  Remarks: Its range south of Cape Cod is doubtful. The sin-
  gle Connecticut record appears to refer to a specimen taken
  from the stomach of a codfish.

Spindle Shells: Family Fusinidae

COMMONLY large, rather spindle-shaped (fusiform) shells. Lip
not thickened, umbilicus wanting. Operculum horny. The spire
is tall, acuminate, and many-whorled. Canal long and straight.
Found chiefly in warm seas.

Genus *Fusinus* Rafinesque 1815

SLENDER SPINDLE   *Fusinus amphiurgus* (Dall)   **Pl. 61**
  Range: Gulf of Mexico.
  Habitat: Deep water.
  Description: About 2 in. high. A tall and slender shell of 9
  or 10 whorls, sutures indented. Volutions convex, decorated
  by sturdy vertical folds and sharp encircling lines. Aperture
  rather small, canal long, slender, and nearly closed. Operculum horny. Color grayish white.

COUE'S SPINDLE   *Fusinus couei* (Petit)   **Pl. 61**
  Range: Gulf of Mexico.
  Habitat: Deep water.
  Description: Height 4 in. A tall and graceful shell of 7 or 8
  whorls, the sutures plainly marked. Sculpture of distinct
  spiral lines, no vertical folds. Aperture small, canal long and
  nearly closed. Inner lip reflected. Color white.

ORNAMENTED SPINDLE   **Pl. 61**
Fusinus eucosmius (Dall)
  Range: Gulf of Mexico.

Habitat: Moderately deep water.
Description: About 2½ in. high. 9 whorls, a tall, sharply pointed spire, and a long thin canal. Volutions convex, sculptured with rounded vertical ridges crossed by wavy encircling lines. Sutures impressed. Aperture small, inner lip reflected, operculum horny. Color orange-white to pure white.

TURNIP SPINDLE *Fusinus timessus* (Dall)   **Pl. 61**
    Range: Gulf of Mexico.
    Habitat: Deep water.
    Description: Height 3 in. A rather stocky spindle-shaped shell of about 9 well-rounded whorls, sutures deeply impressed. Early volutions decorated with strong vertical folds and sharp revolving lines. Aperture moderately small, operculum horny. Outer lip strongly crenulate, inner lip reflected. Canal long, nearly closed. Color white, sometimes tinged with orange-yellow.

Olive Shells: Family Olividae

MEMBERS of this group tend to be cylindrical, with a greatly enlarged body whorl that conceals most of the earlier volutions. The shells are smooth and polished and often brightly colored. Widely distributed in warm and tropical seas.

Genus *Oliva* Bruguière 1789

CARIBBEAN OLIVE   **Pl. 61**
Oliva caribaeensis Dall & Simpson
    Range: Caribbean.
    Habitat: Intertidal.
    Description: Height 1½ in. high. Elongate, short pointed spire, sutures channeled. Columella with strong plications. Smooth and glossy, colored a mottled grayish purple, with obscure darker bands and numerous scattered white specks. Aperture purplish.
    Remarks: This species has been listed as *O. trujilloi* Clench and *O. jamaicensis* Marrat.

NETTED OLIVE   *Oliva reticularis* Lam.   **Pl. 61**
    Range: S. Florida to West Indies.
    Habitat: Intertidal.
    Description: Height 1½ in. About 4 whorls, short spire, sutures plain. Body whorl large, aperture narrow, inner lip with plications toward base. Surface highly polished. No opercu-

lum. Color white or grayish, with pattern of purplish-brown reticulations.
Remarks: A deep-water subspecies, *O. reticularis bollingi* Clench, has been obtained at 200 ft. off Miami, Florida.

LETTERED OLIVE *Oliva sayana* Ravenel   **Pls. 6, 61**
Range: S. Carolina to Florida.
Habitat: Intertidal.
Description: Height 2½ in. Shell strong and solid. Cylindrical, with short pointed spire. 4 or 5 whorls, sutures deeply incised. Aperture long, notched at base, columella reflected at lower end. No operculum. Surface highly polished, bluish gray, variously marked with chestnut and pink.
Remarks: The pattern of dark markings suggest characters or hieroglyphics, hence the popular name of Lettered Olive. It used to be known as *O. litterata* Lam. The Coast Indians made necklaces of them long before white men set foot on American shores. There is a pale yellowish to nearly golden variety that is unspotted, a form eagerly sought by collectors. It has been named *O. sayana citrina* John., Golden Olive; see Plates 6 and 61.

Genus *Olivella* Swainson 1831

RICE DWARF OLIVE *Olivella floralia* (Duclos)   **Pl. 61**
Range: N. Carolina to West Indies.
Habitat: Shallow water.
Description: About ½ in. high. Elongate, 4 or 5 whorls, no shoulders, well-defined sutures. Apex sharp. Aperture rather narrow, lower columella pleated. Small horny operculum. Surface highly polished. Color white or bluish white, a few darker mottlings at the sutures. Apex frequently orange.

MINUTE DWARF OLIVE *Olivella minuta* (Link)   **Pl. 61**
Range: West Indies.
Habitat: Shallow water.
Description: Nearly ½ in. high. A solid shell of 3 or 4 whorls, apex sharply pointed. Aperture elongate, base of columella grooved. Color bluish gray, whitish at base. Generally a white or brown band at sutures.

DWARF OLIVE *Olivella mutica* (Say)   **Pl. 61**
Range: N. Carolina to West Indies.
Habitat: Shallow water.
Description: About ½ in. high. Shell small but solid, about 5 whorls and short pointed spire. Outer lip thin and sharp, inner lip without plications. Small horny operculum. Surface

highly polished, coloring variable. Specimens may be found that range from nearly white to dark chocolate, with or without bands, but the commonest color is yellowish white, with 2 or 3 revolving bands of purplish brown.

WEST INDIAN DWARF OLIVE Pl. 61
Olivella nivea (Gmelin)
Range: Bermuda, Florida, West Indies.
Habitat: Shallow water.
Description: About ¾ in. high. A slender shell of about 5 whorls, sutures plainly marked. Well-developed spire, sharp apex. Surface shiny. Color white, clouded with orange-brown, and commonly with suggestions of brownish dots encircling the shell near base and at shoulders.

TINY DWARF OLIVE   *Olivella perplexa* Olsson Pl. 61
Range: Florida to West Indies.
Habitat: Shallow water.
Description: About ³⁄₁₆ in. high. 4 or 5 whorls, well-developed spire. Sutures deep, aperture elongate, base rather square. Surface highly polished, color white.

CARIBBEAN DWARF OLIVE Pl. 61
Olivella petiolita Duclos
Range: West Indies to S. America.
Habitat: Shallow water.
Description: About ½ in. high. 4 whorls, sutures plainly delineated, apex sharp. Aperture moderately wide. Color whitish, with pattern of irregular wavy vertical lines of purplish brown.

Genus *Jaspidella* Olsson 1956

JASPER DWARF OLIVE Pl. 61
Jaspidella jaspidea (Gmelin)
Range: N. Carolina to West Indies.
Habitat: Shallow water.
Description: About ½ in. high. An oval shell of 4 or 5 whorls, short pointed spire, impressed sutures. Aperture narrow, inner lip weakly plicate toward base, operculum horny and small. Surface highly polished, color pale yellowish gray, with narrow band of chocolate just below sutures.
Remarks: Formerly and sometimes still listed as *Olivella jaspidea* (Gmelin).

Miter Shells: Family Mitridae

THE members of this family on our shores are mostly small shells. In the Pacific and Indian Oceans they are much larger, and often brilliantly colored. The shell is spindle-shaped (fusiform), rather thick and solid, with a sharply pointed spire. Aperture is small, notched in front, and there are several distinct pleats on the columella. Chiefly inhabit warm seas.

Genus *Mitra* Lamarck 1799

SULCATE MITER *Mitra albocincta* C. B. Adams **Pl. 62**
    Range: N. Carolina to West Indies.
    Habitat: Shallow water.
    Description: About 1 in. high. A stoutly fusiform shell, but quite variable in proportions, some specimens being much stouter than others. 7 or 8 whorls with indistinct sutures, pointed apex. Ornamentation consists of revolving ribs and furrows, faintly marked with vertical lines. Aperture narrow, columella with several strong white plications. Color brown, sometimes with paler bands.
    Remarks: Formerly called *M. sulcata* (Gmelin).

BARBADOS MITER *Mitra barbadensis* (Gmelin) **Pl. 62**
    Range: S. Florida to West Indies.
    Habitat: Intertidal.
    Description: Height 1 to 1½ in. About 6 rather flat whorls, sutures not well impressed. Surface moderately smooth, but there are weak revolving lines. Aperture long and narrow, 4 or 5 slanting ridges on columella. Outer lip somewhat thickened. Color yellowish gray, often with whitish flecks.

GULF STREAM MITER **Pl. 62**
Mitra fluviimaris Pilsbry & McGinty
    Range: S. Florida to West Indies.
    Habitat: Moderately deep water.
    Description: Height 1 in. 7 or 8 somewhat flat whorls, sutures indistinct, apex sharp. Aperture small and narrow, 3 or 4 pleats on columella. Quite smooth in texture, color grayish white with some brownish stains with weak revolving lines.

NODULAR MITER *Mitra nodulosa* (Gmelin) **Pl. 62**
    Range: N. Carolina to West Indies.
    Habitat: Shallow water.
    Description: About 1 in. high. Shell elongate, with fairly

sharp apex and 9 or 10 flattish whorls that are slightly shouldered. Sutures distinct. Surface sculptured with raised granules produced by vertical ribs crossed by revolving lines. Aperture quite short, notched at base. 4 pleats on columella. Color pale brown.

Remarks: Formerly listed as *M. granulosa* Lam.

ANTILLEAN MITER Pl. 62
Mitra swainsoni antillensis Dall
 Range: N. Carolina to West Indies.
 Habitat: Moderately shallow water.
 Description: Height 2 to 3 in. when fully grown. A trim, spindle-shaped shell of about 10 flattened whorls, sutures plainly impressed. Surface with revolving lines or ridges that are weakly checked. Aperture narrow, 4 pleats on columella. Color grayish brown.
 Remarks: Typical *M. swainsoni* occurs in S. America.

Genus *Pusia* Swainson 1840

WHITE-LINED MITER Pl. 62
Pusia albocincta (C. B. Adams)
 Range: Florida to West Indies.
 Habitat: Shallow water.
 Description: Height ¾ in. A fusiform shell of about 6 whorls, sutures well indented. Aperture narrow, columella with 4 folds. Sculpture of numerous vertical ribs. Color dark brown, with encircling band of white on each volution.

LITTLE GEM MITER *Pusia gemmata* (Sby.) Pl. 62
 Range: Florida to West Indies.
 Habitat: Shallow water.
 Description: Height ¼ in. About 7 whorls, sutures distinct. Columella with 4 folds. Sculpture of rounded vertical ribs. Color black or dark brown, a whitish band showing on the crests of the ribs.

HANLEY'S MITER *Pusia hanleyi* (Dohrn) Pl. 62
 Range: Florida to West Indies.
 Habitat: Shallow water.
 Description: Height ¼ in. A rather slender shell of about 5 whorls, sutures well impressed. Aperture narrow, columella with 4 folds. Sculpture of rounded vertical ribs. Color brown, with whitish encircling band.

PAINTED MITER *Pusia histrio* (Reeve) Pl. 62
 Range: West Indies.
 Habitat: Moderately deep water.

Description: Height ½ in. A spindle-shaped shell of 5 whorls, sutures impressed, apex sharp. Aperture narrow, columella with 4 folds. Sculpture of robust vertical ribs, about 12 on body whorl. Color whitish, with an orange band.

MAIDEN MITER *Pusia puella* (Reeve) Pl. 62
   Range: S. Florida to West Indies.
   Habitat: Shallow water.
   Description: Height ½ in. A stubby little shell of about 5 whorls. Apex rounded, sutures indistinct. Aperture small and very narrow, weak ridges on columella. Surface smooth, with just a suggestion of spiral threads. Color deep brown or black, commonly with a few encircling dots of white on body whorl.

BEAUTIFUL MITER *Pusia pulchella* (Reeve) Pl. 62
   Range: West Indies.
   Habitat: Shallow water.
   Description: ½ in. high. A rather stubby shell of 5 whorls, sutures well indented. Sculpture of strong vertical ribs and weaker revolving lines. Columella with robust plications. Color reddish brown, with encircling rows of white dots.

Chank Shells: Family Xancidae

LARGE, thick, heavy shells, often ponderous. There are several distinct pleats on the columella. The operculum is clawlike. These snails all are natives of tropical or subtropical seas. In India rare left-handed specimens are regarded as sacred and are mounted in gold and placed on altars. This family has also been called Turbinellidae.

Genus *Xancus* Röding 1798

LAMP SHELL *Xancus angulatus* (Lightf.) Pl. 62
   Range: S. Florida to West Indies.
   Habitat: Shallow water.
   Description: About 5 to 9 in. high. Heavy and ponderous, about 6 whorls with prominent knobs on shoulders. Sutures very distinct. Spire moderately high, apex bluntly rounded. Sculpture of weak revolving lines. Aperture large, descending into a short open canal. Inner lip partly reflected, forming small umbilicus. Operculum horny, clawlike. Columella strongly pleated. Color yellowish white, with brownish periostracum. Interior delicate pink in fresh specimens.
   Remarks: Formerly listed as *Turbinella scolymus* Gmelin.

Vase Shells: Family Vasidae

THESE are generally heavy shells, often ponderous. The outline is like an inverted vase. There are strong plications on the columella, the operculum is clawlike, and usually there is a heavy periostracum. Confined to tropical or subtropical waters.

Genus *Vasum* Röding 1798

SPINY VASE *Vasum capitellus* (Linn.) **Pl. 60**
  Range: West Indies.
  Habitat: Shallow water.
  Description: Height 2 to 3 in. A sturdy vase-shaped shell of about 6 whorls, the shoulders flattened. Sculpture of strong revolving ridges and strongly rounded vertical ribs that produce robust spines at the shoulders. A double row of spines near base of body whorl. Base rolled over to form deep umbilicus. Columella with 3 plications. Color yellowish brown.

VASE SHELL *Vasum muricatum* (Born) **Pl. 60**
  Range: S. Florida to West Indies.
  Habitat: Shallow water.
  Description: About 3 to 4 in. high. Shell rough, heavy, and strong, with a fairly short spire and bluntly pointed apex. 6 to 7 whorls, sculptured with many revolving ribs and ridges, and there is a series of rather sharp nodes on the shoulders, with a double row of spinelike nodes sometimes on lower part of body whorl. Aperture long, narrowing to canal, inner lip with transverse folds at center. Color yellowish brown, with tough brown periostracum.

Volutes: Family Volutidae

THE volutes are attractive, colorful shells and have always been great favorites with collectors, sharing honors with the cones and cowries. These shells exhibit a wide assortment of ornamentation and color. The aperture is notched in front and the columella bears several pleats. The group is noted for having a large, often bulbous initial whorl (protoconch) at the apex. Volutes are well distributed in tropical seas, living chiefly in rather deep water. For the collector who likes to add to his collection by purchasing specimens from dealers, the volutes can be the most costly of all marine shells.

Genus *Voluta* Linnaeus 1758

MUSIC VOLUTE   *Voluta musica* Linn.   **Pl. 7**
   Range: West Indies to S. America.
   Habitat: Moderately deep water.
   Description: Height 2 to 2½ in. Shell strong and solid,
   about 6 whorls, sutures indistinct. Shoulders sloping, spire
   short and capped by a large rounded apex. Sculpture of
   rounded vertical ridges. Outer lip thick, notched below, inner
   lip with 9 or 10 robust pleats. Small horny operculum. Color
   delicate rosy pink, with fine revolving lines in groups of 5 or
   6, occasionally marked with small dark spots.
   Remarks: The popular Music Volute is perhaps as aptly
   named as any shell from any ocean, for its pattern of spotted
   lines does suggest the lines of the staff and notes of written
   music.

GREEN VOLUTE   *Voluta virescens* Sol.   **Pl. 7**
   Range: Cen. and S. America.
   Habitat: Moderately deep water.
   Description: About 2 in. high. 5 or 6 whorls, short pointed
   spire, large body whorl. A series of rounded knobs at shoulders. Aperture long and rather wide, outer lip thickened,
   inner lip with several weak pleats. Surface shiny, color greenish brown, with weak and narrow bands of paler shade that
   are heavily spotted with brown. Inside of aperture bright
   yellow, outer lip strongly barred with deep brown and white.

Genus *Scaphella* Swainson 1832

DOHRN'S VOLUTE   *Scaphella dohrni* (Sby.)   **Pl. 63**
   Range: S. Florida.
   Habitat: Deep water.
   Description: 2 to 3 in. high. A rather slender shell with
   moderately tall spire. 4 or 5 whorls, sutures distinct, apex
   blunt. Aperture elongate, inner lip pleated at base. Color
   gray or creamy white, with revolving rows of squarish chocolate-brown spots.

DUBIOUS VOLUTE   *Scaphella dubia* (Brod.)   **Pl. 63**
   Range: S. Florida.
   Habitat: Deep water.
   Description: Height 2½ in. A slender shell of about 5
   whorls, a moderate spire capped by a rounded apex. Aperture
   long and narrow, weak plications on inner lip. Upper volutions ribbed vertically. Surface smooth, color yellowish brown

with chocolate spots, but the spots are fewer — both in number of spots and number of rows — than in the better-known Dohrn's Volute.

FLORIDA VOLUTE Pl. 63
Scaphella florida Clench & Aguayo
Range: Florida Keys.
Habitat: Deep water.
Description: Height 3 to 4 in. A substantial shell of about 7 whorls, the early ones bearing vertical ribs at shoulder. 3 distinct pleats on inner lip. Aperture moderately wide. Surface smooth, color yellowish gray with encircling rows of rather widely spaced brownish spots.

GOULD'S VOLUTE *Scaphella gouldiana* (Dall) Pl. 63
Range: N. Carolina to West Indies.
Habitat: Deep water.
Description: Height 2 to 3 in. About 5 whorls, spire well developed, sutures impressed. Surface sculptured with vertical knobs at shoulders, and by very minute spiral lines. Aperture long and narrow, outer lip thin and sharp. Color yellowish gray, sometimes pinkish aperture; occasional specimens may be nearly white and others may show indistinct bands of brown.

JUNONIA *Scaphella junonia* (Lam.) Pls. 4, 63
Range: N. Carolina to Gulf of Mexico.
Habitat: Moderately deep water.
Description: Height 3 to 6 in. Shell spindle-shaped, strong and solid, 5 or 6 whorls, sutures distinct. Apex blunt, spire moderate. Outer lip relatively thin, inner lip with 4 oblique pleats on lower part. Canal short, no operculum. Color pinkish white, with slanting rows of squarish spots that may be chocolate-brown or reddish orange.
Remarks: This is one of the prizes in any shell collection. Good specimens have brought as much as $100 in the past, when the species was considered extremely rare. Every season a few examples are washed up on shore during storms, but the shrimp fishermen now bring them in quite regularly.

Genus *Auriniopsis* Clench 1953

KIENER'S VOLUTE *Auriniopsis kieneri* Clench Pl. 63
Range: W. Florida to Texas.
Habitat: Deep water.
Description: Height 6 in. A gracefully slender shell of 7 or 8 whorls. Apex sharp, sutures rather indistinct. Aperture long,

lip thin and sharp. Columella without pleats. Surface smooth, color smoky brown with squarish spots of deep brown or black encircling the shell. Often these spots are in double rows.

Remarks: This is another of the "lost" species that has turned up with the extensive shrimp trawling of the last several years. It is also called *Aurinia kieneri*. The shell was originally described as *Fusus tessellatus* by Louis Kiener in 1840 from a single specimen collected off Mexico.

Nutmegs: Family Cancellariidae

SMALL but solid shells, with a striking cross-ribbed sculpture. The aperture is drawn out, with a short canal at base. The inner lip is strongly plicate, and the outer lip is ribbed within. There is no operculum. These snails are vegetarians and live in warm waters as a rule.

Genus *Cancellaria* Lamarck 1799

NUTMEG *Cancellaria reticulata* (Linn.) **Pls. 6, 60**
Range: N. Carolina to Florida.
Habitat: Shallow water.
Description: About 1 to 1½ in. high. Shell strong and rugged, with 6 or 7 well-rounded whorls, sutures distinct. Surface sculptured with vertical ribs and revolving lines, producing a network of raised lines over the shell. Aperture moderately narrow, canal short. Inner lip with strong oblique pleats. Color bright or pale orange, with weak orange-brown bands.
Remarks: *C. conradiana* Dall is a slightly slimmer shell with stronger sculpture on its upper whorls, and the color is apt to be more whitish; it was described by W. H. Dall in 1890, but most malacologists today believe that it is only a form of the common Nutmeg.

Genus *Trigonostoma* Blainville 1827

AGASSIZ'S NUTMEG *Trigonostoma agassizi* Dall **Pl. 60**
Range: N. Carolina to West Indies.
Habitat: Deep water.
Description: About ½ in. high. 6 whorls, sutures channeled, spire turreted. Aperture moderately wide, parietal wall reflected. Columella grooved. Sculpture of well-spaced revolving ridges and rounded vertical ribs that produce pointed knobs at shoulders. Color yellowish white.

PHILIPPI'S NUTMEG **Pl. 60**
Trigonostoma tenerum (Phil.)
 Range: S. Florida.
 Habitat: Shallow water.
 Description: Height ¾ in. high. A solid and sturdy little shell of about 4 whorls that are sharply flattened at the shoulders and form a winding terrace up the spire. Below this angle the volutions show distinct slanting folds. Aperture fairly large, with short canal at base, outer lip thickened. Color pale yellowish orange, sometimes with a row or two of chocolate spots on body whorl.

Genus *Admete* Kröyer 1842

COUTHOUY'S NUTMEG *Admete couthouyi* (Jay) **Pl. 60**
 Range: Arctic Ocean to Massachusetts; California.
 Habitat: Moderately deep water.
 Description: About ½ in. high. A stout shell of 5 well-rounded whorls, the body whorl relatively large, sutures well impressed. Spire sharply pointed, canal short and open. Sculpture of revolving lines, made nodular at shoulders by wavy vertical ridges. Aperture rather large, outer lip thin and sharp, inner lip deeply arched. Color yellowish brown.

Marginellas:
Family Marginellidae

THESE are small, porcelaneous, highly polished shells found on sandy bottoms in warm seas. Spire short or nearly lacking and the body whorl very large. Aperture narrow and long, the outer lip usually somewhat thickened, the inner lip plicate.

Genus *Marginella* Lamarck 1799

BANDED MARGINELLA **Pl. 64**
Marginella aureocincta Stearns
 Range: N. Carolina to West Indies.
 Habitat: Shallow water.
 Description: About ⅛ in. high. Spindle-shaped, with moderate spire. 5 or 6 whorls, sutures indistinct. Elongate aperture toothed on both lips, with outer lip noticeably thickened. Surface very glossy, color yellowish gray, with 2 pale brownish bands encircling the body whorl.

DENTATE MARGINELLA   Pl. 64
Marginella denticulata Conrad
    Range: N. Carolina to West Indies.
    Habitat: Shallow water.
    Description: Nearly ½ in. high. 4 or 5 whorls, rather high
spire for this group. Aperture narrow, about half the length
of whole shell. 4 strong pleats on columella, outer lip considerably thickened. Surface polished, color yellowish tan.
    Remarks: This shell looks very much like a small dove shell
of the genus *Columbella* (page 200).

MILK MARGINELLA *Marginella lactea* (Kiener)   Pl. 64
    Range: Florida to West Indies.
    Habitat: Shallow water.
    Description: Height ¼ to ½ in. Short spire, aperture nearly
as long as entire shell. Columella with 4 plications near base.
Surface glossy, color pure white.

Genus *Bullata* Jousseaume 1875

TEARDROP MARGINELLA   Pl. 64
Bullata ovuliformis (Orb.)
    Range: S. Florida to West Indies.
    Habitat: Shallow water.
    Description: About ⅛ in. high. A glossy, small shell, rather
globular, with a narrow aperture. Columella with 3 plications, outer lip feebly toothed within. Color pure white.

Genus *Prunum* Herrmannsen 1852

COMMON MARGINELLA   Pl. 64
Prunum apicinum (Menke)
    Range: N. Carolina to Gulf of Mexico and West Indies.
    Habitat: Shallow water.
    Description: Nearly ½ in. high. A solid shell of 3 or 4
whorls, with a low spire and a greatly enlarged body whorl.
Aperture narrow, outer lip thickened, inner lip with 4 pleats.
Surface highly polished, color varies from bright golden yel-
low to orange-brown. Outer lip usually white, sometimes with
a pair of brownish spots.
    Remarks: The most abundant marginellid on our shores.
Very rarely it is sinistral.

BELL MARGINELLA *Prunum bellum* (Conrad)   Pl. 64
    Range: N. Carolina to Florida.
    Habitat: Shallow to deep water.

Description: Height ¼ in. A flat-topped shell of 3 or 4 whorls. Aperture long and narrow, 4 pleats on inner lip. Outer lip thickened but without teeth. Whitish in color and glossy in appearance.

ORANGE MARGINELLA *Prunum carneum* (Storer) **Pl. 64**
Range: S. Florida and West Indies.
Habitat: Shallow water.
Description: Height ¾ in. 3 or 4 whorls, apex rounded, last volution constitutes most of shell. Aperture very narrow, outer lip rolled in and thickened. Inner lip with 4 pleats. Shiny orange, with whitish bands on body whorl.

SPOTTED MARGINELLA *Prunum guttatum* (Dill.) **Pl. 64**
Range: S. Florida to West Indies.
Habitat: Shallow water.
Description: About ¾ in. high. 3 or 4 whorls, low spire, apex bluntly rounded. Aperture long and narrow, outer lip thickened, inner lip with 4 pleats. Color pinkish gray, with numerous whitish dots scattered over the highly polished surface. 2 brown spots on apertural side of outer lip, and 3 or 4 at the edge on other side.
Remarks: This is one of the most colorful and attractive species of the group to be found on our shores.

ROYAL MARGINELLA *Prunum labiatum* (Kiener) **Pl. 64**
Range: Off s. Texas and Mexico.
Habitat: Moderately deep water.
Description: Height 1 in. or more. A robust and solid shell of 3 or 4 whorls. Spire short, top of shell rounded, narrow aperture about as high as whole shell. 4 very prominent pleats on inner lip, and the thickened outer lip has a number of small teeth along its inner margin. Glossy surface pale yellowish gray, with very weak bands of darker shade.

Genus *Persicula* Schumacher 1817

PRINCESS MARGINELLA **Pl. 64**
Persicula catenata (Montagu)
Range: S. Florida to West Indies.
Habitat: Moderately shallow water.
Description: About ¼ in. high. Apex very low, even concave, and usually covered by a callus. Narrow aperture runs full height of shell, outer lip only moderately thickened, with a series of tiny teeth inside. Weak pleats on columella. Color grayish white, with encircling rows of triangular whitish and brownish spots and 2 separated rows of brownish spots.

Remarks: This species is very similar to the Decorated Marginella (below), but is less strongly marked.

LINED MARGINELLA Pl. 64
Persicula interruptalineata (Mühl.)
    Range: West Indies.
    Habitat: Shallow water.
    Description: Nearly ½ in. high. A sturdy shell that is almost all body whorl and without a spire. Aperture narrow, columella with sharp folds below. Color grayish, with horizontal broken (or interrupted) dashes of brown over entire surface. Aperture white.

SNOWFLAKE MARGINELLA Pl. 64
Persicula lavelleeana (Orb.)
    Range: S. Florida to West Indies.
    Habitat: Moderately shallow water.
    Description: Height ⅛ in. Tiny but solid. 3 or 4 whorls, short spire. Aperture moderate, columella with 4 folds, outer lip curled in and with minute teeth. Color pure white.

DECORATED MARGINELLA Pl. 64
Persicula pulcherrima (Gaskoin)
    Range: S. Florida to West Indies.
    Habitat: Shallow water.
    Description: Height ¼ in. An oval shell, flat on top. Outer lip feebly toothed. Surface glossy, color buffy yellow, with encircling rows of broken brown bands, the area between decorated with vertical brownish-yellow lines.
    Remarks: See Remarks under Princess Marginella (above).

Genus *Hyalina* Schumacher 1817

WHITE-LINED MARGINELLA Pl. 64
Hyalina albolineata (Orb.)
    Range: Bermuda, West Indies.
    Habitat: Shallow water.
    Description: Height ¼ in., sometimes slightly more. A slender, highly polished snail of 3 or 4 whorls; the last whorl makes up most of shell. Aperture long and narrow, lower columella with 4 folds. Banded with orange-yellow and white.

ORANGE-BANDED MARGINELLA Pl. 64
Hyalina avena (Val.)
    Range: N. Carolina to West Indies.
    Habitat: Shallow water.
    Description: About ½ in. high. Spire short but pointed, shell

solid and highly polished. 4 or 5 whorls. Long aperture narrow above and wider below. Outer lip thickened, inner lip with 4 distinct pleats. Color creamy white, generally with 2 or 3 pale orange bands.

OAT MARGINELLA *Hyalina avenacea* (Desh.) **Pl. 64**
Range: N. Carolina to West Indies.
Habitat: Shallow water.
Description: About ½ in. high. A slender shell of about 3 whorls, a short pointed spire, and an elongate aperture that widens noticeably at its base. Surface polished, color white or yellowish white.
Remarks: Formerly listed as *Marginella succinea* Conrad.

PALE MARGINELLA *Hyalina pallida* (Donovan) **Pl. 64**
Range: Florida Keys.
Habitat: Shallow water.
Description: Height ½ in. A slender shell much like *H. avenacea* (above). It is thinner in substance, and its spire is proportionally shorter. Surface has a high gloss, color is yellowish tan.

VELIE'S MARGINELLA *Hyalina veliei* (Pilsbry) **Pl. 64**
Range: W. Florida.
Habitat: Shallow water.
Description: About ½ in. high. Shell elongate and thin, moderate spire. About 4 whorls, sutures distinct. Aperture wider at bottom, inner lip with 4 pleats. Highly polished, the color whitish or yellowish.

Cone Shells: Family Conidae

THIS is a large family of many-whorled, cone-shaped snails noted for their variety of colors and patterns. They live among the rocks and corals in tropical seas. This group is unusual among mollusks in that some of its members possess poison glands. The venom passes through a tiny duct to the teeth of the radula, some of which are modified to resemble small harpoons, and serves to benumb the gastropod's prey. None of the cones living in our waters are known to be dangerous, but all living examples should be handled with care. Certain s. Pacific and Indian Ocean species are capable of inflicting serious and even fatal wounds. Many of the cones have a strong periostracum during life, which must be removed to display the shell's colors.

Genus *Conus* Linnaeus 1758

GOLDEN-BANDED CONE   **Pl. 65**
Conus aureofasciatus Rehder & Abbott
   Range: Florida Keys and off Yucatan.
   Habitat: Moderately deep water.
   Description: Height 2 to 3 in. high. 9 or 10 whorls, moderate spire, sharp apex. Aperture long and narrow, notched at suture. Color reddish brown, with several bands of pale yellow.
   Remarks: This shell may turn out to be a subspecies of *C. spurius*.

AUSTIN'S CONE *Conus austini* Rehder & Abbott **Pl. 66**
   Range: Florida Keys to West Indies.
   Habitat: Moderately deep water.
   Description: About 1 to 2 in. high. Roughly 10 whorls, moderately high spire, sharp apex. Sculpture consists of revolving cords that cross over weak vertical lines. Color whitish.
   Remarks: This shell has been considered quite rare, but shrimp fishermen now bring them in rather frequently. *C. clarki* Rehder & Abbott (not illus.) is much like *C. austini* but has the cords at its shoulders strongly beaded.

CARROT CONE *Conus daucus* Hwass **Pl. 65**
   Range: Florida to West Indies.
   Habitat: Moderately deep water.
   Description: About 2 in. high. 9 or 10 whorls, spire short. Solid in build. Revolving lines at base of shell. Color varies from deep orange to bright yellow, some individuals showing a weak central band of yellowish. Top of shell — the tightly coiled spire — often has white blotches on orange background.

ANTILLEAN CONE *Conus dominicanus* Hwass **Pl. 65**
   Range: West Indies.
   Habitat: Moderately deep water.
   Description: About 2 in. high. Strong and solid, 10 or 11 whorls. Spire moderately elevated. Color yellowish tan, with irregular pinkish-white blotches and encircling lines of minute white beads, each outlined and connected by a thread of brown.
   Remarks: This species closely resembles the Crown Cone (below). It is a rare shell and may be found in some collections as *C. cedonulli* Linn. but that species is from the Indo-Pacific.

AGATE CONE *Conus ermineus* Born **Pl. 65**
   Range: West Indies.
   Habitat: Moderately shallow water.

Description: Height 2½ in. About 6 whorls, sutures weak. Spire sloping, apex pointed. Sculpture of faint revolving lines, stronger toward base. Color grayish or bluish white, blotched in an irregular fashion by deep brown.

FLORIDA CONE *Conus floridanus* Gabb **Pl. 65**
Range: N. Carolina to Florida.
Habitat: Shallow water.
Description: About 1½ in. high. 7 or 8 whorls, sutures distinct, spire elevated, apex sharp. Aperture long and narrow, notched at suture. Inner lip bears weak spiral ridges on lower margin. Color buffy yellow, marked with yellowish brown and white, generally in the form of broad bands.

GLORY-OF-THE-ATLANTIC CONE **Pls. 8, 66**
Conus granulatus Linn.
Range: S. Florida to West Indies.
Habitat: Moderately deep water.
Description: Height 1 to 1¾ in. high. A handsome shell of 8 or 9 whorls, slender in build, the shoulders rounded so that the short spire lacks the sharpness of most cones. Surface shows distinct revolving lines. The color varies from orange to bright pink, with encircling rich brown markings. Generally there is a darker area near the middle of the body whorl and another at the shoulder.
Remarks: An uncommon species, much desired by collectors.

JASPER CONE *Conus jaspideus* Gmelin **Pl. 66**
Range: N. Carolina to West Indies and south to Brazil.
Habitat: Shallow water.
Description: About 1 in. high. Shell trim and sturdy, with 10 whorls and a rather prominent spire that is frequently carinated. Apex sharp. Sculpture of evenly spaced spiral lines (especially on lower part of shell) and a few small tubercles. Color gray, with encircling rows of tiny spots of white and brown.
Remarks: Formerly listed as *C. peali* Green.

JULIA'S CONE *Conus juliae* Clench **Pl. 66**
Range: Florida to West Indies.
Habitat: Moderately deep water.
Description: About 1 to 2 in. high. Spire short, shoulders rounded, the whole shell having a substantial appearance. About 8 whorls. Color pinkish, with a fairly broad central band of white, and over all is a pattern of fine brownish lines and dots.
Remarks: Named in honor of Mrs. William J. Clench, wife of the distinguished malacologist of Harvard University.

MAZE'S CONE *Conus mazei* Desh. **Pl. 66**
  Range: Florida Keys and south through West Indies to Brazil.
  Habitat: Deep water.
  Description: About 2 in. high. Unusually slender, with tall pointed spire. About 10 whorls, those of spire often beaded. A few indistinct lines at base of shell. Color yellowish tan, with spiral rows of orange-brown spots.
  Remarks: One of the most attractive American cones, and relatively rare.

MOUSE CONE *Conus mus* Hwass **Pl. 66**
  Range: S. Florida to West Indies.
  Habitat: Shallow water.
  Description: About 1 in. high. A rather stubby shell of 6 or 7 whorls, the spire short and rounded, apex blunt. Surface bears faint revolving lines. Color dull yellowish gray, spotted with reddish brown and generally with a light-colored central band. There may be a row of whitish spots at shoulders.

CROWN CONE *Conus regius* Gmelin **Pl. 65**
  Range: Florida to West Indies and south to Brazil.
  Habitat: Moderately shallow water.
  Description: About 2 in. high, at times slightly more. Shell strong and solid, 7 or 8 whorls. Small spire, apex usually rounded. Surface sculptured with spiral threads or lines, more pronounced on the spire; some of these lines may be beaded. Mottled chocolate-brown and purplish, sometimes more or less banded. Compare Antillean Cone (above) for similarity.
  Remarks: Formerly listed as *C. nebulosus* Hwass.

SENNOTT'S CONE **Pl. 66**
Conus sennottorum Rehder & Abbott
  Range: Gulf of Mexico.
  Habitat: Deep water.
  Description: About 1½ in. high. 9 or 10 whorls, prominent terraced spire, apex sharp. Body whorl large at top, tapering very rapidly to base. Surface smooth, almost polished, color pale bluish gray with revolving rows of yellowish-brown dots.

SOZON'S CONE *Conus sozoni* Bartsch **Pls. 4, 65**
  Range: S. Carolina to Florida and Gulf of Mexico.
  Habitat: Deep water.
  Description: Height 2 to 4 in. A trim, shapely shell of 9 or 10 whorls. Well-developed spire, sharply pointed apex. Surface smooth. Color rich orange, with a pair of conspicuous white bands on body whorl, and over these is a series of distinct revolving lines of interrupted brownish dots.

Remarks: Many collectors would regard this as our handsomest cone. It is uncommon but not as rare as many others.

ATLANTIC ALPHABET CONE Pl. 65
Conus spurius atlanticus Clench
Range: Florida.
Habitat: Moderately shallow water.
Description: Height 2 to 3 in. 9 or 10 whorls, the first few forming a short spire on the otherwise rather flat top. Aperture long and narrow, notched at suture. Operculum horny and very small. Color creamy white, with revolving rows of squarish orange and brown spots and blotches.
Remarks: The irregular markings often resemble letters of the alphabet. The shell was known for years as *C. proteus* Hwass, but that name rightfully belongs to an Indian Ocean species our cone very closely resembles. The typical Alphabet Cone, *C. s. spurius* Gmelin (see Plate 65), lives in the West Indies; its markings are usually arranged in revolving bands, whereas those of the subspecies *atlanticus,* found in Florida, are scattered so that ordinarily no banding is discernible.

STEARNS' CONE *Conus stearnsi* Conrad Pl. 66
Range: N. Carolina to Florida and West Indies.
Habitat: Shallow water.
Description: About 1 in. high. Moderately high spire, tops of whorls concave. Apex sharp. About 8 whorls, sutures well defined. A few spiral lines at base of shell. Color gray, with revolving rows of tiny dots of brown and white.

STIMPSON'S CONE *Conus stimpsoni* Dall Pl. 66
Range: S. Florida.
Habitat: Deep water.
Description: Height about 2 in. 8 or 9 whorls, spire well developed, slightly concave, capped by a sharp apex. Surface smooth, with a few spiral lines at base. Color yellowish or ivory-white, usually with 2 or 3 broad yellowish bands.

WARTY CONE *Conus verrucosus* Hwass Pl. 66
Range: S. Florida to West Indies.
Habitat: Shallow water.
Description: About ¾ in. high. A stocky shell of 10 or 11 whorls, the shoulders abruptly sloping, so that the spire appears angulated. Sculpture of rather stout revolving ribs, with regularly spaced beadlike pustules. Color pinkish gray, somewhat blotched with brown.

VILLEPIN'S CONE *Conus villepini* Fischer & Bern. Pl. 66
Range: Florida Keys to West Indies.
Habitat: Deep water.

Description: Height 1½ in. A rather slender cone of about 10 whorls, the spire well developed and sharply pointed. Each volution slightly concave at top. Surface smooth, with a few spiral lines at base. Color white or pale gray, with irregular vertical splashes of reddish brown.

Remarks: As with many of the cones, there is a strong periostracum in life that quite conceals the shell's true colors.

Auger Shells: Family Terebridae

THESE are slender, elongate, many-whorled shells, confined to warm and tropical seas. There are no pleats on the columella, but the base of the inner lip is twisted. Some members of this family are provided with a mild poison, but no American species is dangerous.

Genus *Terebra* Bruguière 1789

GRAY AUGER   *Terebra cinerea* (Born)   Pl. 67
  Range: S. Florida, West Indies, and south to Brazil.
  Habitat: Shallow water.
  Description: Height 1 to 2 in. About 10 rather flat whorls, the whole shell tapering gradually and regularly to a slender point at apex. Each volution sculptured with numerous fine vertical grooves. Aperture small. Surface shiny, color gray or brown, with a whitish band encircling each volution just below the suture.
  Remarks: See Sallé's Auger (below) for a similar shell.

CONCAVE AUGER   *Terebra concava* Say   Pl. 67
  Range: N. Carolina to Florida.
  Habitat: Shallow water.
  Description: About 1 in. high. A slender shell of some 12 whorls, sutures fairly distinct. There is a line of beads just below the suture on each volution; middle of whorl is concave. Color gray, sometimes with a tinge of yellow.

COMMON AUGER   *Terebra dislocata* Say   Pl. 67
  Range: Virginia to Florida and Gulf of Mexico.
  Habitat: Shallow water.
  Description: About 1 to 1¾ in. high. Shell elongate, tapering gradually to a fine point. About 15 whorls, rather indistinct sutures. Surface decorated with wavy vertical folds and fine spiral grooves. A knobby spiral band encircles shell just below each suture. Aperture small, distinct twist at base of

columella. Operculum horny. Ashy gray to pale brown, sometimes nearly white. See Florida and Sallé's Augers (below) for comparison.

Remarks: The knobby spiral band just under the suture gives this shell the appearance of being composed of alternating large and small whorls. This gastropod is very common as a Pleistocene fossil in Florida and Bermuda.

FLORIDA AUGER   *Terebra floridana* Dall   **Pl. 67**
Range: S. Carolina to Florida.
Habitat: Shallow water.
Description: To 3 in. high, with as many as 20 whorls. Tall and spikelike shell, volutions marked with wavy vertical lines, and there is a distinct raised line running around middle of each whorl. Columella strongly twisted at base. Color yellowish white.
Remarks: This rather uncommon species may be mistaken for the Common Auger, but it can be distinguished by the lack of a "double suture" as well as by its larger size.

SHINY AUGER   *Terebra hastata* (Gmelin)   **Pl. 67**
Range: S. Florida to West Indies.
Habitat: Shallow water.
Description: About 1 to 1½ in. high. 10 to 12 whorls. The shell does not taper as rapidly or regularly as for most of this group but commonly remains nearly the same diameter until near the upper end, where it tapers rather abruptly. Sutures indistinct, each volution bearing fine vertical grooves. Surface shiny, color creamy white, with rather broad bands of pale orange.

BLACK AUGER   *Terebra protexta* Conrad   **Pl. 67**
Range: N. Carolina to West Indies; Texas.
Habitat: Moderately shallow water.
Description: About 1 in. high. Shell narrow and tall, 12 to 14 whorls, sutures fairly distinct. Each volution bears several sharp-edged vertical folds. Aperture small, columella twisted at base, operculum horny. Despite name, color not black but a deep chocolate-brown.

SALLÉ'S AUGER   *Terebra salleana* Desh.   **Pl. 67**
Range: N. Carolina, West Indies, and south to Brazil.
Habitat: Shallow water.
Description: About 1 in. high. This shell looks much like the Gray Auger (above), but it is smaller and slimmer, with fewer vertical ribs to a volution, and the apex is purple rather than white. Dark gray or brown, with a paler band at base of each whorl.

FLAME AUGER *Terebra taurinum* (Sol.) **Pls. 7, 67**
Range: S. Florida, Gulf of Mexico, and West Indies.
Habitat: Moderately shallow water.
Description: Height to 6 in. Our largest auger shell, tall and spikelike, with 25 or more whorls. Lower half of shell rather smooth, the upper volutions show vertical lines. Color yellowish white, streaked with spiral rows of reddish-brown marks.
Remarks: Formerly listed as *T. flammaea* Lam. Once regarded as extremely rare, this shell has long been a collector's item; it is now known to be uncommon rather than rare.

Turret or Slit Shells:
Family Turridae

A VERY large family of gastropods, many of which are small and highly ornate. The general shape is fusiform (spindle-shaped), and the outer lip commonly has a slit, or notch. There are said to be more than 500 genera and subgenera in the family, and thousands of species. Many hundreds occur along our shores, mostly in deep water. Their classification is difficult and often causes argument among the specialists themselves. Only a few examples of the commoner forms can be discussed here.

Genus *Gemmula* Weinkauff 1875

GEM TURRET *Gemmula periscelida* (Dall) **Pl. 67**
Range: N. Carolina to Florida.
Habitat: Deep water.
Description: About 2 in. high. A stoutly fusiform shell of some 10 whorls, elaborately sculptured with spiral cords. A broad band at the shoulders is marked by numerous vertical bladelike ridges. Above shoulders the volutions are channeled. Aperture relatively small, outer lip thin with deep notch at bottom, inner lip reflected on body whorl. Canal moderately long, and open. Color gray, with yellowish periostracum.

Genus *Polystira* Woodring 1928

WHITE GIANT TURRET *Polystira albida* (Perry) **Pl. 68**
Range: S. Florida and Gulf of Mexico.
Habitat: Deep water.
Description: Height 3 to 4 in. A graceful shell of about 8

rounded whorls, sutures distinct. Shape fusiform, spire high, sharply pointed apex. Canal open and rather long. Sculpture consists of sharp-edged revolving ridges and oblique striations between the ridges. Distinct slit toward upperpart of outer lip. Color white. See Delicate Giant Turret (below), for similarity.

Remarks: Formerly listed as *Turris virgo* Lam. This handsome shell was rarely encountered in the past. Shrimp fishermen have turned it up by dozens in recent years.

FLORENCE'S GIANT TURRET Pl. 68
Polystira florencae Bartsch
Range: Puerto Rico.
Habitat: Deep water.
Description: Height 1¼ in. 10 to 12 whorls. An elongate shell with pointed apex and long canal. Sculpture of revolving cords with sharp vertical lines in between. Outer lip notched. Color white, lightly mottled with brown.

DELICATE GIANT TURRET Pl. 68
Polystira tellea (Dall)
Range: S. Florida.
Habitat: Deep water.
Description: Height 2 to 3 in. Up to 10 whorls, sutures distinct. Very much like the White Giant Turret (above), but the sculpture is not as sharp and distinct. Color white or grayish white.

GIANT TURRET *Polystira vibex* (Dall) Pl. 68
Range: Florida Keys.
Habitat: Deep water.
Description: Height 1½ in. A rather slim shell of 9 or 10 whorls. Sculpture of sharp revolving ridges. Aperture quite small, open canal moderately long. Operculum horny. Color white.

Genus *Ancistrosyrinx* Dall 1881

PEBBLED STAR TURRET Pl. 67
Ancistrosyrinx elegans Dall
Range: S. Florida to West Indies.
Habitat: Deep water.
Description: Nearly 2 in. high. 6 or 7 whorls, sometimes more. Well-developed spire, long open canal. Sharply angled volutions produce steeply sloping shoulders ringed with numerous short spines. Surface pebbled by encircling rows of very tiny beads. Color yellowish white.
Remarks: See the Star Turret for comparison.

STAR TURRET *Ancistrosyrinx radiata* Dall   **Pl. 67**
Range: S. Florida to West Indies.
Habitat: Deep water.
Description: About ½ in. high. 6 or 7 whorls, sharply angled to produce steeply sloping shoulders, ringed with short spines that curve upward toward the apex. Spines less numerous but larger than those of Pebbled Star Turret. Surface of shell smooth, often shiny. Canal long and open. Color drab yellowish gray.
Remarks: For delicate beauty of form, there are few shells of any size that equal this diminutive snail.

Genus *Leucosyrinx* Dall 1889

CHANNELED TURRET   **Pl. 68**
Leucosyrinx subgrundifera Dall
Range: N. Carolina to Florida.
Habitat: Deep water.
Description: About 1 in. high. 9 or 10 whorls, each sharply angled at the center, so spire is pagodalike. Aperture small, outer lip deeply notched at suture, canal long and open. Fine lines of growth provide the only sculpture. Color yellowish white.

LONG TURRET *Leucosyrinx tenoceras* Dall   **Pl. 68**
Range: N. Carolina to Florida.
Habitat: Deep water.
Description: About 1 in. high. 9 or 10 whorls, a row of nodules encircling each at the shoulder, the volution above this concave. Aperture small, outer lip thin and deeply notched, canal long and open. Inner lip reflected on last volution. Color yellowish white.

VERRILL'S TURRET *Leucosyrinx verrilli* Dall   **Pl. 68**
Range: N. Carolina to Florida.
Habitat: Deep water.
Description: About 1 in. high. A rather stout shell of 7 or 8 whorls, the sutures indistinct. A row of large nodes at shoulders. Aperture small, canal open and moderately long. Color ashy white.

Genus *Crassispira* Swainson 1840

SPINDLE TURRET *Crassispira alesidota* Dall   **Pl. 69**
Range: Cape Hatteras to Florida.
Habitat: Deep water.
Description: About 1 in. high. Whorl count 10. Slender,

with sharp apex. Sculpture consists of prominent spiral lines that cross over curving vertical ridges. A revolving crease above the shoulder imparts a look of a tiny volution between each major whorl. Aperture elongate, notched above, the canal short and open. Color ashy brown.

BLACK TURRET *Crassispira fuscescens* (Reeve)   **Pl. 69**
Range: S. Florida to West Indies.
Habitat: Moderately shallow water.
Description: About ¾ in. high. A solid shell of 10 or 11 whorls, the sutures set off by an encircling band ridged at center. Sculpture consists of rather sharp vertical ridges and revolving threads. Aperture elongate, notched above. Color dark brown or black.

STRIPED TURRET *Crassispira mesoleuca* Rehder   **Pl. 69**
Range: S. Florida.
Habitat: Shallow water.
Description: About ¾ in. high. Fusiform, with 7 or 8 whorls decorated with vertical ridges that form small knobs at the shoulders; these knobs commonly are white, the rest of the shell dark brown or black. Sutures indistinct, aperture elongate, canal open and short.

OYSTER TURRET *Crassispira ostrearum* (Stearns)   **Pl. 69**
Range: N. Carolina to Florida.
Habitat: Moderately shallow water.
Description: About 1 in. high. Shell strong and solid, composed of 7 or 8 whorls. There is a prominent elevated ridge following the suture line, effectively separating the volutions. Each whorl bears numerous closely set vertical ribs, crossed by revolving threadlike lines. Aperture oval, strongly notched at suture, and opens into a short open canal. Color reddish brown.

SANIBEL TURRET   **Pl. 69**
Crassispira sanibelensis Bartsch & Rehder
Range: W. Florida.
Habitat: Moderately shallow water.
Description: Nearly 1 in. high. About 8 whorls, an elongate spire, and sharp apex. Volutions ornamented with rounded vertical ridges and prominent spiral lines. Revolving ridge at suture not as prominent as in Oyster Turret. Color pale brown or reddish brown.

TAMPA TURRET   **Pl. 69**
Crassispira tampaensis Bartsch & Rehder
Range: W. Florida.

Habitat: Moderately shallow water.
Description: About ¾ in. high. Approximately 8 whorls, decorated with vertical ridges and weak spiral lines. Apex well pointed, aperture small, open canal short. Noticeable ridge at suture line. Color reddish brown.

Genus *Monilispira*
Bartsch & Rehder 1939

WHITE-BANDED DRILLIA   **Pl. 68**
Monilispira albinodata (Reeve)
  Range: S. Florida to West Indies.
  Habitat: Shallow water.
  Description: Height ½ in. A rugged and stubby shell with about 5 whorls, sutures not deeply impressed. Aperture small, notched above, outer lip somewhat thickened, canal short. Surface decorated with revolving knobs, those on shoulders larger and white, the areas between them dark brown or black.

KNOBBY DRILLIA   *Monilispira leucocyma* (Dall)   **Pl. 68**
  Range: S. Florida, Gulf of Mexico, and West Indies.
  Habitat: Shallow water.
  Description: About ½ in. high. A slender shell of 6 or 7 whorls, sutures indistinct. Sculpture consists of a row of knobs at the shoulders, about 12 to a volution. Aperture small and elongate, outer lip rather thick. Color grayish brown, the knobs whitish.

COLLARED DRILLIA   **Pl. 68**
Monilispira monilis Bartsch & Rehder
  Range: S. Florida.
  Habitat: Shallow water.
  Description: Height ¾ in. A slender shell of about 8 whorls, sutures rather indistinct. Volutions bear rugged knobs at the shoulders and weak revolving lines. Aperture small and elongate, canal short. Brown, the encircling knobs paler.

Genus *Drillia* Gray 1838

CYDIA TURRET   *Drillia cydia* Bartsch   **Pl. 69**
  Range: Florida to West Indies.
  Habitat: Moderately shallow water.
  Description: Height ¾ in. A sturdy shell of about 8 whorls, sutures well defined. Aperture narrow, notched above. Parietal wall reflected. Sculpture of strongly rounded vertical ribs. Color white.

Genus *Cerodrillia*
Bartsch & Rehder 1939

CLAPP'S DRILLIA   **Pl. 68**
Cerodrillia clappi Bartsch & Rehder
   Range: W. Florida.
   Habitat: Shallow water.
   Description: About ½ in. high. A slender shell of some 6
whorls, made angular by a series of rather sharp oblique
ridges. Outer lip thickened and notched above. Canal open
and short. Color waxy white, sometimes with a weak band
on body whorl.

THEA DRILLIA   *Cerodrillia thea* (Dall)   **Pl. 68**
   Range: W. Florida.
   Habitat: Shallow water.
   Description: Height ½ in. High-spired, with 6 angled whorls,
sutures only moderately distinct. Each volution has 12 or so
slanting ribs. Aperture fairly large, inner lip reflected on body
whorl. Color brown, the ribs yellowish, aperture dark brown.

Genus *Inodrillia* Bartsch 1943

TALL-SPIRED TURRET   **Pl. 69**
Inodrillia aepynota (Dall)
   Range: N. Carolina to Florida.
   Habitat: Moderately deep water.
   Description: Height ½ in. High-spired, sharp apex. About 7
sharply angled whorls, sutures well impressed. Sculpture con-
sists of rather strong vertical ribs. Aperture somewhat elon-
gate, notched above. Color pinkish white.

MIAMI TURRET   *Inodrillia miama* (Dall)   **Pl. 69**
   Range: E. Florida.
   Habitat: Moderately deep water.
   Description: Height ½ in. Stouter than Tall-spired Turret,
with about 6 whorls, sutures impressed. Decorated with
rather widely spaced vertical ridges. Aperture moderately
elongate, outer lip thin, and open canal short. Color pinkish
white.

Genus *Fenimorea* Bartsch 1934

JANET'S TURRET   *Fenimorea janetae* Bartsch   **Pl. 68**
   Range: Florida to West Indies.
   Habitat: Moderately shallow water.

Description: About 1 in. high. 6 or 7 slightly shouldered whorls, sutures impressed. Sculpture of strongly rounded vertical ribs, about 12 on body whorl. Lip thickened. Color whitish, with pinkish-brown blotches between the ribs, especially on lower portions of volutions.

MOSER'S TURRET *Fenimorea moseri* (Dall) **Pl. 68**
Range: N. Carolina to Florida and Gulf of Mexico.
Habitat: Shallow water.
Description: Height ½ in. An elongate, sturdily built shell of about 10 whorls, sutures distinct. About 12 strongly curving ribs on each volution, plus an encircling ridge just below the suture. Aperture somewhat lengthened, notched above. Color creamy white, with thin yellowish-brown periostracum.

Genus *Kurtziella* Dall 1918

GRAY-PILLARED TURRET **Pl. 69**
Kurtziella atrostyla (Dall)
Range: N. Carolina to Gulf of Mexico and West Indies.
Habitat: Shallow water.
Description: Height ¼ in. A small shell of 6 or 7 whorls, with sharp vertical ribs that produce slight shoulders. Aperture somewhat twisted, outer lip thickened. Color grayish white.

Genus *Lora* Gistel 1848

CANCELLATE LORA **Pl. 70**
Lora cancellata (Mighels & Adams)
Range: Labrador to Massachusetts.
Habitat: Moderately deep water.
Description: Height ½ in. Shell rather slender, with 7 or 8 turreted whorls. Sculpture of numerous vertical ribs (about 20 on body whorl) that are crossed by raised revolving lines, giving the surface a cancellate appearance. Aperture narrow and small. Color pinkish white.
Remarks: The generic status of this group is uncertain at the present time. They were once listed under *Bela* Gray 1847 and later under *Lora* Gistel 1848. Many authorities believe they belong in the genus *Oenopota* Mörch 1852, but until more decisive work is done on them it seems best to leave them in the genus *Lora*.

HARP LORA *Lora harpularia* (Couthouy) **Pl. 70**
Range: Labrador to Rhode Island.

Habitat: Moderately shallow water.
Description: About ¾ in. high. Stoutly elongate, with 6 to 8 whorls flattened somewhat above and forming slightly sloping shoulders. Sutures distinct. Each volution bears numerous oblique rounded ribs crossed by fine revolving lines. Aperture narrow and oval, canal a mere notch. Buffy flesh color.
Remarks: See Remarks under Cancellate Lora.

NOBLE LORA *Lora nobilis* (Möller) **Pl. 70**
Range: Greenland to Maine.
Habitat: Moderately shallow water.
Description: About ¾ in. high. Stoutly fusiform, with some 7 whorls, the shoulders flattened to form a turreted spire. About 15 vertical ribs to a volution, crossed by weak spiral lines. Aperture small and narrow, canal short. Color yellowish white.
Remarks: See Remarks under Cancellate Lora (above).

STAIRCASE LORA *Lora scalaris* (Möller) **Pl. 70**
Range: Greenland to Massachusetts.
Habitat: Deep water.
Description: Nearly 1 in. high. A rather solid shell of 6 or 7 whorls. Volutions angled at shoulders, producing a turreted spire. Apex sharp, sutures distinct. Sculpture of vertical ridges crossed by threadlike lines. Aperture elongate, canal short. Color yellowish brown.
Remarks: See Remarks under Cancellate Lora (above).

Genus *Mangelia* Risso 1826

WAX-COLORED MANGELIA **Pl. 70**
Mangelia cerina Kurtz & Stimpson
Range: Massachusetts to Florida.
Habitat: Moderately deep water.
Description: About ½ in. high. Stoutly elongate, 7 whorls with well-developed shoulders and distinct sutures. Each volution bears a few broad, rounded, vertical folds and ribs, crossed by very fine spiral lines. Aperture small, canal short, no operculum. Color pale yellowish white.

LITTLE WAX-COLORED MANGELIA **Pl. 70**
Mangelia cerinella Dall
Range: N. Carolina to Florida and Texas.
Habitat: Moderately deep water.
Description: About ½ in. high. Slender and spikelike, with about 8 whorls. Sutures distinct, volutions angled slightly at

shoulders. Broad vertical folds, aperture relatively small. Color yellowish white.

DARK MANGELIA   *Mangelia fusca* C. B. Adams   **Pl. 70**
Range: West Indies.
Habitat: Shallow water.
Description: About ¼ in. high. A rather stubby shell of 6 whorls, sutures well impressed. Aperture narrow and small, strongly notched above. Sculpture of prominent vertical ribs. Color reddish brown.

BLACK MANGELIA   *Mangelia melanitica* Dall   **Pl. 70**
Range: West Indies.
Habitat: Moderately shallow water.
Description: Height ¼ in. 6 or 7 whorls, spire moderately turreted. Aperture narrow, deep round notch above. Sculpture of vertical ribs and encircling lines produces a cancellate surface. Color white.

RIBBED MANGELIA   **Pl. 70**
Mangelia plicosa C. B. Adams
Range: Maine to Gulf coast of Florida.
Habitat: Shallow water.
Description: Height ¼ in. A sturdy shell of 7 whorls, sutures quite distinct. Surface displays a network of vertical ribs and nodulous lines. Aperture narrow, outer lip thick, notched at top. Color dark brown.

Genus *Pleurotomella* Verrill 1872

AGASSIZ'S PLEUROTOMELLA   **Pl. 69**
Pleurotomella agassizi Verrill & Smith
Range: Massachusetts to Florida.
Habitat: Deep water.
Description: Height ¾ in. A stoutly built shell of 6 whorls, sutures moderately impressed. A row of slanting ribs is present at the shoulders, and most of each volution is decorated by distinct revolving lines. Aperture somewhat elongate, outer lip thin and sharp. Color yellowish brown.

BAIRD'S PLEUROTOMELLA   **Pl. 69**
Pleurotomella bairdi Verrill & Smith
Range: Off Chesapeake Bay.
Habitat: Deep water.
Description: Height 1 to nearly 2 in. About 7 sharply shouldered whorls leading to an acute apex. Surface bears revolving lines, and there is a row of weak oblique nodules at

the shoulders. Base of shell lengthened to form a moderately long, open canal. Color pale yellowish white.
Remarks: A real giant for its group. It has been dredged from more than 10,200 ft.

JEFFREY'S PLEUROTOMELLA Pl. 69
Pleurotomella jeffreysi Verrill
Range: Off Delaware and Chesapeake Bays.
Habitat: Deep water.
Description: Height 1 to 1½ in. 7 or 8 whorls, sharply angled at the shoulders, so spire is noticeably turreted. Sculpture consists of row of oblique elongated nodules at angle of the shoulder, those on the spire most pronounced. These nodules are continued downward as weak curved ribs that fade out before reaching middle of the whorl, leaving much of the surface relatively smooth. Color lustrous white.

PACKARD'S PLEUROTOMELLA Pl. 69
Pleurotomella packardi Verrill
Range: Bay of Fundy to Cape Cod.
Habitat: Deep water.
Description: Height ½ in. Shell thin and fragile, somewhat translucent. About 9 whorls and an acute, turreted spire. Shape rather stoutly fusiform. Surface sculptured with rounded, partially oblique ribs; spaces in between bear revolving lines. Aperture broad above and elongate below, terminates in a short but wide canal. Pale flesh-colored.

RATHBUN'S PLEUROTOMELLA Pl. 69
Pleurotomella rathbuni (Verrill)
Range: Off Chesapeake Bay.
Habitat: Deep water.
Description: Height ¾ in. About 6 whorls, angled slightly at the shoulders to form a somewhat turreted spire. Sculpture consists of weak revolving and vertical lines, so surface is moderately reticulate. Outer lip thin and sharp, short open canal. Color yellowish white.

Genus *Gymnobela* Verrill 1884

BRIEF TURRET *Gymnobela brevis* Verrill Pl. 69
Range: Nova Scotia to Massachusetts.
Habitat: Deep water.
Description: Height ¼ in. A stout little shell of 5 or 6 rounded whorls; the body whorl constitutes most of the shell, but there is a short turreted spire. Surface sculptured with obscure nodes at shoulders, the nodes continuing down the

sides as broken ribs. Aperture short and broad, outer lip thin and sharp, inner lip reflected on body whorl. Color yellowish white.

SHORT TURRET *Gymnobela curta* Verrill   **Pl. 69**
Range: Off southern New England.
Habitat: Deep water.
Description: Slightly more than ¼ in. high. About 5 whorls, body whorl much enlarged. Surface sculptured with fine vertical and revolving lines of nearly equal size. Sutures deeply impressed, often slightly channeled. Aperture moderately large, with slightly recurved canal. Color grayish or greenish white, somewhat translucent when fresh.

Genus *Glyphoturris* Woodring 1928

FROSTED TURRET   **Pl. 69**
Glyphoturris quadrata rugirima Dall
Range: Florida to West Indies.
Habitat: Moderately shallow water.
Description: Height ¼ in. 6 shouldered whorls. Sculpture of strong vertical ribs and sharp revolving lines. Aperture moderate, outer lip thickened. Color white.
Remarks: *G. rugirima* Dall is a rare shell, occurring from N. Carolina to the Gulf of Mexico.

Genus *Daphnella* Hinds 1844

SLUGLIKE TURRET *Daphnella limacina* Dall   **Pl. 69**
Range: Massachusetts to Florida.
Habitat: Moderately deep water.
Description: Height ½ in. Fusiform, with 5 or 6 whorls, sutures distinct. Apex sharp, aperture elongate-oval. Most of surface smooth and semiglossy, with extremely fine lines of growth, but there is a row of squarish knobs on upperpart of each volution, making a twisting circuit up the spire. Color ivory-white.

VOLUTE TURRET   **Pl. 69**
Daphnella lymneiformis (Kiener)
Range: S. Florida to West Indies.
Habitat: Shallow water.
Description: Height ¼ in. Thin in substance. 7 or 8 whorls, the body whorl constituting about ⅔ of whole shell. Sutures well impressed. Aperture moderate, large below, notched above. Sculpture of distinct revolving lines. Color yellowish white, heavily blotched with orange-brown.

Obelisk Shells: Family Eulimidae

SMALL, high-spired shells, usually polished. The spire is often slightly bent to one side. This is a large family that lives chiefly in warm waters. Many of the species are parasitic on other forms of marine life. Formerly known as the Melanellidae, this group is now considered to be in the Episthobranchia.

Genus *Eulima* Risso 1826

GOLDEN-BANDED EULIMA   **Pl. 70**
Eulima auricincta Abbott
    Range: N. Carolina to West Indies.
    Habitat: Shallow water.
    Description: About ¼ in. high. A slender shell of about 12 flattened whorls, the sutures scarcely discernible. Aperture narrow and elongate, lip thin and sharp. Surface very glossy. Color grayish white, with a thin brownish band encircling each volution.
    Remarks: Formerly listed as *E. acuta* (Sby.)

TWO-BANDED EULIMA *Eulima bifasciata* (Orb.) **Pl. 70**
    Range: S. Florida to West Indies.
    Habitat: Shallow water.
    Description: Height ¼ in. A slender spikelike shell of 10 or 11 flat whorls, sutures quite indistinct. Aperture narrow and elongate. Surface very glossy. Color yellowish white, with a pair of brown bands encircling each volution.

TALL EULIMA *Eulima hypsela* Verrill & Bush **Pl. 70**
    Range: Bermuda.
    Habitat: Moderately shallow water.
    Description: Nearly ½ in. high. A slender spikelike shell of about 12 whorls, sutures indistinct. Aperture relatively small. Surface glossy, color milky white.
    Remarks: The specimen shown on Plate 70 is the holotype.

NARROWMOUTH EULIMA   **Pl. 70**
Eulima stenotrema (Jeffreys)
    Range: Labrador to Massachusetts.
    Habitat: Moderately shallow water.
    Description: About ¼ in. high. Tall and slender, with 8 or 9 whorls, the sutures moderately well impressed. Aperture narrow and elongate, lip thin and sharp. Glassy in appearance and amber in color.

Genus *Balcis* Leach 1847

TWO-LINED BALCIS   *Balcis bilineata* (Alder)   **Pl. 70**
   Range: N. Carolina to West Indies.
   Habitat: Shallow water.
   Description: About ⅛ in. high. 7 or 8 rather flat whorls, with very weak sutures, so that the taper from base to sharply pointed apex is very regular. Aperture rather small, outer lip somewhat thickened. Surface shiny, color milky white.

CONELIKE BALCIS   **Pl. 70**
Balcis conoidea (Kurtz & Stimpson)
   Range: Cape Hatteras to West Indies.
   Habitat: Shallow water.
   Description: Slightly more than ½ in. high. A slender shell of about 12 rather flat whorls, sutures quite indistinct. Shell tapers very regularly to a sharp apex. Aperture oval and small, lip thin. Surface highly polished, color pure white.

HEMPHILL'S BALCIS   *Balcis hemphilli* (Dall)   **Pl. 70**
   Range: Florida
   Habitat: Moderately shallow water.
   Description: Not quite ½ in. high. 9 or 10 flattened whorls, the sutures scarcely indented, so sides of spire are quite regular. Aperture relatively small, inner lip partially reflected on last volution. Color glossy white.

INTERMEDIATE BALCIS   **Pl. 70**
Balcis intermedia (Cantraine)
   Range: New Jersey to West Indies.
   Habitat: Shallow water.
   Description: About ½ in. high. 10 to 13 whorls, tightly coiled and tapering very gradually to sharp apex. Sutures scarcely perceptible. Aperture narrow, lip thin and sharp, operculum horny. Color glossy white, sometimes flecked with yellowish brown.

JAMAICAN BALCIS   **Pl. 70**
Balcis jamaicensis (C. B. Adams)
   Range: Virginia to West Indies.
   Habitat: Shallow water.
   Description: Height ¼ in. About 7 whorls, sutures fairly well indented. Spire not so flattened as for most of this genus. Aperture oval and moderate in size. Surface glossy, color white.

Genus *Aclis* Lovén 1846

STRIATED ACLIS *Aclis striata* Verrill   **Pl. 70**
Range: Nova Scotia to Rhode Island.
Habitat: Deep water.
Description: About ⅛ in. high. A tiny but stocky shell of 4 or 5 well-rounded whorls, sutures well impressed. Aperture oval and rather large, lip thin and sharp. Surface bears revolving striae. Color yellowish white.

WALLER'S ACLIS *Aclis walleri* Jeffreys   **Pl. 70**
Range: Labrador to Massachusetts.
Habitat: Deep water.
Description: Height ¼ in. A shell of about 10 rounded whorls and steeply inclined spire. Sutures clearly defined. Aperture nearly round, lip thin. Color glossy white.

Genus *Niso* Risso 1826

INTERRUPTED NISO *Niso interrupta* Sby.   **Pl. 70**
Range: N. Carolina to Florida.
Habitat: Moderately shallow water.
Description: About ½ in. high. A conical shell, rather wide at base, with 10 to 12 flattened whorls, the taper from base to apex very regular. Sutures indistinct, apex acute. Aperture relatively small. Surface polished, color pale brown, sometimes with patches of darker brown. Sutures marked with a reddish-brown line.

Genus *Stilifer* Broderip 1832

STIMPSON'S STYLUS *Stilifer stimpsoni* Verrill   **Pl. 71**
Range: Massachusetts to Florida.
Habitat: Moderately deep water; on sea urchins.
Description: About ⅛ in. high. An odd little shell of 4 or 5 whorls, the last one forming most of the shell. The apex, or initial chamber, is relatively tall and slender, standing up like a miniature steeple. No operculum. Surface of shell smooth, the color is pale orange-yellow.
Remarks: Found most commonly as a parasite on the sea urchin, living among the many spines of the echinoid.

AWL-SHAPED STYLUS   **Pl. 71**
Stilifer subulatus Brod. & Sby.
Range: West Indies.
Habitat: Moderately deep water.

Description: Height ¾ in. About 8 rounded whorls, the 1st 2 or 3 small and forming a cylindrical apex, sometimes curved to one side. Shell expands rapidly. Sutures plainly marked. Surface smooth, color translucent white.

Pyramid Shells: Family Pyramidellidae

SMALL pyramidal or conical shells, usually white and often polished. Many-whorled. The columella is usually plicate (folded or pleated). They inhabit sandy bottoms, and the family contains a vast number of very small gastropods. Many of them are parasites.

Genus *Pyramidella* Lamarck 1799

SHINING WHITE PYRAM   **Pl. 71**
Pyramidella candida (Mörch)
    Range: N. Carolina to Florida, Gulf of Mexico, and West Indies.
    Habitat: Moderately deep water.
    Description: Slightly more than ¼ in. high. A sturdy little shell of 9 or 10 moderately flattened whorls, sutures slightly channeled and indented. Aperture quite small, columella twisted to form 1 or 2 teeth. Color pure white, surface brightly polished.

GRACEFUL PYRAM   *Pyramidella charissa* (Verrill)   **Pl. 71**
    Range: Massachusetts to New Jersey.
    Habitat: Moderately deep water.
    Description: Height ¼ in. About 10 rather well rounded whorls, sutures quite deeply impressed. Aperture nearly round, outer lip simple and inner lip slightly reflected on body whorl. Color translucent white.

CRENATE PYRAM   *Pyramidella crenulata* Holmes   **Pl. 71**
    Range: S. Carolina to Florida.
    Habitat: Shallow water.
    Description: Height ¾ in. About 12 rather flat whorls, sutures plainly marked and slightly crenulate. Aperture moderate, lip thin. Operculum horny, and notched on one side to fit the plications on columella. Surface shiny, color whitish, with brownish mottlings.

GIANT ATLANTIC PYRAM   **Pl. 71**
Pyramidella dolabrata Lam.
    Range: Florida to West Indies.

Habitat: Shallow water.
Description: About 1 in. high, our largest member of the genus. 9 or 10 rounded whorls, sutures deeply impressed. The taper is rather abrupt, and the shell appears quite plump. Aperture semilunar, with 2 or 3 distinct ridges on columella. Small umbilicus and horny operculum. Surface polished, color milky white, with revolving lines of brown.

BROWN PYRAM *Pyramidella fusca* (C. B. Adams) **Pl. 71**
Range: Prince Edward Island to Florida.
Habitat: Shallow water.
Description: Height ¼ in. A chubby small shell of 7 or 8 whorls, sutures fairly well impressed. Aperture moderately large, oval, with lip thin and sharp. A twist on inner lip produces a tiny umbilicus. Surface lustrous, color pale brown.

POLISHED PYRAM *Pyramidella lissa* (Verrill) **Pl. 71**
Range: Off Cape Hatteras.
Habitat: Deep water.
Description: Height ¼ in. Elongate, with blunt apex. About 7 whorls, sutures rather sharply impressed. Inner lip reflected, aperture somewhat square. Surface only slightly polished, color amber-brown.

BRIGHT PYRAM *Pyramidella lucida* (Verrill) **Pl. 71**
Range: Off New England coast.
Habitat: Deep water.
Description: Height ¾ in. A slender shell of some 15 rather flat whorls, sutures only slightly impressed. Aperture relatively small and oval, lip thin and sharp. No pleats on columella. Surface but faintly polished, color white.
Remarks: This gastropod has been dredged from more than 12,000 ft.

EXTENDED PYRAM **Pl. 71**
Pyramidella producta (C. B. Adams)
Range: Massachusetts to New Jersey.
Habitat: Shallow water.
Description: Height ¼ in. A translucent yellowish-brown shell of only 5 or 6 whorls. Sutures plainly marked, apex blunt. Aperture oval, inner lip with a fold, or twist, well back inside. Surface semipolished.

SMITH'S PYRAM *Pyramidella smithi* (Verrill) **Pl. 71**
Range: Off New England coast.
Habitat: Moderately deep water.
Description: Slightly more than ¼ in. high. A slender shell of about 10 whorls, sutures well impressed. Apex rather

blunt, aperture small, with single fold on columella. Surface highly polished, color milky white.

Genus *Turbonilla* Risso 1826

THE genus *Turbonilla* is a large one, containing many species of very small, high-spired gastropods — all much alike to the casual observer. The 1st (nuclear) whorl is coiled at a right angle to rest of the shell. An enlarged example is illustrated on Plate 71.

BUSH'S TURBONILLE *Turbonilla bushiana* Verrill **Pl. 71**
Range: Vineyard Sound to Long Island Sound.
Habitat: Moderately deep water.
Description: Height nearly ½ in. A slender shell of 10 or 11 rather flat whorls, sutures moderately impressed. Sculpture of weak vertical ribs, about 18 on body whorl. Surface between ribs smooth and polished. Aperture small and oval, columella slightly flattened. Color yellowish white.

DALL'S TURBONILLE *Turbonilla dalli* Bush **Pl. 71**
Range: N. Carolina to Florida and Gulf of Mexico.
Habitat: Moderately shallow water.
Description: About ¼ in. high. 12 well-rounded whorls, sutures sharply indented. About 18 strong and prominent vertical ribs to a volution. Surface also bears very fine revolving lines (striae). Aperture small, color bluish white.

ELEGANT TURBONILLE **Pl. 71**
Turbonilla elegantula Verrill
Range: Off New England coast.
Habitat: Moderately deep water.
Description: About ⅛ in. high. Rather stocky, with 10 whorls, sutures distinct. Sculpture consists of conspicuous vertical ribs, and revolving lines that give the surface a cross-hatched look. Color yellowish white, the shell rather shiny.

HEMPHILL'S TURBONILLE **Pl. 71**
Turbonilla hemphilli Bush
Range: W. Florida.
Habitat: Moderately shallow water.
Description: Nearly ½ in. high. A tall and slender shell of about 15 whorls, sutures distinct. Strong vertical ribs. Aperture small, operculum horny. Color white.

INCISED TURBONILLE *Turbonilla incisa* Bush **Pl. 71**
Range: W. Florida.
Habitat: Moderately shallow water.

Description: About ¼ in. high. Slender and high-spired, with 12 rather flat whorls, sutures distinct. Vertical ribs not as sharp as for most of this group. Color pale brown.

INTERRUPTED TURBONILLE Pl. 71
Turbonilla interrupta Totten
    Range: Maine to Florida and West Indies.
    Habitat: Moderately deep water.
    Description: About ¼ in. high. 9 or 10 rather flat whorls, sutures but little impressed. Vertical ribs small and crowded, some 20 to a volution. Color pale yellow, the shell having a waxy luster.

CLARET TURBONILLE   *Turbonilla punicea* Dall   Pl. 71
    Range: N. Carolina to Florida.
    Habitat: Moderately shallow water.
    Description: Height ¼ in. A sturdy shell of about 12 whorls, sutures moderately impressed. Each volution bears about 16 rounded vertical ribs. Aperture quite small. Color yellowish white.

RATHBUN'S TURBONILLE Pl. 71
Turbonilla rathbuni Verrill & Smith
    Range: Massachusetts to N. Carolina.
    Habitat: Deep water.
    Description: Nearly ½ in. high, one of the largest of this genus. About 12 rounded whorls, sutures well impressed. Vertical ribs rather strong and somewhat crowded. Color white or buffy white.

STRIPED TURBONILLE   *Turbonilla virgata* Dall   Pl. 71
    Range: N. Carolina to Florida.
    Habitat: Moderately deep water.
    Description: Height ¼ in. A slender shell of some 10 to 14 rather flat whorls, sutures plainly marked. Vertical ribs rounded and numerous, aperture small and somewhat elongate. Surface shiny, color yellowish white.
    Remarks: Formerly listed as *T. exarata* Holmes.

Genus *Odostomia* Fleming 1817

DOUBLE-SUTURED ODOSTOME Pl. 72
Odostomia bisuturalis Say
    Range: Gulf of St. Lawrence to Florida.
    Habitat: Shallow water.
    Description: About ⅛ in. high. 7 or 8 whorls, somewhat flattened, sutures well defined. Apex sharp, aperture oval, columella with 2 oblique folds. A deeply impressed line on upper-

part of each volution. Operculum horny. Color dull white, with thin brownish periostracum.

CHANNELED ODOSTOME   Pl. 72
Odostomia canaliculata C. B. Adams
   Range: West Indies.
   Habitat: Shallow water.
   Description: About ⅛ in. high. Whorls flattened, 6 in number, sutures well defined and slightly channeled. Surface smooth, color white.

ANGULATED ODOSTOME   Pl. 72
Odostomia engonia Bush
   Range: N. Carolina to Florida.
   Habitat: Shallow water.
   Description: About ⅕ in. high. A rather slender shell of some 7 whorls, sutures deeply impressed. Volutions somewhat flattened, giving shell a flat-sided appearance. Apex blunt, aperture oval, inner lip partially reflected at base to form tiny umbilicus. Surface smooth, almost polished, color white.

GEMLIKE ODOSTOME   Pl. 72
Odostomia gemmulosa C. B. Adams
   Range: West Indies.
   Habitat: Shallow water.
   Description: Height ⅛ in. About 5 whorls, sutures moderately indented. Prominent nodular ridges encircle volutions. Apex rather blunt, aperture oval, strong pleat on columella. Color white.

INCISED ODOSTOME   *Odostomia impressa* Say   Pl. 72
   Range: Massachusetts to Florida and Gulf of Mexico.
   Habitat: Shallow water.
   Description: Height ⅛ in. Shell elongate, consisting of 6 or 7 rather flat whorls, with channeled sutures. Surface decorated with 3 equally spaced spiral grooves. Aperture oval, outer lip thin and sharp, occasionally flaring a little. Color milky white.

HALF-SMOOTH ODOSTOME   Pl. 72
Odostomia seminuda C. B. Adams
   Range: Prince Edward Island to Florida and Gulf of Mexico.
   Habitat: Shallow water.
   Description: Height ⅛ in. Stoutly conical, with 6 or 7 whorls, sutures distinct. Sculptured with several revolving ridges cut by vertical striations, so surface appears beaded. Apex sharp, aperture oval, with oblique fold at base of columella. Color white.

TRIPARTITE ODOSTOME Pl. 72
Odostomia trifida Totten
  Range: Massachusetts to New Jersey.
  Habitat: Shallow water.
  Description: Height ⅕ in. Rather stoutly conical, with 5 or 6 whorls, sutures well defined. 2 or 3 impressed lines below the suture give the shell appearance of having a multiple suture. Aperture oval, outer lip simple, inner lip with oblique fold. Color pale greenish white.

Small Bubble Shells:
Family Acteonidae

SMALL, cylindrical, solid shells with a short, sharp spire. The inner lip bears a single pleat. Surface usually spirally grooved. Aperture long and narrow.

Genus *Acteon* Montfort 1810

ADAMS' BABY BUBBLE Pl. 72
Acteon punctostriatus (C. B. Adams)
  Range: Massachusetts to West Indies.
  Habitat: Shallow water.
  Description: Nearly ¼ in. high. 3 or 4 whorls, sutures indented. Body whorl somewhat elongate, spire moderately extended. Surface smooth and glossy. Lower portion of last volution bears encircling rows of tiny punctate dots. Aperture large, broadest below; columella shows a twisted fold. Color white.

Helmet Bubble Shells:
Family Ringiculidae

TINY, almost globular shells, but with well-developed pointed spires. The surface generally is polished. They occur from Maine to Cuba and live in fairly deep water.

Genus *Ringicula* Deshayes 1838

VERRILL'S HELMET BUBBLE Pl. 72
Ringicula nitida Verrill
  Range: Gulf of Maine to Florida and Gulf of Mexico.

Habitat: Deep water.
Description: About ³⁄₁₆ in. high. A fat little shell with 4 whorls. Apex bulbous. Surface smooth and polished. Outer lip thickened, inner lip with 2 strong folds. Color white.

STRIATE HELMET BUBBLE Pl. 72
Ringicula semistriata Orb.
Range: N. Carolina to West Indies.
Habitat: Moderately deep water.
Description: About ⅛ in. high. A stout shell of 3 whorls. Body whorl nearly globular, but the short and blunt spire is set off by well-indented sutures. Aperture large, outer lip thickened and swollen at middle, inner lip with 3 folds. Surface polished, color white.

Lined Bubble Shells: Family Hydatinidae

SHELLS oval, inflated, and thin in substance. The surface is smooth, and there is no umbilicus. The spire is involute (rolled inward from each side). The animal is large, generally extending beyond the shell. Distributed widely in warm seas.

Genus *Hydatina* Schumacher 1817

LINED BUBBLE *Hydatina vesicaria* (Sol.) Pl. 72
Range: S. Florida to West Indies.
Habitat: Shallow water.
Description: About 1½ in. high. A globose, well-inflated shell, thin and rather fragile. Spire not concealed as in *Bulla* (below), but top of shell is flattened and shows a small tight spiral. Surface smooth and polished. Color yellowish gray, with numerous thin wavy brown lines encircling the shell; every 4th or 5th line heavier than the others.
Remarks: Formerly listed as *H. physis* (Linn.), which occurs in Japan, India, and Australia and used to be considered conspecific with our snail.

Genus *Micromelo* Pilsbry 1894

MINIATURE MELO *Micromelo undata* (Brug.) Pl. 72
Range: Florida to West Indies.
Habitat: Shallow water.
Description: Height ½ in. Semiglobular, thin and fragile in substance, nearly all body whorl. Aperture wide at bottom

and narrower at top. Surface shiny, decorated with 4 or 5 encircling lines of reddish brown, these lines connected at intervals by curving vertical splashes of reddish hue.

Remarks: If this shell were larger, it might well be one of the most sought after of our East Coast mollusks. Unfortunately it is not very common.

Paper Bubble Shells:
Family Diaphanidae

VERY small globose shells occurring in moderately deep water, from cold to tropical seas. There is a small chinklike umbilicus.

Genus *Diaphana* Brown 1827

ARCTIC BUBBLE   *Diaphana minuta* Brown   Pl. 72
   Range: Greenland to Connecticut.
   Habitat: Moderately deep water.
   Description: About ¼ in. high. A globose shell, the flaring aperture narrow above and broad below. Small umbilicus, covered by reflected inner lip. Surface glossy, color translucent tan.
   Remarks: Formerly listed as *D. globosa* (Lovén).

True Bubble Shells:
Family Bullidae

OVAL in outline, the aperture flaring. Small to fairly large shells, usually rolled up like a scroll. The shell is thin and light. These are carnivorous snails, burrowing in the sands and muds for their prey, chiefly in warm seas.

Genus *Bulla* Linnaeus 1758

COMMON BUBBLE   *Bulla occidentalis* A. Adams   Pl. 72
   Range: N. Carolina to West Indies.
   Habitat: Shallow water.
   Description: About 1 in. high, sometimes more. Shell oval and inflated, spire depressed. Surface smooth and semipolished. Aperture longer than shell, rounded at both ends. A reflected white shield at base of columella. Color pale reddish

gray, mottled with purplish brown; some individuals show traces of banding.

Remarks: These gastropods inhabit grassy mudflats, slimy banks of river mouths, and brackish waters and conceal themselves in mud or under seaweeds while the tide is out, being most active at night. They feed upon small mollusks, which they swallow whole and crush by interior calcareous plates. See Striate Bubble (below) for comparison and classification.

SOLID BUBBLE *Bulla solida* Gmelin   **Pl. 72**
Range: Florida Keys.
Habitat: Shallow water.
Description: Height ¾ in. A rather solid shell, well inflated, and less elongate than Common Bubble. Aperture narrow above, rounding to a wide opening at the base, where the inner lip provides a polished white area. Color mottled gray and brown.

STRIATE BUBBLE *Bulla striata* Brug.   **Pl. 72**
Range: W. Florida to Texas; West Indies.
Habitat: Shallow water.
Description: About 1 to 1½ in. high. Shaped much like the Common Bubble (above), but somewhat more robust in build. Though colors are the same, this species has a series of well-defined spiral striations at the base of its shell.
Remarks: Some authorities believe that this snail and the Common Bubble, *B. occidentalis,* are the same species. If this proves to be a fact, then the mollusk will have to be called *B. striata,* since it was named by Bruguière in 1792; *B. occidentalis* was not named until 1850, by A. Adams.

Glassy Bubble Shells: Family Atyidae

SMALL and fragile shells inhabiting muddy and brackish waters, as a rule, chiefly in warm seas. The animal is too large for its shell, which is partially internal.

Genus *Atys* Montfort 1810

CARIBBEAN GLASSY BUBBLE   **Pl. 72**
Atys caribaea Orb.
Range: S. Florida to West Indies.
Habitat: Shallow to moderately deep water.

Description: About ½ in. high. A graceful species, rather elongate, the outer lip rising well above top of shell. Surface smooth and shiny, with a few obscure lines at both ends of shell. Color pure white.

SHARP'S GLASSY BUBBLE   *Atys sharpi* Vanatta   **Pl. 72**
Range: S. Florida.
Habitat: Moderately deep water.
Description: Height ½ in. A slender species, the aperture not as wide as that of Caribbean Glassy Bubble. There is a small umbilicuslike depression at the top of shell. Color milky white.

Genus *Haminoea*
Turton & Kingston 1830

ANTILLEAN GLASSY BUBBLE   **Pl. 72**
Haminoea antillarum (Orb.)
Range: W. Florida to West Indies.
Habitat: Shallow water.
Description: About ½ in. high. Somewhat globular, aperture widely flaring. Outer lip thin, rising above top of shell. No depression at top. Color greenish yellow, the shell very thin and translucent.

ELEGANT GLASSY BUBBLE   **Pl. 72**
Haminoea elegans (Gray)
Range: S. Florida to West Indies.
Habitat: Shallow water.
Description: About ½ to ¾ in. high. Shell remarkably thin and fragile, and semitransparent. Shape somewhat short and squat, surface sculptured with very minute revolving lines, but the general appearance is smooth and glossy. Color greenish yellow.

SOLITARY GLASSY BUBBLE   **Pl. 72**
Haminoea solitaria (Say)
Range: Massachusetts to N. Carolina.
Habitat: Shallow water.
Description: About ½ in. high. Thin and delicate, well inflated. Aperture wide, small depression at top of shell. Surface shiny, with tiny but sharp spiral lines. Color bluish white.

AMBER GLASSY BUBBLE   **Pl. 72**
Haminoea succinea (Conrad)
Range: Florida to West Indies; Texas.

Habitat: Shallow water.
Description: About ½ in. high. More slender than most of the genus *Haminoea,* the aperture less flaring. Marked depression at top of shell. Surface with microscopic wrinkled lines. Shell substance very thin, translucent amber in color.

Genus *Cylindrobulla* Fischer 1856

BEAU'S GLASSY BUBBLE Pl. 72
Cylindrobulla beaui Fischer
   Range: Florida Keys.
   Habitat: Moderately deep water.
Description: Height ¼ in. Broadly cylindrical, paper-thin in substance. Outer lip rolled over columella and practically closes the aperture, although it is open at bottom and top. Inner lip rather strongly reflected. Surface shiny, color soiled white.

Barrel Bubble Shells: Family Retusidae

SMALL cylindrical bubble shells, with or without a short spire. The sutures are deeply channeled and the inner lip usually bears a single fold. These tiny snails range from cold to tropical seas.

Genus *Retusa* Brown 1827

CHANNELED BARREL BUBBLE Pl. 73
Retusa canaliculata (Say)
   Range: Canada to Mexico.
   Habitat: Shallow water.
Description: Height ⅕ in. Shell cylindrical, with about 5 whorls, the summit of each with a shallow rounded groove. Spire slightly elevated, but body whorl makes up about ⅞ of shell. Outer lip arches slightly forward, inner lip is overspread with a thin plate of enamel. Oblique fold near base. Color dull chalky white.
Remarks: This diminutive snail commonly is found clinging to an old oyster or clam shell, and also on decaying, floating timbers. By gathering a few handfuls of broken fragments found in the coves of a shell beach and running the material through a sieve, one will usually find examples of this gastropod, as well as many other tiny varieties.

CANDÉ'S BARREL BUBBLE *Retusa candei* (Orb.) **Pl. 73**
Range: Cape Hatteras to West Indies.
Habitat: Shallow water.
Description: About ¹⁄₁₆ in. high. Cylindrical, with short spire, the sutures slightly channeled. Aperture elongate, widest below, the inner lip with a distinct fold. Color milky white.

IVORY BARREL BUBBLE **Pl. 73**
Retusa eburnea (Verrill)
Range: Cape Hatteras to Florida.
Habitat: Moderately deep water.
Description: About ¼ in. high. Oval in shape, the outer lip rising slightly above rounded apex of shell. Small chinklike umbilicus. Lip thin and sharp, describing a graceful curve from top to bottom. Surface smooth, color ivory-white.

GOULD'S BARREL BUBBLE **Pl. 73**
Retusa gouldi (Couthouy)
Range: Maine to N. Carolina.
Habitat: Moderately deep water.
Description: Height ¼ in. An oval shell, the aperture wide at bottom. Layer of enamel on inner lip. No spire, top of shell bluntly rounded. Color pale brown.
Remarks: Some modern authors believe that this species belongs in the genus *Cylichna* (p. 269).

ARCTIC BARREL BUBBLE **Pl. 73**
Retusa obtusa (Montagu)
Range: Arctic seas to off N. Carolina.
Habitat: Moderately deep water.
Description: Height ⅛ in. Early whorls form a spire that is flat and scarcely discernible. Aperture elongate, narrow above, broad and rounded below. About 4 whorls, surface smooth. Color dingy white.
Remarks: Specimens are usually obtained from the stomachs of haddock and other bottom-feeding fishes.

Genus *Pyrunculus* Pilsbry 1894

OVATE LITTLE PEAR SHELL **Pl. 73**
Pyrunculus ovatus (Jeffreys)
Range: Newfoundland to Florida.
Habitat: Moderately deep water.
Description: Height ⅛ in. A pyriform snail with its spire buried in a small pit at top of shell. Aperture rises above apex, and almost wraps around upper part of shell, but broadens considerably below. Color brownish gray.

Genus *Rhizorus* Montfort 1810

SPINDLE BUBBLE *Rhizorus oxytatus* Bush **Pl. 73**
   Range: N. Carolina to West Indies.
   Habitat: Moderately deep water.
   Description: About ¼ in. high. Cylindrical, with rounded bottom and sharp spikelike apex. Aperture narrow, widening at base. Surface shiny, color white, with pale yellow periostracum.

Canoe Shells:
Family Scaphandridae

SMALL to fairly large shells, usually rolled up like a scroll. The shell is thin and brittle. These are carnivorous snails, burrowing in the muds and sands for their prey, which are chiefly scaphopods.

Genus *Scaphander* Montfort 1810

NOBLE CANOE SHELL *Scaphander nobilis* Verrill **Pl. 74**
   Range: Massachusetts to Delaware.
   Habitat: Deep water.
   Description: About 2 in. high. Oval in shape, with widely flaring aperture, the outer lip extending well above the rounded top of the shell. Inner lip with narrow band of enamel. Surface smooth, color chalky white, with thin yellowish-brown periostracum.
   Remarks: The largest bubble-type shell found on our coast.

COMMON CANOE SHELL **Pl. 74**
Scaphander punctostriatus (Mighels & Adams)
   Range: Gulf of St. Lawrence to West Indies.
   Habitat: Moderately deep water.
   Description: Nearly 2 in. high. Oval in shape, narrower toward the top. Aperture large, flaring at base. Spire concealed. Surface bears numerous very fine revolving lines (striae), hardly to be seen with the naked eye. Color pale yellowish brown, with thin grayish periostracum.

WATSON'S CANOE SHELL **Pl. 74**
Scaphander watsoni Dall
   Range: N. Carolina to Florida.

Habitat: Deep water.
Description: About 1¼ in. high. Narrow at top, broad at bottom. Aperture flares widely at base. Thin layer of enamel on inner lip. Color yellowish brown; interior white.

Flat-topped Bubble Shells: Family Acteocinidae

SMALL cylindrical shells, generally smooth and glossy. There is little or no evidence of a spire. Aperture long, widening at base. They prefer cold seas as a rule.

Genus *Cylichna* Lovén 1846

BROWN'S BUBBLE *Cylichna alba* (Brown)   Pl. 73
  Range: Greenland to N. Carolina; Europe.
  Habitat: Shallow water.
  Description: Height ¼ in. Spire sunken, so that there is a shallow pit at top of shell. General shape cylindrical, aperture narrow and elongate, widening at base. Surface bears very delicate lines of growth, but appearance is smooth and often shiny. Color white, with rusty-brown periostracum.

AUBER'S BUBBLE *Cylichna auberi* (Orb.)   Pl. 73
  Range: Bermuda, West Indies.
  Habitat: Shallow water.
  Description: About ⅛ in. high. Oval, with flat sides, top flattish. Aperture long, narrow above, widening below. No fold on columella. Color white.

ORBIGNY'S BUBBLE *Cylichna bidentata* (Orb.)   Pl. 73
  Range: N. Carolina to Florida and West Indies.
  Habitat: Moderately shallow water.
  Description: Height ⅛ in. Elongate-oval, rounded at top and bottom. Distinct fold at base of inner lip. Surface smooth, color yellowish white.
  Remarks: See Rice Bubble (below) for comparison.

CONCEALED BUBBLE   Pl. 73
Cylichna occulta (Mighels & Adams)
  Range: Circumpolar; Greenland to Maine.
  Habitat: Moderately shallow water.
  Description: Height ¼ in. Oval in shape, top flat or concave, base of shell somewhat constricted. Prominent layer of

enamel on inner lip. Weak folds on columella. Color dingy white, periostracum yellowish brown.

RICE BUBBLE *Cylichna oryza* (Totten) **Pl. 73**
  Range: Nova Scotia to Connecticut.
  Habitat: Shallow water.
  Description: Less than ⅛ in. high. Very much like Orbigny's Bubble (above), but smaller and more chubby in build. Fold on columella produces a toothlike structure. Surface smooth and shiny, color yellowish gray.

VERRILL'S BUBBLE *Cylichna verrilli* Dall **Pl. 73**
  Range: N. Carolina to West Indies.
  Habitat: Moderately deep water.
  Description: Height ⅛ in. Cylindrical shape, rounded below and somewhat flattened above. Aperture long and narrow, widening at base. Color bluish white, thin yellowish periostracum.

DEPRESSED BUBBLE *Cylichna vortex* Dall **Pl. 73**
  Range: Maine to Chesapeake Bay.
  Habitat: Deep water.
  Description: Height ¼ to ½ in. One of the largest of this group. A bluntly oval shell, almost all body whorl, with a tiny sunken spiral at top. Aperture narrow above, broad below. Shell thin and brittle, smooth, the color white or yellowish white.

Wide-mouthed Bubble Shells: Family Philinidae

SMALL, loosely coiled shells with flaring apertures. The shell is partially internal, concealed by the mantle of the snail. They are chiefly mollusks of cold seas.

Genus *Philine* Ascanius 1772

GIRDLED PAPER BUBBLE **Pl. 73**
Philine cingulata (Sars)
  Range: Nova Scotia to Maine; Norway.
  Habitat: Deep water.
  Description: Less than ¼ in. high. A fragile shell that seems to be made up of but 1 whorl. Aperture flares widely, the upperpart forming a wing well above the rounded top of shell. Surface smooth, color yellowish white.

FINMARKIAN PAPER BUBBLE Pl. 73
Philine finmarchia Sars
Range: Nova Scotia to Massachusetts; Europe.
Habitat: Deep water.
Description: Height ¼ in. Top of shell flattened, aperture a mere chink at that point but widening considerably at base. Surface smooth and shiny, color ivory-white.

FUNNEL-LIKE PAPER BUBBLE Pl. 73
Philine infundibulum Dall
Range: Bermuda; s. Florida to West Indies.
Habitat: Moderately deep water.
Description: Height ¾ in. Shell extremely thin and fragile, in life pretty much concealed by animal itself. The flaring aperture makes up almost the whole shell; the body whorl is surprisingly small. Color translucent white.

FILE PAPER BUBBLE *Philine lima* (Brown) Pl. 73
Range: Arctic Ocean to Massachusetts.
Habitat: Moderately shallow water.
Description: Height ¼ in. Elongate-oval, top partially flattened. Flaring aperture begins below apex. Surface bears microscopic encircling lines but appears quite smooth. Color translucent yellowish white.
Remarks: Formerly listed as *P. lineolata* Couthouy.

QUADRATE PAPER BUBBLE Pl. 73
Philine quadrata (Wood)
Range: Greenland to N. Carolina.
Habitat: Deep water.
Description: About ¼ in. high. Thin-shelled, inflated, aperture flaring widely. Apex deeply excavated. Body whorl small. Color translucent grayish or whitish.

False Limpets:
Family Siphonariidae

THESE snails look very much like the limpets. The shell is roughly circular and conical, with a deep groove on one side that makes a distinct projection, or bulge, on the margin. The animals possess both gills and lungs and spend their time between the tide limits living a partially amphibious life, so they form a connecting link between the purely aquatic snails and the more highly developed air-breathing mollusks. Inside, the horseshoe-shaped muscle scar is open at the side instead of at the end as it is in the true limpets.

Genus *Siphonaria* Sowerby 1824

FALSE LIMPET *Siphonaria alternata* Say   **Pl. 74**
  Range: Bermuda, s. Florida, and Bahamas.
  Habitat: Intertidal.
  Description: About ¾ in. long. Shell rather oval in outline, conical, open at base. A deep siphonal groove on right side makes a noticeable bulge on that margin, so the shell is not symmetrical. Surface decorated with numerous fine ribs that radiate from summit, the ribs varying somewhat in size. Apex commonly eroded. Color brownish, marked with white; interior glossy.

STRIPED FALSE LIMPET   **Pl. 74**
Siphonaria pectinata (Linn.)
  Range: Florida to Texas and West Indies.
  Habitat: Intertidal.
  Description: Length 1 in. Shape oval and conical, apex situated slightly behind center. Surface sculptured with fine radiating lines, all about same size. Strong bulge on right margin. Color pale gray, with brownish lines radiating from summit. Interior glossy.

Genus *Williamia* Monterosato 1884

KREBS' FALSE LIMPET   **Pl. 74**
Williamia krebsi (Mörch)
  Range: S. Florida to West Indies.
  Habitat: Intertidal; on rocks and seaweeds.
  Description: Diam. about ¼ in. Limpetlike. Only moderately arched. Apex close to one end and slightly hooked. Shell very thin in substance, color amber-brown.

Genus *Trimusculus* Schmidt 1832

GOES' FALSE LIMPET *Trimusculus goesi* Hubn.   **Pl. 74**
  Range: West Indies.
  Habitat: Intertidal; on rocks.
  Description: Diam. about ¼ in., sometimes a bit more. Limpetlike in shape, base rounded. Moderately arched, apex about central. Sculpture of distinct radiating lines. Color white.

Salt-marsh Snails:
Family Ellobiidae

THESE salt-marsh snails spend their time out of water to a considerable degree. The shell is spiral, and covered with a horny periostracum. The aperture is elongate, with a strong fold or two on the inner lip.

Genus *Melampus* Montfort 1810

SALT-MARSH SNAIL *Melampus bidentatus* Say   **Pl. 74**
  Range: Nova Scotia to Texas.
  Habitat: Salt marshes.
  Description: Height ½ in. Shell oval, thin, and shining when clean. About 5 whorls, the last one constituting most of shell and others flattened to form a short, blunt spire. Aperture long and narrow, broadest below; inner lip usually covered with white enamel and 2 folds cross the lower part. Deep within the outer lip are several elevated ridges. Color greenish olive. Young specimens are banded with brown, but old shells are often corroded and coated with a muddy deposit. Thin yellowish-brown periostracum.
  Remarks: This is the commonest salt-marsh snail on the Atlantic Coast. It inhabits marshes occasionally flooded by the tide and is never very far from the high-tide mark. When the tide comes in these snails clamber to tops of the saltgrass, as if to avoid getting wet for as long as possible. Formerly listed as *M. lineatus* Say.

COFFEE-BEAN SNAIL *Melampus coffeus* (Linn.)   **Pl. 74**
  Range: Florida to West Indies.
  Habitat: Intertidal.
  Description: Slightly more than ½ in. high. Shell thin but strong, with low spire. 4 or 5 whorls. Outer lip thin and sharp, crenulate within, inner lip with 2 white folds. Color pale chocolate, usually with 3 creamy-white bands on body whorl. Thin but tough grayish periostracum.
  Remarks: This species, as well as the Salt-marsh Snail, forms a very important food supply for wild ducks.

NECKLACE SNAIL *Melampus monile* (Brug.)   **Pl. 74**
  Range: Florida to West Indies.
  Habitat: Salt marshes.
  Description: Height ½ in. An oval shell of 4 or 5 whorls.

Short pointed spire. Outer lip thin and sharp, columella with deep fold at base. Surface polished, color yellowish brown, often with 3 encircling paler bands.

Genus *Pedipes* Férussac 1821

STEPPING SHELL *Pedipes mirabilis* (Mühl.)   **Pl. 74**
Range: Florida to West Indies.
Habitat: Shallow water.
Description: Nearly ½ in. high. 4 rounded whorls, short spire. Sculpture of revolving lines. Aperture wide, with 3 prominent teeth on columella. Outer lip thickened, 1 tooth within. Color yellowish brown.

Genus *Tralia* Gray 1840

EGG MELAMPUS *Tralia ovula* (Brug.)   **Pl. 74**
Range: Bermuda; Florida to West Indies.
Habitat: Salt marshes.
Description: Height ½ in. Oval, with short spire and about 4 whorls. Columella bears 3 teeth. Whitish glaze on parietal wall. Color yellowish brown.

Genus *Detracia* Gray 1840

BULLA MELAMPUS   **Pl. 74**
Detracia bullaoides (Montagu)
Range: Florida to West Indies.
Habitat: Salt marshes.
Description: About ½ in. high. Elongate-oval, 5 or 6 whorls, blunt apex. Aperture small, columella with a deep fold. Color brown, with encircling grayish lines.

Sea Butterflies:
Family Cavolinidae

THE sea butterflies, or pteropods, are creatures of the open sea, arriving on shore only by chance and generally after a violent storm with onshore winds. They exist in huge colonies, spending the day at considerable depths and rising to the surface at night to feed on plankton. Varied in shape, the shells may be globose, triangular, or thin and slender, according to the species—of which there are many known and described.

Genus *Cavolina* Abilgaard 1791

HUMPBACK CAVOLINE *Cavolina gibbosa* Orb. **Pl. 75**
Range: Worldwide.
Habitat: Pelagic; temperate and tropical seas.
Description: Length about ¼ in. Shell rather inflated. Ventral face with concentric ridges at periphery; dorsal face with several distinct radiating folds. Lateral spines short, middle spine stout. Color milky white.

INFLEXED CAVOLINE *Cavolina inflexa* Lesueur **Pl. 75**
Range: Atlantic Ocean.
Habitat: Pelagic; temperate and tropical seas.
Description: About ⅛ in. long. Shell long and compressed. Ventral face relatively smooth; dorsal face with 3 radiating folds. Lateral spines small, middle spine long, stout, and curved upward. Color milky white.

LONG-SNOUT CAVOLINE **Pl. 75**
Cavolina longirostris Lesueur
Range: Off Martha's Vineyard to Gulf of Mexico and South Atlantic.
Habitat: Pelagic; temperate and tropical seas.
Description: About ⅛ in. long. Ventral face rounded and relatively smooth; dorsal face with 3 robust folds and extended in front by a long folded beak. Spines short and truncate. Color milky white.

TRIDENTATE CAVOLINE **Pl. 75**
Cavolina telemus (Linn.)
Range: Worldwide.
Habitat: Pelagic; temperate and tropical seas.
Description: Nearly ¾ in. long — one of the largest forms in this group. Ventral face semiglobular and smooth; dorsal face with strong radiating folds. Spines sharp, lateral spines short, middle spine longer. Color translucent amber.
Remarks: Sometimes listed as *C. tridentata* Forskål.

THREE-SPINED CAVOLINE **Pl. 75**
Cavolina trispinosa Lesueur
Range: Atlantic Ocean.
Habitat: Pelagic; temperate and tropical seas.
Description: Length about ¼ in. Shell compressed, with 3 straight spines, the middle one long and stout. Ventral face shows faint ridge on each side; dorsal face with sharp radiating folds. Color translucent smoky white.

UNCINATE CAVOLINE *Cavolina uncinata* (Rang) **Pl. 75**
Range: Atlantic Ocean.
Habitat: Pelagic; temperate and tropical seas.
Description: About ¼ in. long. Ventral face swollen and smooth; dorsal face with 3 strong radiating folds. Spines along sides compressed and curved slightly backward, middle spine short, stout, and turned upward. Color pale amber.

Genus *Clio* Linnaeus 1767

CUSPIDATE CLIO *Clio cuspidata* Bosc **Pl. 75**
Range: Worldwide.
Habitat: Pelagic; temperate and tropical seas.
Description: Length about ½ in. An extremely fragile opaque-white shell. Surface wrinkled, 3 dorsal folds. Thin and delicate spines along sides.

PYRAMID CLIO *Clio pyramidata* Linn. **Pl. 75**
Range: Worldwide.
Habitat: Pelagic; arctic and temperate seas.
Description: About ½ in. long. Somewhat variable in shape. General form is kitelike, with no spines along sides. Color translucent white.

Genus *Styliola* Lesueur 1825

KEELED CLIO *Styliola subula* Quoy & Gaimard **Pl. 75**
Range: Worldwide.
Habitat: Pelagic; temperate and tropical seas.
Description: Length nearly ½ in. A straight, elongate, conical shell. There is a marked dorsal groove. Surface smooth, color whitish. This is the only species in the genus.

Genus *Creseis* Rang 1828

STRAIGHT NEEDLE PTEROPOD **Pl. 75**
Creseis acicula Rang
Range: Worldwide.
Habitat: Pelagic; temperate and tropical seas.
Description: Nearly ½ in. long. This is merely a long slender cone that tapers to a sharp point, the larger end open. Surface smooth and shiny, the color white or yellowish white.

Genus *Herse* Gistel 1848

CIGAR PTEROPOD *Herse columnella* (Rang)   **Pl. 75**
 Range: Worldwide.
 Habitat: Pelagic; temperate and tropical seas.
 Description: About ½ in. long. Shell cylindrical and bottle-shaped, rounded at one end, open at other. Cross section almost circular. Surface smooth, color white. The only species in the genus.

Sea Hares: Family Aplysidae

SEA HARES, or sea pigeons, bear little resemblance to any snail. The shell is completely internal, buried away in the soft parts of the back, and is really no more than a thin, triangular, shelly plate. The animal may be as long as 10 in. and is soft-bodied; the mantle forms broad swimming lobes and the narrowed front end bears a pair of earlike tentacles that make it look a bit like the head of a rabbit. Colors frequently brilliant, with greens, purples, and bright reds in irregular patches of blotches or spots. Mollusk secretes an entirely harmless deep purple fluid.

Genus *Aplysia* Linnaeus 1767

WILLCOX'S SEA HARE *Aplysia willcoxi* Heilprin   **Pl. 75**
 Range: Massachusetts to both coasts of Florida.
 Habitat: Temperate and tropical seas.
 Description: Length of animal about 8 in.; color mottled brown, with round yellowish markings on inner margin of lobes. Shell about 2 in. long and internal; thin and rather flat; color glossy white on the attached side, with a thin yellowish periostracum on other.
 Remarks: Formerly in the genus *Tethys* Linn.

III

Amphineurans

THESE ARE the chitons, or coat-of-mail shells — mollusks that usually live in rocky situations close to shore. They are nocturnal in habits, feeding chiefly upon decaying vegetation at night and spending the daytime clinging to the undersides of stones and dead shells, safely out of sight. They are not exclusively herbivorous, however, and many will feed upon animal matter to some extent. All members of this class are marine.

The chiton shell is composed of 8 separate but overlapping plates (valves). This forms a sort of shield, which when empty and turned over suggests a small boat. The girdle is the leathery skin in which the plates are imbedded, and it generally extends beyond them — all the way around — to form a thin border. This girdle may be smooth or granular, or it may be covered with scales or tufts of hair. Frequently it has the appearance of snakeskin.

The chiton crawls about on a muscular creeping organ, the foot, which is really the whole underneath surface of the animal. It can cling to a wave-washed rock with surprising tenacity, and it is not easy to remove a specimen without injury. Slipping a thin knifeblade under one is the best way to dislodge it. In death they commonly curl up like a pillbug, but they can be relaxed by being soaked in water, then spread out on a thin board, such as a lath, and secured by wrapping with strong thread.

Place the tied-down chitons, board and all, in a container of some kind (2 quart fruit jars do nicely) and covered them with either a 10 percent solution of Formalin or a 70 percent solution of alcohol. If a specimen is large you can dig out most of the fleshy part before fastening it to the board. After soaking for about 2 weeks, the chitons can be taken out and allowed to dry slowly in a shady place. Then they may be removed from the boards and are ready for the collection, although there will be some shrinkage of the girdle. There also will be some loss of color, so the chitons rarely make as attractive cabinet specimens as do the snails and clams.

Some collectors remove the 8 plates of the shell and clean and reassemble them. The resulting specimens are things of beauty, exhibiting an unexpected range of colors from pure white to rich blue-green, with splashes of yellow, rose, or orange. The attractiveness of the collection is enhanced by displaying a few cleaned plates in the tray along with the complete specimens. To prepare them, simply boil the chiton for about 5 minutes, then remove the plates and place them in pure Clorox for ½ hour. This softens any adhering tissue so that its removal with a stiff brush is easy. After drying the plates on a paper towel, glue them together in the proper sequence. When completely dry and set, they can be wiped lightly with mineral oil to which a small amount of neat's-foot oil has been added; then they are ready for your cabinet.

Chitons: Family Lepidochitonidae

Genus *Tonicella* Carpenter 1873

MOTTLED CHITON   *Tonicella marmorea* (Fabr.)   **Pl. 76**
   Range: Circumpolar; Arctic Ocean to Massachusetts.
   Habitat: Moderately shallow water.
   Description: About 1 to 1½ in. long. A solid and robust species, oval in shape. Plates keeled; each bears a slight projection at the rear. Color reddish brown, marked with angular whitish lines. Girdle is without scales, is leathery in texture, tan, and sometimes speckled with red.

Chitons: Family Cryptoplacidae

Genus *Acanthochitona* Gray 1821

PYGMY CHITON   **Pl. 76**
Acanthochitona pygmaea Pilsbry
   Range: Florida to West Indies.
   Habitat: Shallow water.
   Description: About ½ in. long. Rather elongate, with a high, somewhat arched back. Plates microscopically dotted. Girdle leathery, covering major portion of plates, margins decorated with minute tufts of hair. Color may be yellowish, brownish, or mottled, sometimes with greenish.

Chitons: Family Ischnochitonidae

Genus *Calloplax* Thiele 1909

RIO JANEIRO CHITON   Pl. 76
Calloplax janeirensis (Gray)
    Range: Florida Keys to S. America.
    Habitat: Shallow water.
    Description: About ¾ in. long. Strongly sculptured, each plate bearing a pair of heavy ribs along its front margin that radiate from the center to the sides; rest of plate deeply scored by vertical grooves. A fanlike display of ribs at each end of shell. Girdle shows some fine silky hairs. Color greenish brown, dotted in places with red; interior of plates white.

Genus *Chaetopleura* Shuttleworth 1853

BEE CHITON   *Chaetopleura apiculata* (Say)   Pl. 76
    Range: Massachusetts to Florida.
    Habitat: Shallow water.
    Description: About ¾ in. long. Oval, the plates heavily scored with longitudinal grooves. Edges marked with rows of tiny elevated dots. Color grayish or pale chestnut.

Genus *Ischnochiton* Gray 1847

WHITE CHITON   *Ischnochiton albus* (Linn.)   Pl. 76
    Range: Circumpolar; Greenland to Massachusetts; Aleutian Islands in n. Pacific.
    Habitat: Shallow water.
    Description: About ½ in. long. Somewhat elongate, plates weakly marked with growth lines. Narrow girdle sandpapery. In life the color is bluish black, but this rubs off easily, and cabinet specimens vary from pale cream to nearly white.

SLENDER CHITON   *Ischnochiton floridanus* Pilsbry   Pl. 76
    Range: S. Florida.
    Habitat: Shallow water.
    Description: About 1½ in. long. Slender, plates broad and rough, with heavy but not keeled longitudinal lines. Girdle rather wide, smooth, and delicately marked with black and gray. Color of back whitish or pale green, mottled with gray.

RED CHITON *Ischnochiton ruber* (Linn.) **Pl. 76**
Range: Arctic Ocean to Connecticut; Alaska south to Sitka.
Habitat: Shallow water.
Description: About ¾ in. long. Oval shape, plates strongly keeled, bearing elevated lines of growth. Color pale brick-red, with a few lines of deeper red. Girdle reddish brown and set with elongate scales. Animal itself bright red.

Chitons: Family Chitonidae

Genus *Chiton* Linnaeus 1758

MARBLED CHITON *Chiton marmoratus* Gmelin **Pl. 76**
Range: S. Florida to West Indies.
Habitat: Shallow water.
Description: Length 2 to 3 in. Broadly oval. Surface quite smooth, almost polished. Plates without ridges or pustules, presenting a silky appearance. Color variable, ranging from olive-gray to greenish black, often with longitudinal streaks of paler shade. Girdle broad, resembles snakeskin, alternately banded with greenish white and dark brown. Interior of plates blue-green.

SQUAMOSE CHITON *Chiton squamosus* Linn. **Pl. 76**
Range: Florida to West Indies.
Habitat: Shallow water.
Description: Length 2 to 3 in. Oval, strong and solid. Plates with sharp longitudinal lines at center, sides with transverse wavy lines. Color greenish gray, with brownish streaks. Girdle broad, strongly scaled, gray-green, with irregular patches of darker green.

TUBERCULATE CHITON **Pl. 76**
Chiton tuberculatus Linn.
Range: Florida to West Indies.
Habitat: Shallow water.
Description: Length 2 to 3½ in. Broadly oval, plates strongly keeled. Each plate divided into a broad triangle toward front, and 2 wedge-shaped triangles at sides. Main triangle marked with longitudinal grooves, side triangles bear transverse wavy ridges. Front and back plates pimpled with small tubercles. Color olive-green, clouded and speckled with greenish. Conspicuous girdle heavily scaled; pale green, with squarish patches of black or dark green, suggesting a border of snakeskin.

IV

Scaphopods

AN EXTERNAL, tubular shell, open at both ends and somewhat curved, covers the animal in this class. Shape of the shell is responsible for the popular names of tusk shell and tooth shell which are generally applied to members of this group. Found only in marine waters, where they range from just below the low-tide mark to depths of several hundred fathoms, they show a preference for clean sand, but are sometimes found living on muddy bottoms. The typical shell is long and tapering, cylindrical, gently curved, and usually pure white, although some tropical species may be pale green or pink. The surface may be smooth and glossy, dull and chalklike, or ribbed longitudinally. They live in the sand with the small end uppermost; from the larger end protrudes the foot, a tough elastic organ admirably suited for burrowing. In fact, the name scaphopod means "plow-footed."

In primitive times the shells were used for money and adornment by various Indian tribes, particularly from our Northwest, where the tubular shells were cut into short sections for beads.

Tusk Shells: Family Dentaliidae

Genus *Dentalium* Linnaeus 1758

SHINING TUSK SHELL   Pl. 74
Dentalium callipeplum Dall
  Range: Gulf of Mexico.
  Habitat: Shallow water.
  Description: Length 2 in. Strongly and regularly curved; the taper is very gradual. Surface perfectly smooth and well polished. Color creamy white.

IVORY TUSK SHELL *Dentalium eboreum* Conrad   Pl. 74
  Range: N. Carolina to West Indies.
  Habitat: Shallow water.
  Description: About 2 in. long. Thin and delicate, gently curved. Surface smooth and polished, with a few raised lines

at the small end. Color white or ivory-white, sometimes flushed with pinkish or yellowish. Apical slit narrow and on the convex side.

VERRILL'S TUSK SHELL   Pl. 74
Dentalium meridionale verrilli Henderson
Range: Newfoundland to Cape Hatteras.
Habitat: Deep water.
Description: Length 4 to 5 in. A strong and robust shell, its small (upper) end slightly curved, but most of the shell relatively straight. In cross section it is nearly round. Surface smooth, color grayish or slaty brown; interior bluish white.
Remarks: Typical *D. meridionale* Pilsbry & Sharp occurs in deep water in the Caribbean and south to Brazil.

RIBBED TUSK SHELL   Pl. 74
Dentalium occidentale Stimpson
Range: Newfoundland to Cape Hatteras.
Habitat: Moderately deep water.
Description: About 1 in. long. A rather solid shell decorated with about 16 well-defined longitudinal ribs. Surface not polished, color a soiled white, sometimes with tinges of ivory.

TEXAS TUSK SHELL   *Dentalium texasianum* Phil.   Pl. 74
Range: N. Carolina to Texas.
Habitat: Shallow water.
Description: About 1 in. long. Surface bears weak longitudinal ribs, the spaces between them rather flat. In cross section the shell is hexagonal. Color dull grayish white.

Swollen Tusk Shells:
Family Siphonodentaliidae

Genus *Cadulus* Philippi 1844

THESE tusk shells differ from the genus *Dentalium* by being short and generally swollen in the middle.

CAROLINA CADULUS   *Cadulus carolinensis* Bush   Pl. 74
Range: N. Carolina to Florida and Texas.
Habitat: Moderately deep water.
Description: About ¼ in. long. Shell rather solid but slightly swollen at middle. In cross section it is round. Small end bears 4 tiny slits. Surface highly polished, color milky white.

GREAT CADULUS *Cadulus grandis* Verrill   **Pl. 74**
Range: Nova Scotia to Cape Hatteras.
Habitat: Deep water.
Description: Length ¾ in. Shell sturdy but slightly curved. Somewhat swollen toward larger end. Small end (apex) with tiny slits. Cross section roundish. Surface well polished, color pure white.
Remarks: The shell illustrated on Plate 74 is the one actually used for the description of this species by A. E. Verrill in 1884. It is the holotype.

V

Cephalopods

THE CEPHALOPODS are highly specialized mollusks, keen of vision and swift in action. The head is armed with a parrotlike beak and surrounded by long flexible tentacles studded with sucking disks. Besides the octopuses, this class includes the beautiful Pearly or Chambered Nautilus of the w. Pacific and the Indian Ocean. The class is abundantly represented in the past by a bewildering variety of nautiloids, belemnites, and ammonites known only as fossils, some of the latter from the Cretaceous with a diameter of more than 3 ft. Two members of the shell-bearing cephalopods may be found on our shores, in addition to several kinds of squid lacking external shells.

Rams' Horns: Family Spirulidae

Genus *Spirula* Lamarck 1799

RAM'S HORN *Spirula spirula* (Linn.) Pl. 75
Range: Worldwide.
Habitat: Pelagic.
Description: Diam. about 1 in. Shell thin and rather fragile, flatly but loosely coiled, the coils not in contact. Shell divided into many chambers by concave partitions (septa). Last partition shows a small hole (opening of the siphuncle — a structure connecting the chambers). Color pure white.
Remarks: This is an internal shell, located at the rear of a small octopuslike mollusk, and it may serve as a balancing device or may enable the creature to rise or sink at will. The small snow-white shells are washed ashore from Cape Cod to Florida (rather rarely in north but not uncommonly in south) and can be picked up on beaches all around the world in warm and temperate seas; the complete mollusk is almost never seen, and then only when taken by dredging or trawling in deep water.

Paper Argonauts:
Family Argonautidae

Genus *Argonauta* Linnaeus 1758

PAPER ARGONAUT   *Argonauta argo* Linn.   **Pl. 75**
Range: Worldwide.
Habitat: Pelagic.
Description: Diam. to 8 in. Shell somewhat flattened, forming a roundish spiral; thin and brittle, not chambered. Exterior bears a double keel, and is decorated with knobs and swollen lines. Color milky white, tinged with brown at the keels.
Remarks: There are no muscular attachments to the shell, and the animal may discard it at any time. This puzzled the early naturalists, for it was contrary to all known varieties of shellfish; the general belief was that the cephalopod lived in the shell of some mysterious "sea snail" after the manner of a hermit crab. Another puzzling fact was that all specimens found were females. About 1840 it was discovered that the female has 2 "arms" broadly expanded to form a pair of web-like appendages and that these organs secrete the shell, not the mantle as in the usual mollusk. The shell is begun when the female is only a few weeks old; eventually it will serve as a case for her eggs. The male argonout is much smaller than the female and has no shell at all, so it in understandable that for many years the life history of this mollusk was so imperfectly known.

Empty shells are occasionally washed up on the beaches of Florida, and rarely as far north as New Jersey. This species has been named *A. americana* Dall, but today most scientists regard it as conspecific with the European and Pacific forms, and they are all listed as *A. argo* Linn.

Appendixes
Index

Glossary of Conchological Terms

Acuminate — tapering to a point.

Acute — sharply pointed.

Animal — the fleshy part of the mollusk.

Annulated — marked with rings.

Anterior end — end toward which the beaks point in a bivalve shell; posterior end is opposite.

Aperture — the entrance or opening of the shell.

Apex — the tip of the spire in snail shells.

Apophysis (pl. **apophyses**) — a shelly brace or peglike structure located inside under each beak to support muscles; also called myophore.

Axis — the central structure (column) of a spiral shell.

Base — *snails,* the extremity opposite the apex; *clams,* the margin opposite the beaks.

Beaks — the initial, or earliest, part of a bivalve shell; umbones.

Bivalve — a shell with 2 valves (or shell parts).

Body whorl — the last whorl of a snail shell.

Byssus — a series of threadlike filaments that serve to anchor the bivalve to some support.

Calcareous — composed of lime.

Callum — thin calcareous covering of the gape in some clams.

Callus — a calcareous deposit, such as enamel.

Canal — usually a somewhat narrow prolongation of the lip of the aperture containing the siphons in many snails.

Cancellate — having longitudinal and horizontal lines or ribs that cross each other.

Carinate — with a keel-like, elevated ridge.

Cartilage — an internal, elastic substance found in bivalves which controls the opening of the valves.

Chondrophore — an internal spoonlike projection on the hinge of some bivalves.

Columella — the pillar around which the whorls form their spiral circuit.

Compressed — having the valves close together.

Concentric — applied to curved ridges or lines on a bivalve shell which form arcs with the beaks at the center.

Conic — shaped like a cone.
Coronate — crowned.
Costate — ribbed.
Crenulate — notched or scalloped.
Cuspidate — prickly pointed.

Deck — a shelly plate under the beaks; also called a platform.
Decussated — having lines cross at right angles.
Denticulate — toothed.
Dextral — turning from left to right; right-handed.
Discoidal — having the whorls coiled in one plane.
Dorsal — belonging to the back.

Ear — *see* Wing.
Epidermis — *see* Periostracum.
Equivalve — when both valves are the same size and shape.
Escutcheon — an elongated depression behind the beaks in a bivalve shell.

Foot — muscular extension of the body used in locomotion.
Fusiform — spindle-shaped.

Gaping — having the valves only partially closed.
Gastropod — a snail (or slug).
Genus — a separate group of species, distinguished from all others by certain permanent marks called generic characters.
Globose — rounded like a globe or ball.
Granulated — covered with minute grains or beads.
Growth lines — lines on the surface of a shell indicating rest periods during growth; also called growth laminae.

Hinge — where the valves of a bivalve are joined.

Inequivalve — having valves that differ in size or shape.
Inner lip — portion of aperture adjacent to the axis, or pillar.
Involute — rolled inward from each side, as in *Cypraea*.

Keel — a flattened ridge, usually at the shoulder or periphery.

Lamellibranchia — an older name for pelecypods (bivalves).
Lanceolate — shaped like a lance.
Ligament — a cartilage that connects the valves.
Linear — relating to a line or lines; very narrow.
Lineated — marked with lines.
Lips — the margins of the aperture.
Lirate — resembling fine incised lines.
Littoral — the tidal zone.

Lunule — a depressed area, usually heart-shaped, in front of the beaks in many bivalve shells.
Lyrate — shaped like a lyre.

Maculate — splashed or spotted.
Mantle — a membranous flap or outer covering of the soft parts of a mollusk; it secretes the material that forms the shell.
Margin — the edges of the shell.
Mesoplax — an accessory plate.
Mouth — the aperture of a snail shell.
Myophore — *see* Apophysis.

Nacre — the pearly layer of certain shells.
Nodose — having tubercles, or knobs.
Nodule — a knoblike projection.
Nucleus — the initial nuclear whorls.

Operculum — a plate or door that closes the aperture in many snail shells.
Orbicular — round or circular.
Outer lip — the outer edge of the aperture.
Ovate — egg-shaped or oval.

Pallets — a pair of simple or compound chitinous or calcareous structures to close the tube when the siphons are withdrawn.
Pallial line — a groove or channel near the inner base of a bivalve shell, where the mantle is made fast to the lower part of the shell.
Pallial sinus — a notch or embayment at the posterior end of pallial line where the retracted siphons are held.
Parietal wall — the inner lip area.
Pelagic — inhabiting the open sea.
Periostracum — the noncalcareous covering on many shells; sometimes wrongly called the epidermis.
Peristome — edge of the aperture.
Platform — *see* Deck.
Pleat — applied to folds on the columella.
Plicate — folded or pleated.
Posterior end — opposite anterior; see Anterior end.
Produced — elongated.
Protoconch — the initial whorls of a snail shell; the nucleus.
Pyriform — pear-shaped.

Quadrate — rectangular-shaped.

Radial — pertaining to a ray or rays from a common center.
Radiating — applied to ribs or lines that start at the beak and extend fanwise to the margins.

Radula — the "tongue," or dental apparatus, of gastropods.

Resilium — internal cartilage in bivalve hinge; causes shell to spring open when muscles relax.

Reticulate — crossed, like network.

Rostrate — beaked.

Rostrum — a beak, or handlike extension of shell.

Rugose — rough or wrinkled.

Scabrous — rough.

Sculpture — the ornamentation of a shell.

Semilunar — half-moon-shaped.

Septum (pl. **septa**)—the platform or deck; in plural usage it can mean partitions.

Sinistral — turning from right to left; left-handed.

Sinuate — excavated.

Sinus — a deep cut.

Siphon — the organ through which water enters or leaves the mantle cavity.

Siphuncle — a membranous tubular extension of the mantle which runs through chambers of *Nautilus*.

Species — the subdivision of a genus, distinguished from all others of the genus by certain permanent marks called specific characters.

Spire — the upper whorls, from the apex to the body whorl.

Striae — very fine lines.

Subglobular — not quite globose.

Sulcus (pl. **sulci**) — groove or furrow.

Suture — the spiral line of the spire, where one whorl touches another.

Teeth — the pointed protuberances at the hinge of a bivalve shell; in snails, the toothlike structures in the aperture.

Trochiform — top-shaped.

Truncate — having the end cut off squarely.

Tubercle — a knob.

Turbinate — top-shaped.

Turreted — having tops of the whorls flattened.

Umbilicus — a small hollow at the base of the shell in snails; visible from below.

Umbo (pl. **umbones**) — beak, or prominent part, of bivalve shells above the hinge.

Univalve — a shell composed of a single piece, as a snail.

Valve — one part of a bivalve shell; one of the 8 plates that make up the dorsal shield of a chiton.

Varicose — bearing 1 or more varices.

Varix (pl. **varices**) — an earlier margin (a kind of rib) of the aperture; a prominent raised longitudinal rib on surface of a snail shell caused by a periodic thickening of the outer lip during rest periods in shell growth.

Ventral — of or pertaining to the underside.

Ventricose — swollen or rounded out.

Volutions — the distinct turns of the spire; also called whorls.

Whorls — *see* Volutions.

Wing — a somewhat triangular projection or expansion of the shell of a bivalve, either in the plane of the hinge or extending above it; also known as an "ear."

List of Authors

Abbott — R. Tucker Abbott
Abilgaard — Peter Abilgaard
Adams, A. — Arthur Adams
Adams, C. B. — Charles B. Adams
Adams, H. — Henry Adams
Aguayo — Carlos G. Aguayo
Alder — Joshua Alder
Angas — George F. Angas
Anton — Hermann E. Anton
Ascanius — Peder Ascanius

Baker — Frank C. Baker
Bartsch — Paul Bartsch
Beck — Heinrich H. Beck
Bellardi — Luigi Bellardi
Benson — William Henry Benson
Bergh — Ludwig S. Rudolph Bergh (Rudolph Bergh)
Bern. — A. C. Bernardi
Bertin — Victor Bertin
Bivona — Antonio Bivona
Blain. — Henri M. D. de Blainville
Born — Ignaz von Born (Ignatius Born)
Bosc — Louis A. G. Bosc
Bourg. — Jules R. Bourguignat
Brocchi — Giovanni B. Brocchi
Brod. — William J. Broderip
Bronn — Heinrich G. Bronn
Brown — Thomas Brown
Brünnich — Martin T. Brünnich
Brug. — Jean G. Bruguière
Bucquoy — Eugène Bucquoy
Bush — Katharine J. Bush

Cantraine — François J. Cantraine

Carpenter — Philip P. Carpenter
Chemnitz — Johann H. Chemnitz
Clench — William J. Clench
Conrad — Timothy A. Conrad
Couthouy — Joseph P. Couthouy
Crosse — J. C. Hippolyte Crosse

Da Costa — E. Mendes da Costa
Dall — William H. Dall
Daudin — François M. Daudin
Dautzenberg — Philippe Dautzenberg
Defrance — Jacques M. L. Defrance
DeKay — James E. DeKay
Desh. — Gérard P. Deshayes
Desj. — Max Desjardin
Desmoulins — Charles Desmoulins
Dill. — Lewis W. Dillwyn
Dohrn — Anton Dohrn
Dollfus — Gustave F. Dollfus
Donovan — Edward Donovan
Duclos — P. L. Duclos
Duméril — André M. C. Duméril
Dunker — Wilhelm B. R. H. Dunker

Emerson — William K. Emerson
Esch. — Johann F. Eschscholtz

Fabr. — Otto Fabricius
Far. — Isabel P. Farfante
Férussac — Jean Baptiste Louis d'Audebard de Férussac
Fischer — Paul Fischer
Fleming — John Fleming

Fleuriau — Fleuriau de Bellevue
Forbes — Edward Forbes
Forskål — Peter Forskål
Friele — Herman Friele

Gabb — William M. Gabb
Gaimard — Joseph P. Gaimard
Gaskoin — John S. Gaskoin
Gistel — Johannes Gistel
Gmelin — Johann F. Gmelin
Gould — Augustus A. Gould
Gray — John Edward Gray
Green — Jacob Green
Gregorio — Antonio de Gregorio
Guild. — Lansdowne Guilding
Guppy — R. J. L. Guppy

Hanley — Sylvanus C. T. Hanley
Harb. — Anne Harbison
Hartman — William D. Hartman
Heilprin — Angelo Heilprin
Helb. — G. S. Helbling
Henderson — John B. Henderson
Hermann — Jean Hermann
Herrmannsen — August N. Herrmannsen
Hidalgo — Joaquin G. Hildalgo
Hinds — Richard B. Hinds
Holmes — Francis S. Holmes
Holten — Joannes S. Holten
Hubn. — Bengt Hubendick
Hwass — Christian H. Hwass

Iredale — Tom Iredale
Issel — Arturo Issel

Jay — John Clarkson Jay
Jeffreys — John Gwynn Jeffreys
Johnson — Charles W. Johnson
Jonas — J. H. Jonas
Jousseaume — Felix P. Jousseaume
Jukes-Brown — Alfred J. Jukes-Brown

Keen — A. Myra Keen

Kiener — Louis C. Kiener
Kingston — J. F. Kingston
Kobelt — Wilhelm Kobelt
Koenen — Adolf von Koenen
Kröyer — Henrik Kröyer
Kurtz — John D. Kurtz

Lam. — Jean Baptiste Lamarck
Lea, H. C. — Henry C. Lea
Leach — William E. Leach
Leche — Wilhelm Leche
Lesson — René P. Lesson
Lesueur — Charles A. Lesueur
Lightf. — John Lightfoot
Link — Heinrich F. Link
Linn. — Carolus Linnaeus
Linsley — James H. Linsley
Lovén — Sven Lovén
Lowe — R. T. Lowe

McGinty — Thomas L. McGinty
Marrat — Frederick P. Marrat
Marryat — F. Marryat
Martens — Carl Eduard von Martens
Maton — William G. Maton
Maury — Carlotta J. Maury
Meek — Fielding B. Meek
Ménard — Ménard de la Groye
Menke — Karl T. Menke
Meus. — Friedrich C. Meuschen
Michaud — André L. G. Michaud
Mighels — Jesse W. Mighels
Modeer — Adolph Modeer
Möller — H. P. C. Möller
Mörch — Otto A. L. Mörch
Moll — Friedrich Moll
Montagu — George Montagu
Monterosato — Tommaso di Maria Allery, Marchese di Monterosato
Montfort — Denys de Montfort
Mühl. — Megerle von Mühlfeld
Müller — Otto F. Müller

Nardo — Giovanni D. Nardo
Nyst — Pierre Henri Nyst (H. Nyst)

Odhner — Nils Hj. Odhner
Olivi — Giuseppi A. Olivi
Olsson — Axel A. Olsson
Orb. — Alcide D. d'Orbigny

Partsch — Paul Partsch
Perry — George Perry
Petit — Petit de la Saussaye
Pfr. — Louis Pfeiffer
Phil. — Rudolf A. Philippi
Phipps — Constantine J. Phipps
Pilsbry — Henry A. Pilsbry
Poiret — Jean L. M. Poiret
Poli — Giuseppe S. Poli
Potiez — V. L. Victor Potiez
Prime — Temple Prime
Puffer — Elton L. Puffer
Pulley — Thomas E. Pulley
Pulteney — Richard Pulteney

Quoy — Jean René C. Quoy

Raf. — Constantine S. Rafinesque
Rang — Paul K. Sander L. Rang
Rav. — Edmund Ravenel
Récluz — Constant A. Récluz
Reeve — Lovell A. Reeve
Rehder — Harald A. Rehder
Richards — Horace G. Richards
Risso — J. Antoine Risso
Rob. — Robert Robertson
Roch — Felix Roch
Röding — Peter F. Röding (or Roeding)
Roemer — Eduard Roemer (or Römer)
Russell — Henry D. Russell

Sacco — Frederico Sacco
Salis — Carl U. von Salis
Salisbury — Albert E. Salisbury
Sars — Georg Ossian Sars
Sasso — Agostino Sasso (or Sassi)
Say — Thomas Say
Sby. — George B. Sowerby

Scacchi — Arcangelo Scacchi
Schenck — Hubert G. Schenck
Schilder — Franz A. Schilder
Schmidt — Friedrich C. Schmidt
Schum. — Christian F. Schumacher
Schwengel — Jeanne S. Schwengel
Scop. — Giovanni A. Scopoli
Seguenza — Giuseppe Seguenza
Sharp — Benjamin Sharp
Shaw — George Shaw
Shutt. — Robert J. Shuttleworth
Simpson — Charles T. Simpson
Smith, E. A. — Edgar A. Smith
Smith, I. — Irving Smith
Smith, M. — Maxwell Smith
Smith, S. — Sanderson Smith
Sol. — Daniel C. Solander
Spengler — Lorenz Spengler
Stearns — Robert E. C. Stearns
Steens. — Johannes Jepetus S. Steenstrup
Stewart — Ralph B. Stewart
Stimpson — William Stimpson
Storer — David Humphreys Storer
Stutchbury — Samuel Stutchbury
Swain. — William Swainson

Thiele — Johannes Thiele
Torell — Otto M. Torell
Totten — Joseph G. Totten
Tuomey — Michael Tuomey
Tryon — George W. Tryon
Turner — Ruth D. Turner
Turton — William Turton

Val. — Achille Valenciennes
Vanatta — Edward G. Vanatta
Verrill — Addison E. Verrill

Wagner — Anton J. Wagner
Waldheim — Gotthelf Fischer von Waldheim
Watson — Hugh Watson

APPENDIX C

Bibliography

Abbott, R. Tucker. American Seashells. New York: Van Nostrand, 1954.

Aldrich, Bertha, and Ethel Snyder. Florida Sea Shells. Boston: Houghton Mifflin, 1936.

Amos, William H. The Life of the Seashore. New York: McGraw-Hill in cooperation with World Book Encyclopedia, 1966.

Arnold, Augusta. The Sea-beach at Ebb-tide. New York: Century, 1903.

Carson, Rachel. The Edge of the Sea. Boston: Houghton Mifflin, 1955.

Dance, S. Peter. Shell Collecting: An Illustrated History. Berkeley: Univ. Calif. Press, 1966.

Gould, Augustus A. Report on the Invertebrata of Massachusetts. 2nd ed., edited by W. G. Binney. Boston: Wright and Potter, 1870.

Hausman, Leon A. Beginner's Guide to Seashore Life. New York: Putnam, 1949.

Jacobson, M. K., and W. K. Emerson. Shells of the New York City Area. Larchmont, N.Y.: Argonaut, 1961.

———. Shells from Cape Cod to Cape May, with Special Reference to the New York City Area. New York: Dover, 1972.

Johnson, Charles W. Fauna of New England: List of the Mollusca. Occas. Papers Boston Society of Natural History, Vol. 7, No. 13 (1915).

———. List of the Marine Mollusca of the Atlantic Coast from Labrador to Texas. Proc. Boston Society of Natural History, Vol. 40, No. 1 (1934).

Johnstone, Kathleen Yerger. Sea Treasure: A Guide to Shell Collecting. Boston: Houghton Mifflin, 1957.

La Rocque, Aurele. Catalogue of the Recent Molluscs of Canada. Natl. Museum of Canada Bull. No. 129. Ottawa, 1953.

Miner, Roy Waldo. Field Book of Seashore Life. New York: Putnam, 1950.

Morris, Percy A. Nature Study at the Seashore. New York: Ronald Press, 1962.

Morton, J. E. Molluscs. 4th ed. London: Hutchinson, 1967.

Nowell-Usticke, Gordon W. A Checklist of the Marine Shells of St. Croix, U.S. Virgin Islands. Burlington, Vt.: Lane Press, 1959.

Perry, Louise M., and Jeanne S. Schwengel. Marine Shells of the Western Coast of Florida. Ithaca, N.Y.: Paleontological Research Institution, 1955.

Pratt, Henry Sherring. A Manual of the Common Invertebrate Animals. Rev. ed. Philadelphia: Blakiston, 1935.

Rios, Eliézes C. Coastal Brazilian Seashells. Rio Grande do Sol, Brazil: Museu Oceanográfico de Rio Grande, 1970.

Rogers, Julia E. The Shell Book. New York: Doubleday, Page, 1908. Reprinted 1951 by C. T. Branford, Boston, with names brought up to date in the appendix by Harald A. Rehder.

Smith, Maxwell. East Coast Marine Shells. Ann Arbor: Edwards, 1937.

Turner, Ruth D. A Survey and Illustrated Catalogue of the Teredinidae. Cambridge: Museum of Comparative Zoölogy, Harvard University, 1966.

Verrill, Addison E., and S. T. Smith. Report upon the Invertebrate Animals of Vineyard Sound and Adjacent Waters. Washington: Government Printing Office, 1874.

Verrill, A. Hyatt. Shell Collector's Handbook. New York: Putnam, 1950.

Vilas, Curtis N. and N. R. Florida Marine Shells. 2nd ed. Indianapolis: Bobbs-Merrill, 1952.

Van Nostrand's Standard Catalog of Shells. 2nd ed., edited by Robert J. L. Wagner and R. Tucker Abbott. Princeton: Van Nostrand, 1964.

Warmke, Germaine L., and R. Tucker Abbott. Caribbean Seashells. Narberth, Pa.: Livingston, 1961.

Webb, Walter F. Handbook for Shell Collectors. 8th ed. St. Petersburg, Fla.: Privately printed, 1948.

Zim, Herbert S., and Lester Ingle. Seashores. New York: Simon and Schuster, 1955.

IN ADDITION

How to Collect Shells: A publication written by dozens of experts in their fields. Subjects covered include shore collecting, reef collecting, dredging, collecting mollusks from fish, land collecting, freshwater collecting, and many others. Obtainable from the American Malacological Union (see The Nautilus, below).

The Nautilus: A quarterly journal written for and by conchologists. Technical and semipopular articles. The official organ of the American Malacological Union. Address: Delaware Museum of Natural History, Greenville, Delaware 19807.

Johnsonia: Monographs of the marine mollusca of the western Atlantic. Vols. 1-4, William J. Clench, Editor. Museum of Comparative Zoology, Harvard University, Cambridge, Mass. 02138. Obtainable on a subscription basis.

APPENDIX D

Some Eastern Shell Clubs

COMPLETE ADDRESSES are given below for those clubs with permanent addresses. All other addresses may be obtained from the Secretary of the American Malacological Union, Mrs. Robert Hubbard, 3957 Marlow Court, Seaford, New York 11783.

Argonaut Trail Shell Club, Eau Gallie, Fla.
Boston Malacological Club, Museum of Comparative Zoology, Harvard University, Cambridge, Mass. 02138
Brazoria County Shell Club, Lake Jackson, Texas
Broward Shell Club, Fort Lauderdale, Fla.
Chicago Shell Club, Field Museum of Natural History, East Roosevelt Road, Chicago, Ill. 60605
Coastal Bend Shell Club, Corpus Christi, Texas
Conchology Section, Buffalo Society of Natural History, Buffalo, N.Y. 14211
Conchology Group, Outdoor Nature Club, Houston, Texas
Connecticut Shell Club, Peabody Museum of Natural History, Yale University, New Haven, Conn. 06501
Connecticut Valley Shell Club, Museum of Natural History, Springfield, Mass. 01103
Fort Myers Shell Club, Fort Myers, Fla.
Galveston Shell Club, Galveston, Texas
Jacksonville Shell Club, Jacksonville, Fla.
Lower Keys Shell Club, Marathon, Fla.
Miami Malacological Society, Miami, Fla.
Naples Shell Club, Naples, Fla.
National Capital Shell Club, Washington, D.C.
New York Shell Club, American Museum of Natural History, Central Park West, New York, N.Y. 10024
North Carolina Shell Club, Wrightsville Beach, N.C.
Palm Beach County Shell Club, West Palm Beach, Fla.
Philadelphia Shell Club, Academy of Natural Sciences, 19th and Benjamin Franklin Parkway, Philadelphia, Pa. 19103
Pittsburgh Shell Club, Carnegie Museum, Pittsburgh, Pa. 15213
Rochester Shell and Shore Club, Rochester, N.Y.
St. Petersburg Shell Club, St. Petersburg, Fla.
Sanibel-Captiva Shell Club, Captiva Island, Fla.
Sarasota Shell Club, Sarasota, Fla.
South Florida Shell Club, Museum of Science and Natural History, Miami, Fla. 33129
South Padre Island Shell Club, Port Isabel, Texas

Index

Only chapters containing the main species descriptions are covered by this index. Numbers in lightface roman type indicate the main text page; **boldface** type indicates the plate numbers of the illustrations. In general, the shells are to be found under their family classifications. However, if the English common name of a species has as the last part of its name the genus name, the shell will be found only under the anglicized genus name rather than under the family name when this differs from the genus name. Macomas and strigillas will therefore appear under the anglicized name of their genera rather than under the entry "Tellin(s)," which is their family name.

Barrel Bubble (contd.)
Ivory, 267, **Pl. 73**
Barrel Bubble Shells, 266. *See also* Bubble, Spindle; *and* Pear Shell, Ovate Little
Basket Clam(s), 90
Caribbean, 91, **Pl. 31**
Common, 91, **Pl. 31**
Equal-valved, 91, **Pl. 31**
Fat, 91, **Pl. 31**
Snub-nosed, 91, **Pl. 31**
Swift's, 92, **Pl. 31**
See also Paramya, Ovate
Bathyarca, 13
glomerula, 13, **Pl. 11**
pectunculoides, 13, **Pl. 11**
Batillaria, 148
minima, 148, **Pl. 43**
Bean Clams, 80. *See also* Coquina
Bela. See under *Lora cancellata,* 248
Bentharca, 14
profundicola, 14, **Pl. 11**
Benthonella, 137
gaza, 137, **Pl. 40**
Benthonella, Treasure, 137, **Pl. 40**
Beringius, 209
ossiani, 209, **Pl. 57**
Bittersweet
American, 15, **Pl. 12**
Comb, 16, **Pl. 12**
Grooved, 16, **Pl. 12**
Lined, 16, **Pl. 12**
Bittersweet Shells, 15
Bittium
Alternate, 149, **Pl. 43**
Variable, 150 **Pl. 43**
Bittium, 149
alternatum, 149, **Pl. 43**
varium, 150, **Pl. 43**
Bleeding Tooth, 127, **Pls. 4, 39**
Blind Limpet(s), 115
Northern, 115, **Pl. 37**
Boat Shell. *See* Slipper Shell, Common, 164, **Pl. 45**
Bonnet
Baby, 181, **Pl. 6**
Polished Scotch, 180, **Pl. 8**
Royal, 180, **Pl. 49**
Scotch, 180, **Pl. 49**
Boreotrophon, 196
clathratum, 196, **Pl. 52**
clathratum rostratum, 196, **Pl. 52**
craticulatum, 196, **Pl. 52**
gunneri, 196, **Pl. 52**
scalariformis, 196, **Pl. 52**
Borer
Atlantic Wood, 95, **Pl. 32**
Truncate, 94, **Pl. 32**

Botula, 21
fusca, 21, **Pl. 13**
Brachidontes, 18. *See also under* Mussel, Ribbed, 17
citrinus, 18, **Pl. 12**
exustus, 18, **Pl. 12**
recurvus, 18, **Pl. 12**
Bubble
Adams' Baby, 261, **Pl. 72**
Arctic, 263, **Pl. 72**
Auber's, 269, **Pl. 73**
Brown's, 269, **Pl. 73**
Common, 263, **Pl. 72**
Concealed, 269, **Pl. 73**
Depressed, 270, **Pl. 73**
Lined, 262, **Pl. 72**
Orbigny's, 269, **Pl. 73**
Rice, 270, **Pl. 73**
Solid, 264, **Pl. 72**
Spindle, 268, **Pl. 73**
Striate, 264, **Pl. 72**
Verrill's, 270, **Pl. 73**
See also Barrel, Glassy, Helmet, and Paper Bubbles
Bubble Shells
Barrel, 266
Flat-topped, 269
Glassy, 264
Helmet, 261
Lined, 262
Paper, 263
Small, 261
True, 263
Wide-mouthed, 270
Buccinidae, 206
Buccinum, 206
abyssorum, 206, **Pl. 55**
cyaneum, 207, **Pl. 55**
donovani, 207, **Pl. 55**
groenlandicum. See *B. cyaneum,* 207, **Pl. 55**
tenue, 207, **Pl. 55**
totteni, 207, **Pl. 55**
undatum, 207, **Pl. 55**
Bulla, 263
occidentalis, 263, **Pl. 72.** See also under *B. striata,* 264
solida, 264, **Pl. 72**
striata, 264, **Pl. 72**
Bullata, 232
ovuliformis, 232, **Pl. 64**
Bullidae, 263
Bursa, 186
corrugata, 186, **Pl. 51**
cruentatum. See *B. thomae,* 187, **Pl. 51**
cubaniana, 186, **Pl. 51**
granularis. See *B. cubaniana,* 186, **Pl. 51**

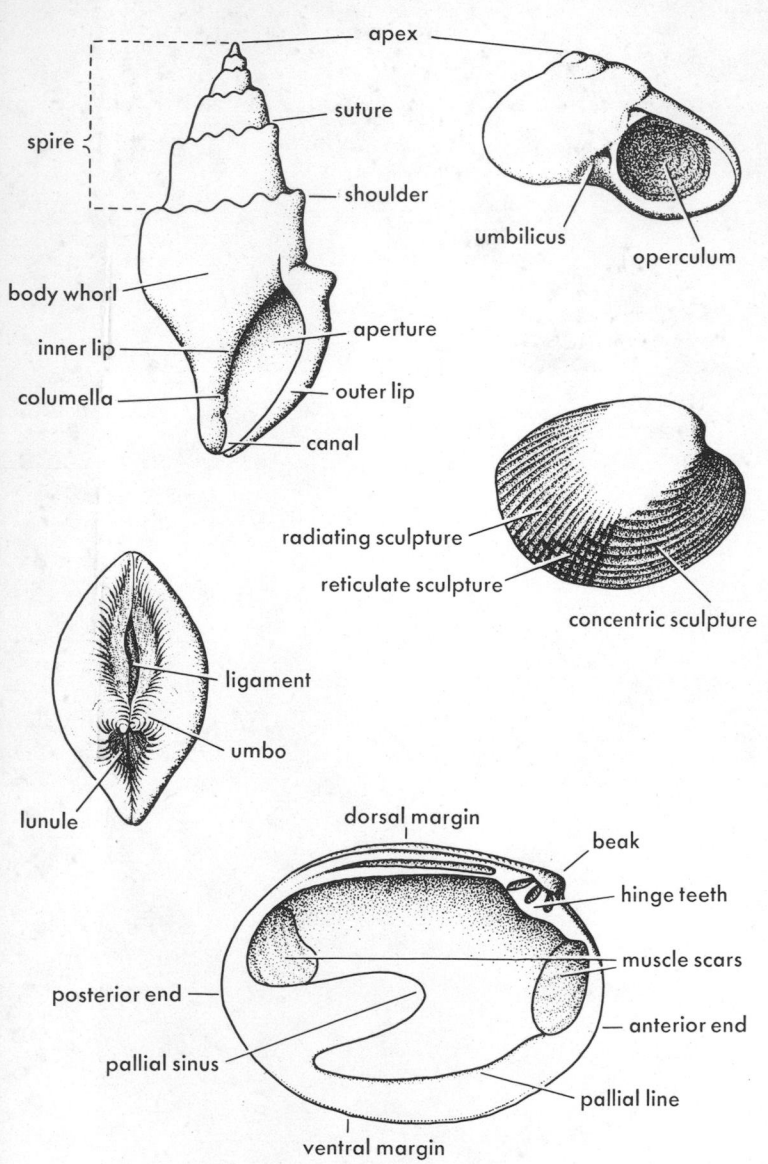

TERMINOLOGY OF UNIVALVE AND BIVALVE SHELLS